# Numerical Methods in Rock Mechanics

# Numerical Methods in Rock Mechanics

**G. N. Pande**

*Department of Civil Engineering*
*University College Swansea*
*UK*

**G. Beer**

*Division of Geomechanics*
*Commonwealth Scientific and Industrial Research Organisation*
*Australia*

**J. R. Williams**

*Department of Civil Engineering*
*Massachusetts Institute of Technology*
*USA*

JOHN WILEY & SONS LTD
Chichester · New York · Brisbane · Toronto · Singapore

*Other Wiley Editorial Offices*

John Wiley & Sons, Inc., 605 Third Avenue,
New York, NY 10158-0012, USA

Jacaranda Wiley Ltd, G.P.O. Box 859, Brisbane,
Queensland 4001, Australia

John Wiley & Sons (Canada) Ltd, 22 Worcester Road,
Rexdale, Ontario M9W 1L1, Canada

John Wiley & Sons (SEA) Pte Ltd, 37 Jalan Pemimpin 05-04,
Block B, Union Industrial Building, Singapore 2057

***Library of Congress Cataloging-in-Publication Data:***
Pande, G. N.
Numerical methods in rock mechanics / G. Pande, G. Beer, J. Williams.
p. cm.
Includes bibliographical references.
ISBN 0 471 92021 5
1. Rock mechanics—Mathematics. 2. Numerical analysis. I. Beer, G. (Geornot) II. Williams, J. (John) III. Title.
TA706.P34 1990
624.1'5132—dc20 89–24826
CIP

***British Library Cataloguing in Publication Data:***
Pande, G. N.
Numerical methods in rock mechanics.
1. Rocks. Mechanics. Analysis. Numerical methods
I. Title II. Beer, G. III. Williams, J.
624.1'5132'072

ISBN 0471 92021 5

Typeset by Photo·graphics, Honiton, Devon
Printed and bound in Great Britain by Biddles Ltd, Guildford

# Contents

# Preface

In recent years numerical methods have continued to expand and diversify into all the major fields of scientific and engineering studies. They provide a viable alternative to physical models that can be expensive and time consuming. This book deals with the application of numerical methods in the area of rock mechanics—an interdisciplinary subject, knowledge of which is required in activities such as tunnelling, mining, and construction of dams and underground caverns to locate powerhouses, store fuels and waste materials etc. It is aimed at graduate students of civil and mining engineering, as well as professional engineers who are being increasingly called upon to use numerical methods to support and check their designs. The book has been written with emphasis on practical applications and they have been included wherever appropriate. It deals with the three most commonly used methods: the finite element method, the boundary element method and the discrete element method. No attempt has been made to compare these methods as their advantages and disadvantages are problem-dependent.

Chapter 1 gives an introduction to rock mechanics. It highlights the essential features of the behaviour of rocks in comparison to soils; it is assumed that the reader has knowledge of soil mechanics which is a standard course at the undergraduate level. Chapter 2 deals with the theory of elasto-plasticity and elasto-viscoplasticity. The knowledge of these theories is essential for modelling of the constitutive behaviour of jointed rock masses. Chapters 3 and 4 cover mechanical behaviour of intact rocks, rock joints and jointed rock masses. Simple analytical and numerical models are discussed and these are used later as 'building blocks' for a comprehensive constitutive model of the behaviour of jointed rock masses discussed in Chapter 7. The finite element method is discussed in Chapter 5. It covers quasi-three-dimensional analysis which is useful in many rock mechanics applications. Chapter 6 discusses joint elements and infinite elements with emphasis on modelling of discrete major discontinuities such as fault and shear zones. Chapter 7 gives various models of behaviour of jointed rock masses of varying complexities. Multi-laminate models described in this chapter have received wide acceptance in the rock mechanics community and many consulting engineers throughout the world now use these models. Chapter 8 gives examples of application of the finite element method to problems such as tunnels, slopes, powerhouse cavities etc. Some of the applications have been contributed by the engineers in consulting companies such as Lahmeyer International, and British Petroleum Research.

Chapters 9 and 10 deal with boundary element methods and their applications. Direct and indirect methods are discussed together with some advanced methods where *finite element shape functions are used for the boundary elements.* Applications of boundary element methods to problems of mining and tunnelling are given in Chapter 10. In this chapter it is also shown that for some problems a combination with the finite element method is beneficial. A future outlook of this fast developing method concludes the chapters on boundary elements.

Chapter 11 deals with the discrete element method applied to rigid bodies. Basic concepts of the method, rigid body dynamics and techniques of contact detection are discussed. The chapter also gives examples of applications. Chapter 12 describes the discrete element method for deformable bodies. A number of examples of applications are given.

The first eight chapters have been written by the first author, Chapters 9 and 10 by the second and Chapters 11 and 12 by the third author.

G. N. Pande
Swansea G. Beer
J. R. Williams

# Acknowledgements

The first author wishes to express his gratitude to Prof. O. C. Zienkiewicz from whom he has learnt many aspects of numerical methods. He is thankful to a number of individuals with whom he has had collaboration at some time or another during the past fifteen years. These include R. Dungar, C. M. Gerrard, W. Haas, D. J. Naylor, S. Pietruszczak, H. F. Schweiger, S. Semprich, W. Wittke, and K. G. Sharma. He also thanks K. Honish and D. L. Kern of Lahmeyer International, and N. C. Koutsebeloulis of British Petroleum Research Centre for their contributions to Chapter 8. Finally he sincerely thanks Geeta, his wife, who constantly encouraged him and Anil, his son, who put up with his absence on Saturdays. This book would not have been completed without polite and diplomatic reminders from Ellen Taylor of John Wiley & Sons, and recriminations from his co-authors. He thanks Ellen, Gernot and John.

The second author wishes to thank the Mining Research Department of Mount Isa Mines for supporting, over a number of years, research on boundary element methods and for supplying the example using the displacement discontinuity program NFOLD.

He is thankful to Max Lee of James Askew and Co, Australia for his determined effort of generating the mesh for the Mt Charlotte analysis in the days before graphical preprocessing was available. He wishes to acknowledge Kalgoorlie Mining Associates for permission to publish results of the analysis.

JKMRC (Julius Krutschnitt Mineral Research Centre) deserves credit for the analysis of excavation at the Elura Mine, Australia.

He is also thankful to PASMINCO Broken Hill Operations, in particular Dr Volker Tillman for supporting boundary element analyses for mine design and for permission to publish the results of such analyses. Stuart Lowe prepared the mesh for the example presented here using FEMCAD.

Finally, he would like to thank the Chief of the Division of Geomechanics, CSIRO for the support of this venture. Particular thanks are due to Dell Dunn of the Long Pocket Laboratories staff who without complaint typed the manuscript, including the sometimes complicated equations.

The third author would like to thank Grant Horking and Graham Mustoe for their knowledge and companionship during the great adventure of Applied Mechanics, Inc., and his wife Rita, who made even the bad times good.

# Note

Readers wishing to obtain software for numerical methods described in this book should write to:

**Dr G.N. Pande**
*Department of Civil Engineering*
*University College of Swansea*
*Engineering Building*
*Singleton Park*
*Swansea SA2 8PP*
*UK*

For Finite Element Software

**Dr G. Beer**
CSIRO
*Division of Geomechanics*
*Private Bag No. 3*
*PO, Indoorpilly*
*Queensland 4068*
*Australia*

For Boundary Element Software

**Dr J. R. Williams**
*15 Clark Court*
*Brookline*
*MA 02146*
*USA*

For Discrete Element Software

# 1 Introduction

## 1.1 INTRODUCTION

Man has used natural rock caverns for shelter from the times immemorial. Weapons made of rocks were used in the stone age when the art of shaping rocks as useful tools flourished. Rock was the primary building material until the beginning of this century. Rock Mechanics has evolved as an interdisciplinary subject combining mainly geology and mechanics during the early thirties of this century. It is perhaps worth noting that rock mechanics developed at a much slower pace than soil mechanics. The apparent reason is that rocks were generally considered more competent than soils and engineers faced a lesser number of problems relating to rock foundations or structures.

There are a variety of engineering applications where the knowledge of rock mechanics is vital. The modern rock mechanics encompasses the behaviour of rocks on a wide range of scales—from a fraction of a millimetre to many kilometres.

For example, the deformation of mineral grains in rocks and variations in the porosity of rock are used as indicators in petroleum and gas exploration. Behaviour of rock specimens is useful for the determination of its suitability as a construction material and in the studies requiring the knowledge of fracturing and fragmentation of rocks. Problems of foundations, slopes, underground openings, etc., all require knowledge of rock mechanics. Studies of crustal adjustments on the earth's surface and movement on fault planes are situations where many kilometres of rock strata have to be considered.

Rock mechanics is the branch of engineering concerned with the properties of rocks and application of this knowledge to engineering problems. It has gained considerable importance in recent years due to a number of factors. These are: (1) In spite of apparent economic recession in many countries, there is an increase in the constructional activity in the world as a whole. (2) The general awareness of the public about the environment has also led the engineers to accept more constraints in their choice and selection of the construction sites for large projects. (3) A few catastrophic failures like the Malpasset dam in 1959 have focused the attention of engineers on the importance of rock mechanics. (4) With the general enhancements in mining techniques, deeper deposits of minerals are gradually becoming economically viable. This development has given considerable impetus to the study of rock mechanics.

Yet another development in the last decade or so also has been responsible for considerable interest in rock mechanics. Large underground caverns have considerable advantages over above ground structures for the storage of fuels like liquid petroleum gas, and other materials. Sweden already has man-made caverns of capacities in excess of one million cubic metres. Many more are planned or are being constructed throughout Europe and the USA. Underground storage of radioactive nuclear waste is being considered in many countries. It is planned to place canisters of nuclear waste in specially constructed cavities. This has generated interest in the long term properties of rocks at high temperatures.

Quest for renewable energy sources has led to geothermal projects where hot rocks at depth can be used as an energy source by pumping water and extracting heat. Many geothermal projects are in the experimental stage and their success to some extent depends on cost effective methods of 'hydraulic fracturing' of rocks to produce a network of channels for the water or other fluids to flow.

Man started with living underground in caves. It may so happen that eventually the caves are indeed better suited than high-rise tower blocks. Only a few metres below the ground level the vagaries of weather die down and the temperatures hardly change, if at all. It is quite conceivable to have housing requiring almost zero input of net energy. Thus, living underground may become quite an attractive proposition. Although a lot remains to be done in research and development for houses underground to become generally acceptable, one could not be sure that this might be the order of things to come in the twenty-first century.

## 1.2 ROCKS VERSUS SOILS

Soil mechanics has been traditionally taught at the undergraduate level in all courses of civil engineering. Rock mechanics, on the other hand, finds place only in few courses at the undergraduate level and that too in courses on mining engineering. Since the readers are familiar with soils, it is perhaps useful to compare and contrast the essential features of rock behaviour with that of soils.

We restrict ourselves here only to those aspects which would influence the numerical modelling of the two materials. While dealing with problems relating to soils, it is usually assumed that they are isotropic, i.e. they have the same deformational characteristics in all directions. This assumption hardly applies to rocks which are generally anisotropic. A soil mass can be treated as a continuum while there are many situations in rocks where it will be inappropriate to treat rock as continuum. There has been a long debate among rock mechanicians whether to regard rock as continuum or discontinuum. While soils can be regarded as homogenous, rocks seldom are homogenous. They have joints, fissures, faults and in general discontinuities, the spacing of which can vary from a few centimetres to many metres.

The second important difference between rock masses and soil masses lies in the nature of *in situ* stresses. The ratio of *in situ* horizontal stress ($\sigma_x'$ ) to vertical stress ($\sigma_y'$ ) (effective stresses are considered) is denoted by $k_0$. If the soil is treated as a linearly elastic isotropic material, it can be shown that $k_0$ is given by

$$k_0 = \frac{\sigma_{x'}}{\sigma_{y'}} = \frac{\sigma_{z'}}{\sigma_{y'}} = \frac{\nu}{1 - \nu} \tag{1.1}$$

where $\nu$ is the Poisson's ratio and $\sigma_x'$ , $\sigma_y'$ and $\sigma_z'$ are components of stress, $\sigma_y'$ coinciding with the vertical axis. The value of $k_0$ is quite important in geotechnical designs and ranges from 0.1 to 1.00 for most soil situations.

However, for rocks a value of $k_0$ much in excess of 1.0 is quite common and is found to be of the order of 2–5 in many parts of Western Europe. This is because rock strata are capable of carrying tectonic stresses which have been generated during the formation of rocks and subsequent tectonic movements of the earth's crust.

The third important difference lies in the nature of engineering problems. Most soil mechanics problems are two dimensional, i.e. embankments, trenches, slopes, foundations, etc. On the other hand, most rock mechanics problems are three-dimensional and a two-dimensional representation is at best a crude approximation. For slopes and tunnels, even though the dimension perpendicular to the plane of analysis may be large, the orientation and geometry of joints is unlikely to be suitable for two-dimensional idealization. In many situations, the axes of *in situ* principal stresses will also not coincide with axes of the tunnel or excavation and a fully three-dimensional solution will be required.

## 1.3 NUMERICAL METHODS IN ROCK ENGINEERING

A number of numerical methods of analysis have been developed over the past three decades. They have become popular due to rapid advancements in computer technology and its availability to engineers. Before the advent of computers, the rock structures were designed largely based on rules of thumb, experience and a trial and error procedure. Rules of thumb are invariably based on the past experience of the designer. They usually tend to be oversafe and are basically applicable to the situations similar to the ones for which they were developed. Engineers of today are, many a time, faced with problems for which no past experience is available. It is also difficult to 'teach' past experience. The civil or mining engineering construction is usually a 'one off' situation every time. The increased consciousness amongst the public regarding safety and economy has led the engineers to seek more rational solutions to the problems in rock mechanics related to civil and mining engineering.

Analytical or 'closed form' solutions are available for simpler situations or can be developed. However, they can in most cases be developed assuming rock as a linear elastic material which is a very drastic simplification. Numerical methods have, therefore, become very popular for solving problems in rock mechanics.

A number of numerical methods are available for solving problems of load deformation. By the term load-deformation problem, we mean a problem in which a rock mass of arbitrary shape (this includes openings of arbitrary shape) is subjected to loads due to self weight, external forces, *in situ* stresses, temperature changes, fluid pressure, prestressing, dynamic forces, etc. and we seek to find the deformation, strains and stresses throughout the rock mass.

Younger readers having a perhaps more rigorous background of theoretical mechanics will recognize the load-deformation problem as a general boundary value problem. Figure 1.1 shows schematically a vertical section of a three-dimensional rock mass (of the shape of Gower† potato), deformation or possibly collapse, of which we are interested in finding out when it is subjected to a general set of loads. A solution of this problem, i.e. deformations, strains, stresses throughout the rock mass must satisfy the following:

(a) Equilibrium
(b) Strain compatibility
(c) Stress–strain relations of the rock mass
(d) Boundary considerations of tractions (forces) and deformations (conditions of fixity).

All numerical methods satisfy the conditions (a), (b) and (d) in almost a routine manner. Stress–strain relations for rock masses (c) is a wide subject in itself and perhaps most crucial on which the usefulness of the solution depends. We shall see this topic in detail in Chapter 7.

There are mainly three numerical methods which have been used in the problems of rock mechanics. They are (1) The Finite Element Method (FEM), (2) The Boundary Element Method (BEM) and (3) The Discrete Element Method (DEM).

† Gower is a region near Swansea which is designated an area of outstanding natural beauty in Britain.

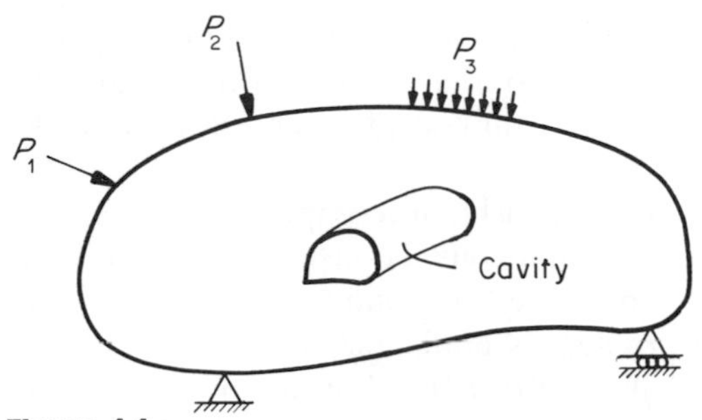


**Figure 1.1**
A body of an arbitrary shape subjected to a set of loads ($P_1$, $P_2$, $P_3$ etc.) and fixity conditions

All methods are approximate methods, i.e. we get an approximate solution to the problem. All methods have their advantages and disadvantages. FEM and BEM can be used for problems other than that of load-deformation, viz. seepage though rocks, consolidation due to pumping, heat conduction, etc.

As an introduction, we shall discuss the three methods in an overall and general manner so that an engineer can appreciate the essential differences in the various methods and their possible advantages and disadvantages.

### 1.3.1 The Finite Element Method

This is the most popular method in engineering sciences. It has been applied to a large number of problems in widely different fields. Its popularity, particularly for load-deformation problems, largely depends on the fact that it is very appealing to engineers. They are able to relate it to a large extent to the background of structural mechanics as the physical meaning of the steps of calculations are relatively transparent. A large part of the finite element program can remain as a 'black box' to the user and even a beginner can obtain interesting results with minimal effort. It does not mean that the method is easy and no experience is required in solving engineering problems of practical importance. On the contrary, to make use of the full potential of the method and interpret the results of the calculation, considerable expertise is required.

The method essentially involves dividing the body in smaller 'elements' of various shapes (triangles or rectangles in two-dimensional cases and tetrahedrons or 'bricks' in three-dimensional cases) held together at the 'nodes' which are corners of elements. The more the number of elements used to model the problem, the better approximation to the solution is obtained. Displacements at the nodes are treated as unknowns and are calculated. Stresses are calculated at one or more points inside each of the elements. Each element can have different material properties.

The major disadvantage of the method is that considerable effort is required in preparing data for a problem. This is particularly crucial in three-dimensional problems and has led to 'mesh generation' programs. These programs produce (to a large extent) the input data required for the Finite Element program. Still considerable effort is needed in 'starting up' the problem.

The method is also expensive in computer time.

A large set of simultaneous equations (several hundreds to several thousands) have to be solved to obtain solutions. The computer time goes up further if the problem is nonlinear, i.e. stress–strain relationship is not linear—which usually is the case. For a nonlinear problem, the sets of simultaneous equations are required to be solved a number of times.

Inspite of the above disadvantages, FEM has been extremely popular with geotechnical engineers. Its strength lies in its generality and flexibility to handle all types of loads, sequences of construction, installation of supports, etc.

Chapters 5 and 6 describe the method in more detail.

### 1.3.2 The Boundary Element Method

This method is becoming increasingly popular. It lacks the generality and flexibility of the FEM. It is not so easily understandable and requires a higher level of understanding of mathematical complexities.

In this method only the surface of the rock mass to be analysed needs to be discretized, i.e. divided into smaller patches. Thus, for two-dimensional situations line elements at the boundary represent the problem, while for fully three-dimensional problems, surface elements are required. The data preparation here is relatively simple. However, the computer program is not so transparent. Whenever there is a change of material properties, the surface defining the separation has to be discretized. Thus, if there are a number of layers of different materials, data preparation can still become complex. BEM appears to be a very efficient method for homogenous, linear elastic problems, particularly in three dimensions. For complex nonlinear material laws with a number of sets of materials, advantages of the method are considerably diminished. The matrices of equations arising in this method are not banded and symmetric as for FEM but are fully populated. Thus, though the number of equations to be solved is considerably reduced, computation time does not reduce in the same proportion.

The method makes use of certain closed form relations of what may be called 'elementary' problems. These solutions frequently contain trigonometric and logarithmic terms which slow down the computations.

Recognizing the advantages and disadvantages of the two methods, viz. FEM and BEM, many researchers have combined the two methods. This is coupled FEM/BEM method in which for a certain region (usually close to an opening or some other feature of interest) FE discretization is used, while for other regions BE discretization is adopted. BEM and coupled

FEM/BEM methods are discussed in Chapter 9. Examples of applications are given in Chapter 10.

### 1.3.3 The Discrete Element Method

This method is based on treating the rock mass as a discontinuum rather than continuum, as in the case of Finite Element and Boundary Element Methods. When loads are applied, the changes in contact forces are traced with time. In the earlier versions of the method rigid spherical balls or discs were used as elements. The equations of dynamic equilibrium for each element are repeatedly solved till the laws of contacts and boundary conditions are satisfied.

In the recent versions of the method, the elements can be of arbitrary shape as in the Finite Element Method. They can also be deformable. Complex constitutive laws can also be used. The elements can split up based on the assumed fracture criterion during the calculation process without any external intervention. Thus, the method is extremely powerful.

There are, however, several drawbacks. Firstly, the parameters required for the description of material behaviour are required to be chosen quite carefully in addition to certain additional parameters like the damping of the system. Computation time required to solve even simple problems can be excessive. At present, the method appears to be extremely useful in explaining the deformation and failure of rock masses qualitatively and provides a valuable insight into the failure mechanism. More experience is, however, required for it to be an acceptable tool of analysis in practice.

The Discrete Element Method has been presented in Chapter 11, together with examples of its applications in Chapter 12.

## Sources of information on numerical methods applied to rock mechanics

*Journals*

*Numerical and Analytical Methods in Geomechanics*, Chichester: Wiley & Sons (bi-monthly).

*Computers and Geotechnics*, Barking, Essex: Elsevier Applied Science (bi-monthly).

*Rock Mechanics and Mining Sciences*, Oxford: Pergamon Press (bi-monthly).

*Books*

Zienkiewicz, O. C. (1978) *The Finite Element Method in Engineering Science*, McGraw-Hill.

Naylor, D. J., and Pande, G. N. (1981) *Finite Element Method in Geotechnical Engineering*, Pineridge Press.

Crouch, S. L., and Starfield, A. M. (1983) 'Boundary Element Methods in Solid Mechanics', Allen & Unwin.

Brady, B. H. G., and Brown, E. T. (1985). *Rock Mechanics for Underground Mining*, London: Allen & Unwin.

*Proceedings*

*International Conferences on Numerical Methods in Geomechanics* (ICONMIG), recent proceedings are: Third, Aachen, 1979; Fourth, Edmonton, 1982; Fifth, Nagoya, Japan, 1985; Sixth, Innsbruck, Austria, 1988. All proceedings are published by A. A. Balkema, Rotterdam.

*International Symposia on Numerical Models in Geomechanics* (NUMOG); First, Zurich, 1982, A. A. Balkema, Rotterdam; Second, Ghent, 1986, M. Jackson, Redruth, Cornwall, England; Third, Niagara Falls, 1989, Elsevier Applied Science Publisher, Barking, Essex, England.

# 2 Elasto-Plasticity and Elasto-Viscoplasticity

## 2.1 INTRODUCTION

A number of powerful and versatile numerical methods are available to the engineer today. The advent of fast and friendly personal computers has made it possible to solve complex problems which, in the past, could not be solved without mainframe machines and without considerable effort in terms of time and resources. The speed of computation has increased by a factor of ten every five years in the past twenty years. This trend is likely to continue at least till the end of this century. It is therefore apparent that numerical techniques will play an increasingly crucial role in the solution of engineering problems.

The main task of the engineer in applying the numerical methods to the problems of rock mechanics is in quantifying the mechanical response of rocks, rock joints, rock masses and rock support systems. This requires the background knowledge of models of standard material behaviour such as elastic, elasto-plastic and elasto-viscoplastic. Theory of elasticity is covered in most undergraduate courses. The theory and its applications to rock mechanics are also covered in texts by Jeager and Cook (1976), Goodman (1980), Hoek and Brown (1980), etc. However, the theories of plasticity and viscoplasticity are not covered in most undergraduate courses.

In this chapter, the above theories will be covered. They will be dealt with at an elementary level. More importance is attached to the concepts involved than the mathematics, except where essential.

## 2.2 THEORY OF PLASTICITY

We shall discuss here only the mathematical theories. There are other theories known as 'physical theories' which are mainly of interest to material scientists. The mathematical theories lead to 'phenomenological laws' of material behaviour, i.e. the relationships between stress and strain on an average basis.

### 2.2.1 Historical background

It is interesting to know a little about the history of a subject before learning it. Theory of plasticity has been around for more than two centuries. Coulomb studied the plastic behaviour of soils in 1773. It was almost a hundred years later, in 1864, that Tresca presented his yield criterion. One of the earliest contributions on the mathematical theory of plasticity was by Saint Venant in 1870.

One of the significant contributions to the theory of plasticity was made by Von Mises who, in 1913, proposed his yield criterion. Prandtl, in 1920, produced the classical solutions of a plane strain punch. Levy, Hencky and Reuss were all actively involved in research relating to the plastic behaviour of metals. It was in 1931 that Nadai published his book on 'plasticity' and he looked at the geological materials as well. He applied the theory to the eruption of volcanoes and revised and updated his book in 1950. During and immediately after the second world war, the theory had an added impetus due to its obvious applications to the development of weaponry. Hill's book was published in 1950 and Prager and Hodge's book in 1951. Drucker and Prager applied the theory to soils in the fifties and sixties. Cam clay model, a model for the behaviour of saturated clays based on the theory of plasticity was developed in Cambridge by Roscoe, Schofield and Worth and their co-workers in the sixties.

In the last decade or so, due to the advances in numerical methods and computing, there has been an exponential growth in the models of the behaviour of engineering materials which are largely based on the theory of plasticity or viscoplasticity.

### 2.2.2 Uniaxial behaviour

For many materials, it is convenient to study the behaviour on a uniaxial sample such as a bar. On this sample, we have only one component of stress $\sigma$, which gives rise to a deformation of the bar. From the deformation, it is possible to calculate the strain ($\epsilon$). For a linearly elastic material, stress is proportional to strain as shown in Figure 2.1(a). If the stress is removed, the bar returns to its original length and the strain becomes zero. Thus on the stress–strain diagram of 2.1(a) the arrows indicate the loading and unloading behaviour. Obviously, elastic behaviour of engineering materials is observed only under a limited range of stresses and strains. Figure 2.1(b) shows the stress–strain relationship of an elastic/ideally plastic material.

The behaviour of the material is elastic up to a stress, $Y_0$, known as the 'yield stress'. At the stress $Y_0$, the strains in the bar become un-bounded,

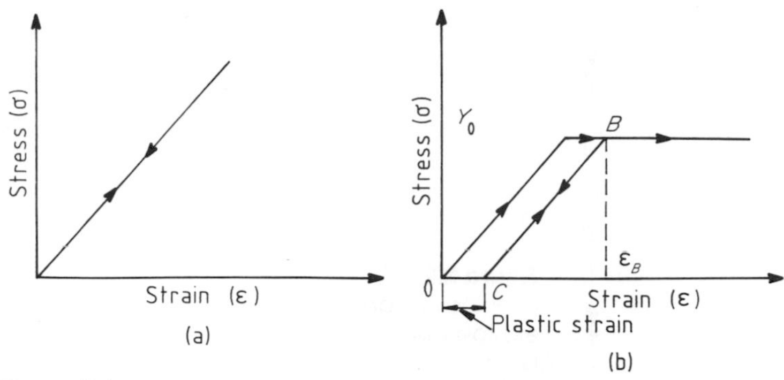


**Figure 2.1**
Idealized stress–strain response. (a) Elastic; (b) elastic/ideally plastic

i.e. the bar deforms indefinitely and thus fails. The stress $Y_0$, is therefore 'failure stress'. There is a difference between the terms 'yield stress' and 'failure stress' but we shall ignore this for the moment.

In general, there are two different types of tests that can be performed on any material. They are 'stress-controlled' or 'strain-controlled' tests. As the name suggests, in a stress-controlled test, strains are monitored for specified stresses or increments of stresses. On the other hand, in a strain-controlled test, strains (or displacements) are specified and stresses are monitored. For an elastic/ideally plastic material, stress can not exceed $Y_0$ in a stress controlled test. However, in a strain-controlled test, any strain can be specified. For example, if a strain $\epsilon_B$ is imposed on the material, the corresponding stress will still be $Y_0$. Suppose, if we reduce the strains, the stress follows the path indicated by $BC$. At the point $C$, the stress is zero but there is a strain; this strain is permanent in the sense that even if we leave the bar free (without any stress), the bar is in a deformed state and has 'plastic' strains.

On reloading the bar from $C$, we reach the point $B$, elastically, i.e. if we unloaded just before the point $B$, the stress path follows $BC$. In a deformation process, all the elastic strains can be recovered by unloading while plastic strain can not be recovered by unloading. There is no loss of energy in an elastic material while there is a dissipation of energy in causing the plastic strains.

### 2.2.3 Rheological analogue of an elasto-plastic material

Rheological analogues are commonly used in indicating the behaviour of materials. These are formed by mechanical components such as 'springs', 'dashports' and 'sliders'. A spring represents elastic behaviour. The rheological model of elastoplastic behaviour is represented by a spring and a slider in series as shown in Figure 2.2. The slider represents the plastic behaviour as it does not move unless the stress on it exceeds a threshold

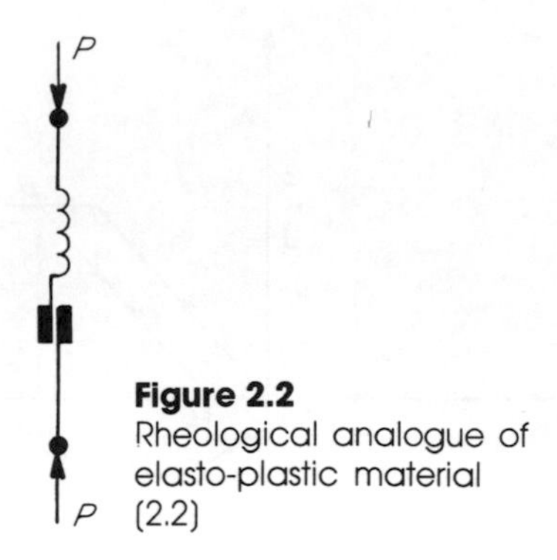


**Figure 2.2**
Rheological analogue of elasto-plastic material (2.2)

limit. Furthermore, the slider does not return to its original position if the stress is removed.

The rheological model also suggests that total strains ($\epsilon$) in the material are composed of two parts, viz. elastic strains ($\epsilon^e$) in the spring (which are recoverable on unloading) and plastic strains ($\epsilon^p$) in the slider (which are not recoverable on unloading). Thus

$$\epsilon = \epsilon^e + \epsilon^p \tag{2.1}$$

Equation (2.1) represents scalar quantities—strains in the uniaxial direction. It is generalized by assuming that it holds good for all components of strain in a multiaxial situation, i.e.

$$\boldsymbol{\epsilon} = \boldsymbol{\epsilon}^e + \boldsymbol{\epsilon}^p \tag{2.2}$$

where $\boldsymbol{\epsilon} = [\epsilon_x, \ \epsilon_y, \ \epsilon_z, \ \gamma_{xy}, \ \gamma_{yz}, \ \gamma_{zx}]^T$ represents the vector of strain components. $T$ represents a transpose and superscripts $e$ and $p$ denote elastic and plastic components, respectively. Equation (2.2) thus represents six equations; additivity of elastic and plastic parts to give the 'total' for each of the six components of strains.

Equation (2.2) is an assumption and is commonly referred to as 'additivity postulate'.

### 2.2.4 A heuristic problem

We shall first discuss a heuristic problem shown in Figure 2.3. Although it is a simple problem it has all the features of a complex problem and is useful in explaining a number of concepts related to the theory of plasticity.

Two parallel bars of ideally elastic plastic material have the same cross-section (1 mm$^2$), same length (1000 mm) and same Young's modulus (10 000 kN/mm$^2$). They differ only in the value of yield stress which are 200 kN/mm$^2$ for bar $A$ and 40 kN/mm$^2$ for bar $B$. They are connected together

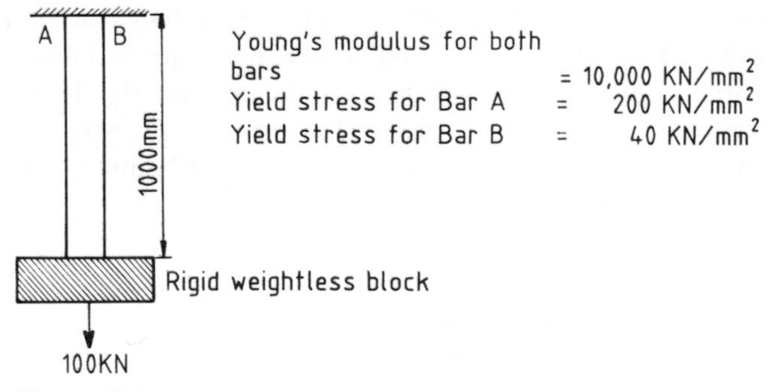


**Figure 2.3**
The two-bar problem

through a weightless rigid block. We would like to answer the following questions:

(a) What is the elongation of the assembly when a load of 100 kN is applied on the rigid block?
(b) What is the elongation when the load is removed? Are there any stresses in the bar?
(c) What happens when a load of 250 kN is applied on the assembly?

*Solution*

(a) Since bar $B$ can take a maximum stress of 40 kN/mm$^2$, the maximum load it can carry is 40 kN.

Load taken by bar $A$ = 100 − 40 = 60 kN. (This is the requirement of equilibrium.) The bar $A$ remains elastic and its deformation can be calculated as

$$\delta_A = \frac{60 \times 1000}{1 \times 10\,000} = 6 \text{ mm}$$

Since the two bars are connected together their deformations ($\delta$) are equal, i.e.

$$\delta_A = \delta_B$$

Hence

$$\delta_B = 6 \text{ mm}$$

Note that the elastic deformation of this bar is 4 mm and 2 mm is the plastic deformation.

(b) If the load is removed, bar $A$ tries to come back to its original length but is prevented by the bar $B$ which has a permanent deformation of 2 mm. Simple equilibrium calculations indicate that the length of the assembly will be 1001 mm with bar $A$ having a tensile stress of 10 kN/mm$^2$ and bar $B$ having a compressive stress of 10 kN/mm$^2$. Equilibrium is satisfied with zero external load.

(c) Maximum force which can be taken by bar $A$ = 200 kN.
Maximum force which can be taken by bar $B$ = 40 kN.
Maximum force which can be taken by the assembly is thus 240 kN. If a force of 250 kN is applied both bars will extend indefinitely and the assembly will collapse. In other words, unconstrained flow will take place.

From the above example the following observations/conclusions can be made for a general boundary value problem:

(1) Whenever there is 'plastic flow' in an elasto–plastic body, the body does not return to its original configuration on removal of the load/loads.
(2) In indeterminate structures (2D and 3D problems are indeterminate), on removal of load (which has caused plastic flow) a 'residual' state of stress is set up. 'Residual stresses' are self equilibriating.
(3) A body collapses if plastic zone progresses in such a way as to intersect the boundary of the body and transform it into a mechanism (see Figure 2.4).
(4) The stresses/deformations/strains under a given set of loads are dependent on the past loading history. This is obvious from the fact that if the bar assembly was reloaded after unloading, its response is not the same as on the first loading.
(5) The collapse load (240 kN) is not influenced by elastic parameters of the body.

### 2.2.5 Basic ingredients of the theory of plasticity

There are four basic ingredients of the theory of plasticity. These are:

(a) *Stress–strain relationship prior to yielding:* It is generally assumed that the behaviour is linear elastic prior to yielding but it is not essential and alternative assumptions such as nonlinear elasticity can as well be made.
(b) *Yield criterion:* yield criterion is a generalization of the yield point in the uniaxial case. It is a scalar function of stresses, i.e.

$$F(\boldsymbol{\sigma}) = 0 \tag{2.3}$$

where $F$ represents a function and $\boldsymbol{\sigma}$ is the vector of stresses. The components of the stress vector will change with a change of coordinate

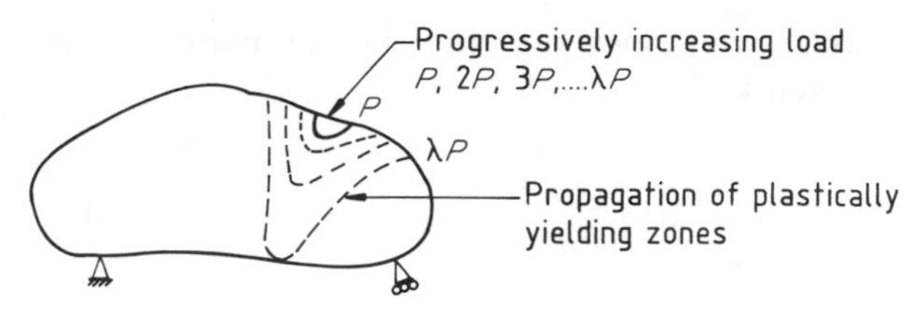


**Figure 2.4**
A three-dimensional body of elasto-plastic material subjected to progressively increasing load

axes to say $\boldsymbol{\sigma}^*$. But the yielding of material should not depend on the choice of the axes of reference if the material is isotropic, i.e.

$$F(\boldsymbol{\sigma}) = F(\boldsymbol{\sigma}^*) = 0 \tag{2.4}$$

This is possible only if $F$ is a function of stress invariants and not stress components. Principal stresses are a set of stress invariants of the stress tensor. Thus, equation (2.3) for isotropic materials will be

$$F(\sigma_1, \sigma_2, \sigma_3) = 0 \tag{2.5}$$

where $\sigma_1$, $\sigma_2$, $\sigma_3$ are the principal stresses.

A yield function represented by equation (2.5) can be plotted in the stress space as a surface which is known as 'yield surface'. The yield surface denotes that any stress situation which lies within the surface, i.e.

$$F(\sigma_1, \sigma_2, \sigma_3) < 0 \tag{2.6}$$

represents elastic behaviour, while any stress situation which lies on the surface, i.e.

$$F(\sigma_1, \sigma_2, \sigma_3) = 0 \tag{2.7}$$

represents plastic behaviour. Stress states outside the yield surface, i.e.

$$F(\sigma_1, \sigma_2, \sigma_3) > 0$$

are not admissible by the theory. This can be verified in the uniaxial diagram 2.1(b) where stress greater than the value of yield stress $Y_0$ is not possible.

(c) *Flow rule:* In the uniaxial example, it was assumed that the plastic strain increments took place in the direction of the stress applied. In a general multidimensional situation, all components of plastic strain will exist.

The flow rule determines the direction of plastic strain increment vector. It is given by:

$$\mathrm{d}\boldsymbol{\epsilon}^p = \mathrm{d}\lambda \frac{\partial Q(\boldsymbol{\sigma})}{\partial \boldsymbol{\sigma}} \tag{2.8}$$

where

$$\mathrm{d}\boldsymbol{\epsilon}^p = [\mathrm{d}\epsilon_x^p, \mathrm{d}\epsilon_y^p, \mathrm{d}\epsilon_z^p, \mathrm{d}\gamma_{xy}^p, \mathrm{d}\gamma_{yz}^p, \mathrm{d}\gamma_{zx}^p]^T$$

is the vector of plastic strain increments ('d' is used to denote infinitesimal increments of various quantities throughout), $\mathrm{d}\lambda$ is a positive proportionality constant (it is not a material parameter) and $Q(\boldsymbol{\sigma})$ = constant is a 'plastic potential function' which is explained below.

Plastic potential function is a scalar function of stresses. Equation (2.8) implies that the plastic strain increment vector $\mathrm{d}\boldsymbol{\sigma}^p$ is normal to the plastic potential surface plotted in the stress space. Thus the direction of plastic strain increment is related to the plastic potential function. It is possible to assume that plastic potential function is the same as the yield function ($Q \equiv F$) and in that case the plastic strain increment vector will be associated with the yield surface. A flow rule of this type is called an 'associated' flow rule and $\mathrm{d}\boldsymbol{\epsilon}^p$ is given by

$$\mathrm{d}\boldsymbol{\epsilon}^p = \mathrm{d}\lambda \frac{\partial F(\boldsymbol{\sigma})}{\partial \boldsymbol{\sigma}} \tag{2.9}$$

When $Q \not\equiv F$ ($Q$ is not the same as $F$), we have a 'nonassociated' flow rule. The flow rule has important implications in the behaviour of geological materials like rocks and soils. The yield functions of these materials are dependent on the mean stress and an associated flow rule leads to dilatancy (plastic volumetric strains on shearing) which are generally far in excess to those measured in experiments. By adopting a suitable plastic potential function it is possible to reduce dilatancy and model it to match closer to the experimentally observed values.

In Figure 2.5, principal plastic strain increments ($\mathrm{d}\epsilon_1^p, \mathrm{d}\epsilon_2^p, \mathrm{d}\epsilon_3^p$) are plotted on the corresponding principal stress axes. Yield and plastic potential surfaces are indicated. The plastic potential function has to pass through the current stress point on the yield surface. This is the reason for writing plastic potential surface as $Q(\boldsymbol{\sigma})$ = constant. On this diagram the direction (but not the magnitude) of plastic strain increments can be marked as a vector normal to the plastic potential function.

(d) *Hardening rule:* The hardening rule specifies as to how the yield function changes, if at all, with the accumulation of plastic strains. If it does not change at all we have ideally elasto-plastic material or perfectly plastic

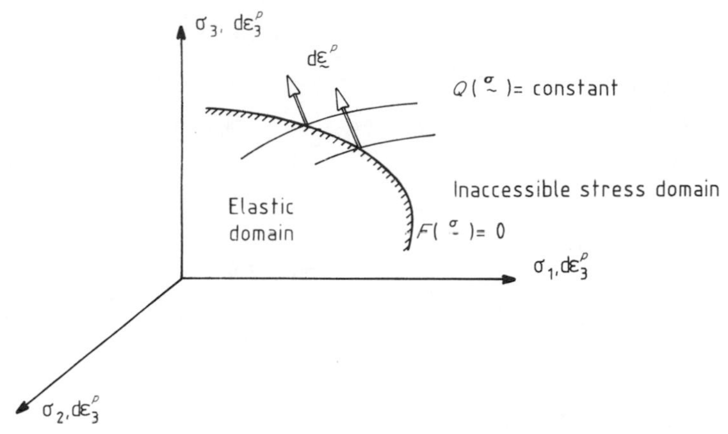


**Figure 2.5**
Yield and plastic potential surfaces in principal stress space

material. If the yield function changes with plastic strains, then it should be a function of stresses as well as plastic strains, i.e.

$$F = F(\boldsymbol{\sigma}, \boldsymbol{\epsilon}^p) = 0 \tag{2.10}$$

Using the arguments given in (b) above, regarding isotropy of the material behaviour, the yield function should be a function of invariants of plastic strain tensor and not individual components. It is convenient to define a hardening parameter ($k$) as a scalar function of principal plastic strains ($\epsilon_1^p$, $\epsilon_2^p$, $\epsilon_3^p$) or some other invariant measure of plastic strains.

$$k = k(\epsilon_1^p, \epsilon_2^p, \epsilon_3^p) \tag{2.11}$$

This leads to equation (2.10) being modified to

$$F = F(\sigma_1, \sigma_2, \sigma_3, k) = 0 \tag{2.12}$$

Figure 2.6 shows an idealized uniaxial stress strain curve of strain hardening material. After yielding at a stress of $Y_0$, the stress and strain increase. Finally at a stress of $Y_f$, the strains increase indefinitely. If the material is unloaded at any point (say $D$) in the strain hardening region, permanent strain denoted by $OE$ takes place. On reloading, the behaviour is elastic along $ED$ as if the yield point had shifted to point $D$ (stress $Y$) due to plastic strain $OE$. In the multiaxial case, the yield surface moves from its original position to a new position corresponding to the stress as shown in Figure 2.7.

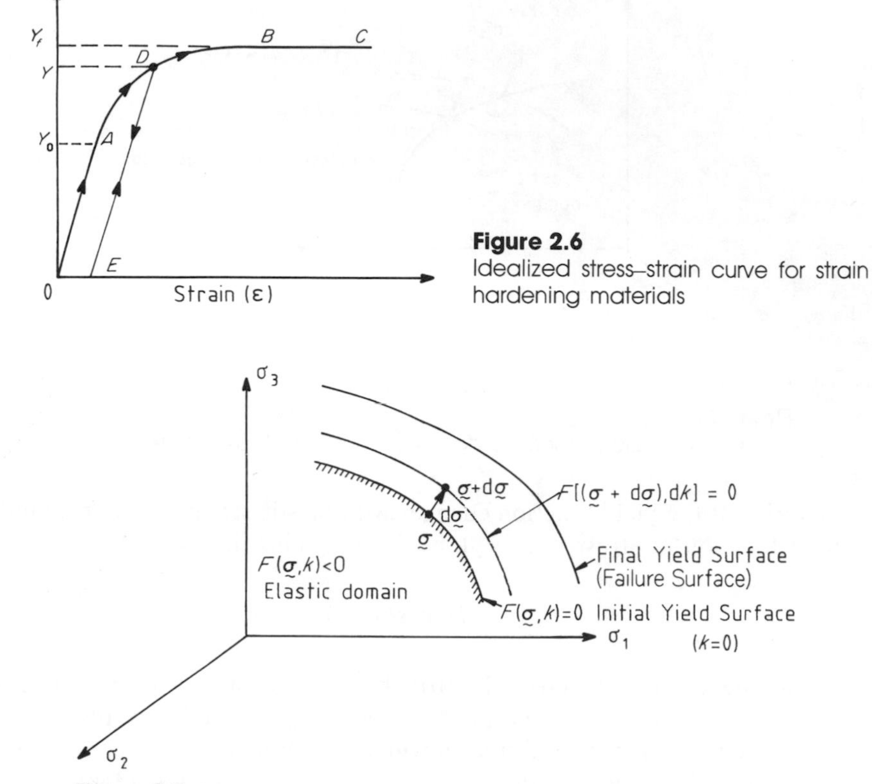


**Figure 2.6**
Idealized stress–strain curve for strain hardening materials

**Figure 2.7**
Initial and final yield surfaces for strain hardening materials—consistent evolution of yield surface

### 2.2.6 Complete stress–strain relation of an elasto-plastic material

The ultimate aim of the theory of plasticity is to establish a relationship between stresses and strains in the form

$$\mathrm{d}\boldsymbol{\sigma} = D_{ep}\, \mathrm{d}\boldsymbol{\epsilon} \tag{2.13}$$

where $D_{ep}$ is a $6 \times 6$ matrix whose coefficients are stress (or strain) dependent. $\mathrm{d}\boldsymbol{\sigma}$ and $\mathrm{d}\boldsymbol{\epsilon}$ are increments of stress and total strain vectors respectively. The concepts developed in Section 2.2.5 above are unified in this section to establish a relationship of the form of equation (2.13).

The additivity postulate gives us

$$\boldsymbol{\epsilon} = \boldsymbol{\epsilon}^e + \boldsymbol{\epsilon}^p \qquad \text{(2.2) bis}$$

This can be written for increments of strain as

$$d\boldsymbol{\epsilon} = d\boldsymbol{\epsilon}^e + d\boldsymbol{\epsilon}^p \tag{2.14}$$

Increments of stress are related to the increments of elastic strains through the usual elasticity matrix ($D_e$) containing terms dependent on elastic moduli and Poisson's ratio. Thus,

$$d\boldsymbol{\sigma} = D_e \, d\boldsymbol{\epsilon}^e \tag{2.15a}$$

or

$$d\boldsymbol{\epsilon}^e = D_e^{-1} \, d\boldsymbol{\sigma} \tag{2.15b}$$

Substituting (2.14) in (2.15a) gives

$$d\boldsymbol{\sigma} = D_e[d\boldsymbol{\epsilon} - d\boldsymbol{\epsilon}^p] \tag{2.17}$$

The flow rule is stated by equation (2.8)

$$d\boldsymbol{\epsilon}^p = d\lambda \frac{\partial Q(\boldsymbol{\sigma})}{\partial \boldsymbol{\sigma}} \tag{2.9) bis}$$

or

$$d\boldsymbol{\epsilon}^p = d\lambda \frac{\partial Q}{\partial \boldsymbol{\sigma}} \tag{2.18}$$

where $Q(\boldsymbol{\sigma})$ has been abbreviated to $Q$.

The yield function is written as

$$F(\boldsymbol{\sigma}, k) = 0 \tag{2.19}$$

Note that in equation (2.19) $\boldsymbol{\sigma}$ has been kept as the chosen variable and not principal stresses or the invariants of stresses. This is to keep the derivation general, to be applicable to isotropic as well as anisotropic materials.

We now need to introduce an important concept known as 'consistency condition'. In Figure 2.7 if the stress changes from $\boldsymbol{\sigma}$ to $\boldsymbol{\sigma} + d\boldsymbol{\sigma}$, leading to certain plastic strain increments $d\boldsymbol{\epsilon}^p$, which in turn leads to change in the value of $k$ to $k + dk$. The new yield surface,

$$F(\boldsymbol{\sigma} + d\boldsymbol{\sigma}, k + dk) = 0 \tag{2.20}$$

must be such that the stress $\boldsymbol{\sigma} + d\boldsymbol{\sigma}$ lies on it. In other words, the evolution of the yield surface should be consistent with the stress increment $d\sigma$.

Subtracting equation (2.19) from equation (2.20) means

$$F(\boldsymbol{\sigma} + \mathrm{d}\boldsymbol{\sigma}, k + \mathrm{d}k) - F(\boldsymbol{\sigma}, k) = 0 \tag{2.21}$$

or

$$\mathrm{d}F(\boldsymbol{\sigma}, k) = 0$$

which, on expanding by the chain rule, can be written as

$$\left(\frac{\partial F}{\partial \boldsymbol{\sigma}}\right)^T \mathrm{d}\boldsymbol{\sigma} + \frac{\partial F}{\partial k}\left(\frac{\mathrm{d}k}{\partial \boldsymbol{\epsilon}^p}\right)^T \mathrm{d}\boldsymbol{\epsilon}^p = 0 \tag{2.22}$$

where $F(\boldsymbol{\sigma}, k)$ has been abbreviated to $F$.

Equation (2.22) is the consistency condition. It ensures that, when plastic straining takes place in the material, the stress always lies on the current yield surface. It is noted that equation (2.22) represents only one equation as each term of the equation is a scalar quantity.

Substituting equation (2.18) in (2.17) leads to

$$\mathrm{d}\boldsymbol{\sigma} = D_e\left[\mathrm{d}\boldsymbol{\epsilon} - \mathrm{d}\lambda\frac{\partial Q}{\partial \boldsymbol{\sigma}}\right] \tag{2.23a}$$

which on rearranging of terms can be written as

$$\mathrm{d}\boldsymbol{\epsilon} = D_e^{-1}\,\mathrm{d}\boldsymbol{\sigma} + \mathrm{d}\lambda\frac{\partial Q}{\partial \boldsymbol{\sigma}} \tag{2.23b}$$

Substituting (2.18) in equation (2.22) leads to

$$\left(\frac{\partial F}{\partial \boldsymbol{\sigma}}\right)^T \mathrm{d}\boldsymbol{\sigma} + \frac{\partial F}{\partial k}\cdot\left(\frac{\partial k}{\partial \boldsymbol{\epsilon}^p}\right)^T \mathrm{d}\lambda\frac{\partial Q}{\partial \boldsymbol{\sigma}} = 0 \tag{2.24}$$

Defining a hardening function ($H$) as

$$H = -\frac{\partial F}{\partial k}\left(\frac{\partial k}{\partial \boldsymbol{\epsilon}^p}\right)^T\frac{\partial Q}{\partial \boldsymbol{\sigma}}, \tag{2.25}$$

substitution of which in equation (2.24) leads to

$$\left(\frac{\partial F}{\partial \boldsymbol{\sigma}}\right)^T \mathrm{d}\boldsymbol{\sigma} - H\,\mathrm{d}\lambda = 0 \tag{2.26}$$

Premultiplying equation (2.23b) with $(\partial F/\partial \boldsymbol{\sigma})^T D_e$ and noting that $D_e \cdot D_e^{-1}$ is an identity matrix, we have

$$\left(\frac{\partial F}{\partial \boldsymbol{\sigma}}\right)^T D_e \, \mathrm{d}\boldsymbol{\epsilon} = \left(\frac{\partial F}{\partial \boldsymbol{\sigma}}\right)^T \mathrm{d}\boldsymbol{\sigma} + \mathrm{d}\lambda \left(\frac{\partial F}{\partial \boldsymbol{\sigma}}\right)^T D_e \frac{\partial Q}{\partial \boldsymbol{\sigma}} \tag{2.27}$$

Substituting $(\partial F/\partial \boldsymbol{\sigma})^T \, \mathrm{d}\boldsymbol{\sigma}$ from equation (2.26) leads to $\mathrm{d}\lambda$ being calculated as

$$\mathrm{d}\lambda = \frac{1}{\beta} \cdot \left(\frac{\partial F}{\partial \boldsymbol{\sigma}}\right)^T \mathrm{D}_e \, \mathrm{d}\boldsymbol{\epsilon} \tag{2.28}$$

where

$$\beta = H + \left(\frac{\partial F}{\partial \boldsymbol{\sigma}}\right)^T D_e \frac{\partial Q}{\partial \boldsymbol{\sigma}}$$

Substituting $\mathrm{d}\lambda$ from equation (2.28) in equation (2.23a) gives the required relationship between $\mathrm{d}\boldsymbol{\sigma}$ and $\mathrm{d}\boldsymbol{\epsilon}$.

The matrix $D_{ep}$ is then:

$$D_{ep} = D_e - \frac{1}{\beta} \cdot D_e \frac{\partial Q}{\partial \boldsymbol{\sigma}} \cdot \left(\frac{\partial F}{\partial \boldsymbol{\sigma}}\right)^T D_e^T \tag{2.29}$$

### 2.2.7 Some comments on $D_{ep}$ matrix

$D_{ep}$ matrix derived in the previous section plays a crucial role in nonlinear analysis. It is the difference of two matrices; elastic matrix ($D_e$) and what may be termed as plastic matrix ($D_p$) represented by

$$D_p = \frac{1}{\beta} D_e \frac{\partial Q}{\partial \boldsymbol{\sigma}} \left(\frac{\partial F}{\partial \boldsymbol{\sigma}}\right)^T D_e^T \tag{2.30}$$

It follows from Maxwell–Betti reciprocal theorem (Timishenko and Goodier, 1951) that $D_e$ matrix is always symmetric (even for anisotropic materials). However, in general, only when $Q \equiv F$, $D_p$ takes the form

$$D_p = \frac{1}{\beta} \cdot D_e \frac{\partial F}{\partial \boldsymbol{\sigma}} \left(\frac{\partial F}{\partial \boldsymbol{\sigma}}\right)^T D_e^T \tag{2.31}$$

and is symmetric. In other words the elastoplastic matrix is symmetric only for an associated flow rule. This has important consequences in the strategies of computation. A nonsymmetric $D_{ep}$ leads to nonsymmetric stiffness equations in the Finite and Boundary Element Methods and solution of these equations becomes more expensive in terms of computer time.

Pande and Pietruszezak (1986) have suggested a method of keeping the $D_{ep}$ matrix symmetric even when the flow rule is nonassociated, thus avoiding the expense in computation time.

In the case of ideal plasticity, the yield function is independent of $k$ and consequently,

$$\frac{\partial F}{\partial k} = 0$$

which leads to

$$\beta = \left(\frac{\partial F}{\partial \boldsymbol{\sigma}}\right)^T D_e \left(\frac{\partial Q}{\partial \boldsymbol{\sigma}}\right) \tag{2.32}$$

It is noted that $\beta$ is not zero even for ideal plasticity.

## 2.3 THEORY OF ELASTO-VISCOPLASTICITY

Theory of elasto-plasticity discussed in the last section does not have a 'time' element. In other words, the plastic strains are generated instantaneously. In the theory of elasto-viscoplasticity, plastic strains, called viscoplastic strains, are assumed to accrue with time. This is quite an appealing aspect of theory and is in line with experience and practice in rock mechanics. After all, monitoring of rock structures by instrumentation is based on the assumption that displacements, strains and stresses will tend to a steady state with time. It is not surprising that theory has been extensively used in the analysis of jointed rock masses.

### 2.3.1 Rheological analogue of elasto-viscoplasticity

Figure 2.8 shows the rheological analogue of an elasto-viscoplastic material. It consists of a spring which is in series with a viscoplastic unit formed by a slider and a viscous dashpot in parallel. The instantaneous response of the material is purely elastic. If the slider fails, the plastic strains symbolized by it take place with time as the viscous dashpot does not respond instantaneously. It is inherent in the theory that stresses can temporarily go outside the yield surface. In this situation ($F > 0$), a viscoplastic strain rate ($\dot{\boldsymbol{\epsilon}}^{vp}$) is developed which is specified by a flow equation. The concepts of yield function, flow rule and hardening rule developed for the theory of elasto-plasticity are valid for the theory of elasto-viscoplasticity as well.

### 2.3.2 Flow equation

The equation which specifies rates of viscoplastic strains can be written as:

$$\dot{\boldsymbol{\epsilon}}^{vp} = \mu \langle \Phi(F) \rangle \frac{\partial Q}{\partial \boldsymbol{\sigma}} \tag{2.33}$$

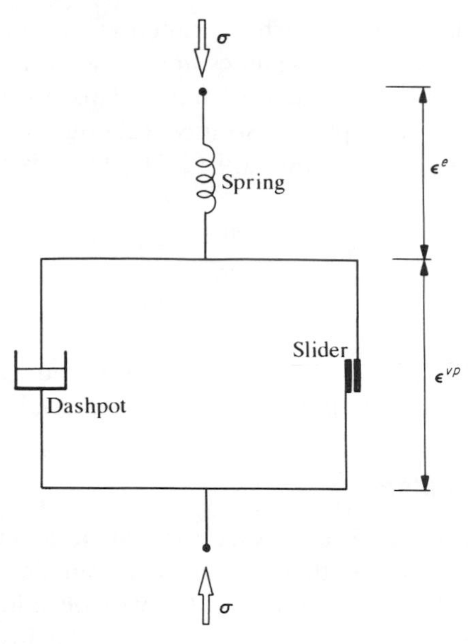


**Figure 2.8**
Rheological analogue of elasto-viscoplasticity

where $\mu$ is the so called 'fluidity parameter'. It is a material constant involving units of 1/time. It has to be identified from experimental tests. $\Phi(F)$ denotes a monotonic function of the yield function ($F$). The meaning of brackets $\langle\ \rangle$ is as follows:

$$\langle\ \rangle = (\cdot) \quad \text{if } F > 0$$

$$\langle\cdot\rangle = 0 \quad \text{if F} \leqslant 0$$

Equation (2.33) represents two sets of equations. If $F \leqslant 0$, equation (2.33) reduces to

$$\dot{\boldsymbol{\epsilon}}^{vp} = 0$$

implying that rates of viscoplastic strains are zero when the stress plotted in stress space lies inside or on the yield surface. Thus, viscoplastic strain rates arise only when stress situation is outside the yield surface. $Q$ is the plastic potential function discussed in Section 2.2.5.

Equation (2.33) is based on the theory of viscoplasticity proposed by Perzyna (1966). There are other theories but they have not been found useful in rock mechanics. Even in equation (2.33), $\Phi(F)$ is simply chosen as $F$ in most cases. Thus, equation (2.33) is simplified to

$$\dot{\boldsymbol{\epsilon}}^{vp} = \mu \langle F \rangle \frac{\partial Q}{\partial \boldsymbol{\sigma}} \tag{2.34}$$

The fluidity parameter $\mu$ can be a function of time and/or viscoplastic strains. Again in most rock mechanics applications, it has been adopted as a constant with an arbitrary value of unity. This distorts the time scale of the development of viscoplastic strains although the sequence of their development is maintained. Equation (2.34), thus, is further simplified to:

$$\dot{\boldsymbol{\epsilon}}^{vp} = F \cdot \frac{\partial Q}{\partial \boldsymbol{\sigma}} \quad \text{if } F > 0 \tag{2.35}$$

$$\dot{\boldsymbol{\epsilon}}^{vp} = 0 \quad \text{if F} \leqslant 0$$

We shall now solve the two-bar problem of Section 2.2.4 again assuming that the bars are made of elasto-viscoplastic material.

### 2.3.3 Two-bar problem

The details of the problem are exactly the same as given in Section 2.2.4 except for the assumption that the bars are made of elasto-viscoplastic material. Simplified flow equation (2.35) will be adopted for solving the problem. At time $T = 0$ (reckoning it from the instant of loading) the response of the system is elastic. Therefore,

$$\sigma_A = \sigma_B = 50 \text{ kN/mm}^2$$

$$\delta_A = \delta_B = 5 \text{ mm}$$

Yield function for

$$\text{bar } A(F_A) = \sigma_A - \sigma_y^A = 50 - 200 = -150 \text{ kN/mm}^2.$$

Since $F_A$ is negative, $\dot{\epsilon}_A^p\text{p} = 0$.
Yield function for

$$\text{bar } B(F_B) = \sigma_B - \sigma_y^B = 50 - 40 = 10 \text{ kN/mm}^2$$

Since $F_B$ is positive, nonzero plastic strain rates arise according to equation (2.35).

For one-dimensional case, as in this problem, equation (2.35) further reduces to

$$\dot{\epsilon}^p = F \quad \text{if } F > 0$$

and

$$\dot{\epsilon}^p = 0 \quad \text{if } F \leqslant 0$$

Thus

$$\dot{\epsilon}_B^p = 10$$
$$\dot{\epsilon}_A^p = 0$$

where $\dot{\epsilon}_A^p$ and $\dot{\epsilon}_B^p$ are rates of viscoplastic strains of bars $A$ and $B$, respectively.

Consider the situation at time $T = 0.0001$ units ($\Delta T = 0.0001$). The increment of viscoplastic strains ($\Delta\epsilon^p$) during this time interval is given by

$$\Delta\epsilon_B^p = \dot{\epsilon}_B^p \times 0.0001$$
$$\Delta\epsilon_B^p = 10 \times 0.0001 = 0.001$$

Therefore, plastic elongation of bar $B = 0.001 \times 1000$ mm $= 1.0$ mm.

Figure 2.9(a) shows the deformed shape of the assembly of two bars at time $t = 0$. Equal and opposite forces ($P$) will act on the bar (Figure 2.9(b)) to bring the assembly to a compatible length of 1005.5 mm at time $T = 0.0001$ units with stresses in bars as

$$\sigma_A = 55 \text{ kN/mm}^2$$
$$\sigma_B = 45 \text{ kN/mm}^2$$

Now, if we take another time step of length $\Delta T = 0.0001$ units so that $T = 0.0002$ units, then the rates of viscoplastic strains for the two bars are given by:

$$\dot{\epsilon}_A^p = 0$$
$$\dot{\epsilon}_B^p = F_B = 45 - 40 = 5$$
$$\Delta\epsilon_B^p = 5 \times 0.0001$$
$$= 0.0005$$

Using the same arguments as in the previous time step

Plastic elongation of bar $B$ (if it was free) $= 0.0005 \times 1000 = 0.5$ mm
Length of bars at $T = 0.0002$ $\quad 1005.5 + 0.25 = 1005.75$
At $T = 0.0002$ units, $\quad \sigma_A = 55 + 2.5 = 57.5$ kN/mm$^2$
$\sigma_B = 45 - 2.5 = 42.5$ kN/mm$^2$

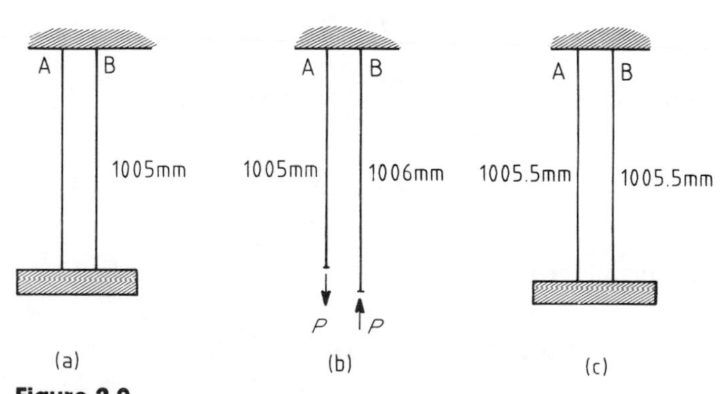


**Figure 2.9**
Two-bar assembly. (a) At time $T = 0$; (b) lengths if bars were free; (c) at time $T = 0.001$ units

If further time steps are continued, after some time $\sigma_A \rightarrow 60$ kN/mm$^2$, $\sigma_B \rightarrow 40$ kN/mm$^2$ and the length of the assembly (1) approaches 1006 mm.

It is noted that the final answer of the problem is the same as obtained in Section 2.2.4, albeit the final deformed shape and stresses are attained after $T \simeq 0.001$ units of time.

In the above problem, the value of $\mu$ was chosen arbitrarily as unity. If a real value was assigned, true history of deformation with time can be traced. In this way the theory of elasto-viscoplasticity can be used to model time dependent behaviour of rocks and rock masses.

### 2.3.4 Complete equations of elasto-viscoplasticity

We can now write the complete equations of elasto-viscoplasticity, identifying how increments of strain are related to increments of stress, or vice versa.

The additivity postulate of strains written in incremental form gives

$$\mathrm{d}\boldsymbol{\epsilon} = \mathrm{d}\boldsymbol{\epsilon}^{e} + \mathrm{d}\boldsymbol{\epsilon}^{vp} \tag{2.36}$$

where $\mathrm{d}\boldsymbol{\epsilon}$, $\mathrm{d}\boldsymbol{\epsilon}^{e}$ and $\mathrm{d}\boldsymbol{\epsilon}^{vp}$ denote increments of total, elastic and viscoplastic strains.

Assuming that $\dot{\boldsymbol{\epsilon}}^{vp}$ is constant over an infinitesimal time increment of $\mathrm{d}t$, we can write

$$\mathrm{d}\boldsymbol{\epsilon}^{vp} = \dot{\boldsymbol{\epsilon}}^{vp}\,\mathrm{d}t \tag{2.37}$$

The flow equation in the simplified form is given by

$$\dot{\boldsymbol{\epsilon}}^{vp} = \mu\,\langle F\rangle \frac{\partial Q}{\partial \boldsymbol{\sigma}} \tag{2.34 bis}$$

Stresses at any stage are related to elastic strains. Therefore,

$$\mathrm{d}\epsilon^{e} = D_e^{-1}\,\mathrm{d}\boldsymbol{\sigma} \tag{2.15b bis}$$

Substituting equations (2.37), (2.34) and (2.15b) in equation (2.36) leads to

$$\mathrm{d}\boldsymbol{\epsilon} = D_e^{-1}\,\mathrm{d}\boldsymbol{\sigma} + \mu\,\langle F\rangle \frac{\partial Q}{\partial \boldsymbol{\sigma}} \cdot \mathrm{d}t \tag{2.38}$$

which on rearranging gives

$$\mathrm{d}\boldsymbol{\sigma} = D_e\,\mathrm{d}\boldsymbol{\epsilon} - \mu\,\langle F\rangle\, D_e \frac{\partial Q}{\partial \boldsymbol{\sigma}}\,\mathrm{d}t \tag{2.39}$$

### 2.3.5 Comparison of theories of elasto-plasticity and elasto-viscoplasticity

Since elasto-viscoplasticity is a relatively new theory and many readers may be less familiar with it, it is useful to compare and contrast its features with the more well established theory of elasto-plasticity. Table 2.1 shows differences and common features of the two theories.

### 2.3.6 Advantages of the theory of elasto-viscoplasticity over the theory of elasto-plasticity

There are many advantages in adopting elasto-viscoplasticity as a model for simulation of mechanical behaviour of rocks and rock masses.

Firstly, the theory is capable of taking into account the time dependence behaviour of rocks and rock masses. Even if the relevant material parameter '$\mu$' is not identified from experiments and a pseudo viscoplastic model with an arbitrarily chosen value of $\mu$ is adopted, a sequence of stress/deformation history is obtained. This allows the study of problems like delayed placement of lining and other rock support systems like rock bolts, etc.

Secondly, the numerical implementation of elasto-viscoplastic model is simpler than that of elasto-plastic model. Moreover, for a predefined load,

**Table 2.1**
Comparison of the theories of elastoplasticity and elasto-viscoplasticity

| Feature | Elasto-plasticity | Elasto-viscoplasticity |
|---|---|---|
| Time dependence | This is time independent theory. Elastic and plastic strains develop instantaneously. | This is a time dependent theory. Initial response is elastic, followed by plastic straining with time. |
| Stress situations | $F < 0$ elastic<br>$F = 0$ plastic strains take place<br>$F > 0$ inadmissible | $F \leqslant 0$ elastic<br>$F > 0$ admissible and time dependent plastic strains take place |
| Load increments | All loads must be applied in small increments since $F$ $>$ is not admissible. | Loads need not be applied in small increments. However, if viscoplasticity is used to obtain elasto-plastic solution, load increments must be small as in theory of elasto-plasticity. In general the rate of loading influences the response. |
| Other details | The concepts of yield function, flow rule and hardening rule apply to both theories. | |

even if the solution does not converge, it provides useful answers as the correct solution is being progressively and sequentially approached.

Finally, nonassociated flow rules and strain-softening features which frequently have to be modelled, are handled quite simply in elasto-viscoplasticity.

## REFERENCES


Goodman, R. E. (1980). *Introduction to Rock Mechanics*. New York: J. Wiley & Sons.

Hoek, E., and Brown, E. T. (1980). *Underground Excavation in Rock*, London Institution of Mining and Metallurgy.

Hill, R. (1971). *The Mathematical Theory of Plasticity*, Oxford University Press.

Jaeger, J., and Cook, N. G. W. (1976). *Fundamentals of Rock Mechanics*, 2nd edn, London: Chapman & Hall.

Martin, J. B. (1975). *Plasticity: Fundamental and General Results*, The MIT Press.

Pande, G. N., and Pietruszczak, S. (1986). 'Symmetric tangential stiffness for nonassociated flow rule.' *Computers and Geotechnics*, **2**, 89–99.

Perzyna, P. (1966). 'Fundamental problems in viscoplasticity.' *Advances in Applied Mechanics*, **9**, 243–377.

Timishenko, S., and Goodier, J. N. (1951). *Theory of Elasticity*, McGraw-Hill.

Salençon, J. (1985). *Application of the Theory of Plasticity in Soil Mechanics*, J. Wiley & Sons.

# 3 Mechanical Properties of Intact Rocks and Rock Joints—Physical Behaviour and Numerical Modelling

## 3.1 INTRODUCTION

Strength and deformability of rocks and rock joints have been the subject of numerous experimental investigations. The subject is also covered by numerous texts on rock mechanics, such as Jeager and Cook (1976), Goodman (1980), Hoek and Brown (1980), Farmer (1983), Brady and Brown (1985), etc.

The aim of this chapter is to introduce to the reader the general concepts so far as they relate to the application of numerical methods. We shall first characterize typical behaviour of intact rocks followed by mathematical models based on the theories of elasto-plasticity and/or elasto-viscoplasticity. We shall repeat the same sequence for rock joints. The mathematical models for intact rocks and rock joints are the 'building blocks' for the final constitutive relations (stress–strain laws) for the jointed rock mass which are presented in Chapter 7.

## 3.2 INTACT ROCK VERSUS ROCK MASS

At this stage, it is essential to distinguish between the terms 'intact rock' and 'rock mass'. For the purpose of design of engineering structures, 'rock mass' implies rock on a scale of tens to hundreds of metres. Rock mass, in general, includes joints, fissures, shear bands and discontinuities. Intact rock, on the other hand, implies rock on a scale of a few centimetres which are free from joints, fissures and other discontinuities. A specimen of rock tested in the laboratory is, thus, a specimen of 'intact rock'. It can be argued that even a few centimetres long specimens contain discontinuities, but these discontinuities are on a micromechanical level and can be ignored.

Behaviour of intact rocks is quite conveniently studied in the laboratory. For the study of rock mass, representative samples several metres in size will be required.

## 3.3 DEFORMABILITY AND STRENGTH OF INTACT ROCKS

### 3.3.1 Deformability of intact rocks

Most intact rocks behave elastically under the range of stresses experienced in engineering and mining structures. However, in many cases, the behaviour is nonlinear. Figure 3.1 shows typical uniaxial stress–strain ($\sigma$, $\epsilon$) curves obtained from experiments on cylindrical specimens. The arrows indicate the path of loading/unloading. From these tests it is possible to obtain an approximate value of modulus of elasticity of intact rock ($E_i$). If the stress strain curve is non linear (type II), it is customary to adopt a value of tangential elastic modulus at a stress equal to half the failure stress ($f_{cu}$) shown in the figure. It is assumed that $E_i$ in tension is the same as in compression.

For a linear isotropic elasticity model to fit the description of rocks, one more parameter, viz. Poisson's ratio ($\mu_i$) is also required. This can be obtained if measurements of radial strain on the cylindrical samples are also made. Theoretically, if linear isotropic elasticity model is applicable, the sample should deform uniformly. If it does not, average value of radial strain will have to be adopted to compute Poisson's ratio. It is noted that $\mu_i$ lies between 0 and 0.5.

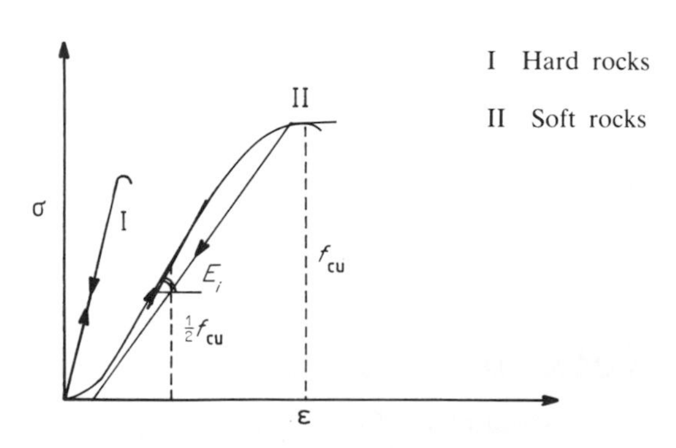


**Figure 3.1**
Typical uniaxial stress–strain curves of intact rocks in compression

If the intact rock is sedimentary, it would have bedding planes giving it a laminated structure. Due to these features, the stress strain curve in the directions parallel and perpendicular to the laminates will be different. This will lead to different values of $E_i$ and $\mu_i$ in the two directions. These types of rock can be described by the transversely isotropic material behaviour which is characterized by five material constants: $E_i^s$, $E_i^n$, $\mu_i^s$, $\mu_i^n$ and $G$, where superscripts $s$ and $n$ refer to the properties parallel and normal to the direction of the bedding planes. $G$ is the shear modulus and is independent of $E_i$ and $\mu_i$ in the parallel and perpendicular directions.

The ratio $E_i^n/E_i^s$ is usually less than 2. There are many problems associated with obtaining reliable values of $E_i^n$ and $E_i^s$. The anisotropy of elastic properties affects the stress distribution around underground openings. It also influences the design of strengthening measures such as rock bolts.

### 3.3.2 Strength of intact rocks

Uniaxial compression tests are the simplest to perform on intact rocks. Crushing strength of rocks can be readily obtained from these tests. Other tests used to determine the strength of rocks are: uniaxial tensile, flexural and shear strength tests. Triaxial tests are also performed but they are less common. It is not easy to test rocks in direct uniaxial tension. Indirect tests such as 'bending tests' and 'splitting tests' are devised to obtain tensile strength of intact rocks.

The strength of most intact rocks is highly variable. The numerical analyst has to keep this in mind and adopt values that will be representative. This requires experience. He has also to adopt a numerical model based on the type of strength data available.

If the framework of strain hardening plasticity is to be adopted, information on yield stresses as well as failure stresses is required in different types of tests. On the other hand, if the assumptions of ideal plasticity are made, failure strengths only are required. In a more sophisticated description of intact rock, post failure stress–strain response in various tests will also be required. In the absence of detailed information a variety of assumptions are made. For example, residual strength after failure can be assumed as a fraction of the peak strength or as an extreme case, assumed to be zero. Figure 3.2 shows typical stress–strain curves for hard rocks like granite and soft rocks like mudstone and coal.

## 3.4 CONSTITUTIVE MODELS FOR INTACT ROCK

For a complete description of behaviour of rock in the framework of plasticity theory, we need to prescribe the basic ingredients discussed in Section 2.2.5. They are: (a) stress–strain relations prior to failure; (b) failure function; (c) flow rule. If a strain-hardening model is to be adopted, yield

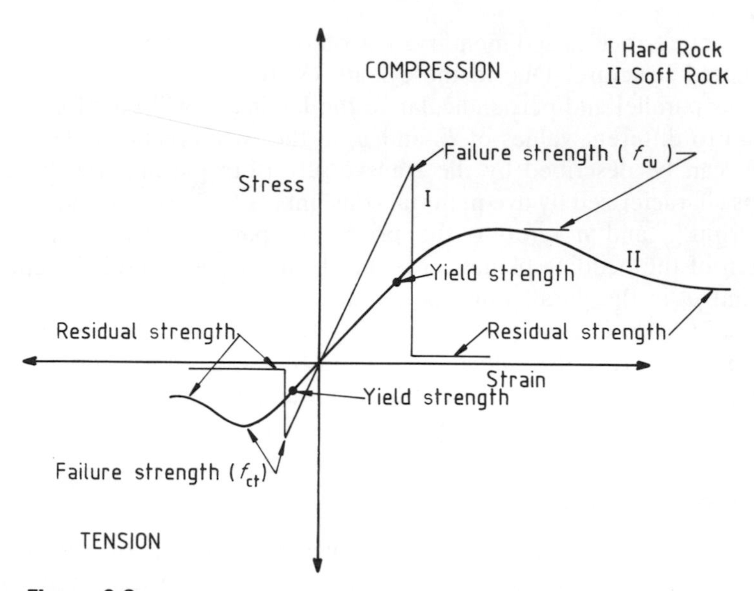


**Figure 3.2**
Typical stress–strain curves for hard and soft rocks

function and hardening functions should also be prescribed. Additionally, a strain-softening function has to be specified if post failure behaviour is also to be modelled.

Plastic yielding/failure of intact rocks is not important in surface excavations, tunnels and cavities which are located at shallow depths. Here, the sets of rock joints which have considerably lower strength, yield-inhibiting yield and failure of intact rock. In deep mines, however, high stresses combined with the existence of relatively fewer joint sets, can cause yielding and failure of intact rock.

Before we discuss the specific forms of various yield/failure functions which can be used in numerical methods for problems in rock engineering, it is important to distinguish between 'total' and 'effective' stresses.

Rocks are porous material and in many practical situations, are saturated with water.

In saturated rocks, the total stress applied on the rock is partly supported by the intergranular stress known as the 'effective stress' and partly by an alround stress in the pores known as the pore water pressure. It is the effective stress that governs yielding, failure and dilatancy of rocks. The effective stress principle states

$$\boldsymbol{\sigma} = \boldsymbol{\sigma}' + \mathbf{m}u \tag{3.1}$$

where

$\boldsymbol{\sigma}^T = (\sigma_x, \sigma_y, \sigma_z, \tau_{xy}, \tau_{yz}, \tau_{zx}]$ is the vector of total stresses,

$$\boldsymbol{\sigma}'^T = [\sigma'_x, \sigma'_y, \sigma'_z, \tau'_{xy}, \tau'_{yz}, \tau'_{zx}] \quad \text{is the vector of effective stresses,}$$
$$\mathbf{m}^T = [1, 1, 1, 0, 0, 0]$$

and $u$ is the pore water pressure (a scalar quantity).

A compression positive criterion has been used in writing equation (3.1) and will be used throughout the book. In the following, all constitutive laws are written in terms of effective stresses ($\boldsymbol{\sigma}'$), but prime has been dropped from the notations for convenience.

### 3.4.1 Mohr–Coulomb yield/failure criterion

This well-known criterion has been extensively used in engineering practice. It states that failure takes place if the magnitude of shear stress ($\tau$) on the failure plane is equal to the value given by the following relationship:

$$|\tau| = \sigma_n \tan \phi_0 + C_0 \tag{3.2}$$

where $|\ |$ denotes the absolute value, $\sigma_n$ is the normal stress on the failure plane and $\phi_0$ and $C_0$ are material constants for intact rock. Suppose, a number of triaxial tests are carried out on intact rock specimens at different cell pressures ($\sigma_3$). If failure takes place at stresses $\sigma_1$, $\sigma_2 = \sigma_3$, where $\sigma_1$ is the major principal stress (it is the vertical axial stress in the triaxial test), a number of Mohr circles can be drawn (Figure 3.3), one corresponding to each test. If a straight line envelope is drawn to the Mohr circles, $\phi_0$ and $C_0$ can be identified. It is noted that the failure envelope is a straight line is only an assumption of this theory and in practice a curved envelope may be obtained, particularly if the range of cell pressures at which tests are carried out is large.

Equation (3.2) can be written in the form of a failure function ($F$) as

$$F = |\tau| - \sigma_n \tan \phi_0 - C_0 - 0 \tag{3.3}$$

Equation (3.3) is not convenient for numerical methods as it involves first finding out the orientation of the failure plane. To obtain a more convenient form, we look at the relationship between principal stresses at failure. From the geometry of Mohr circles, it can be shown (see most books on soil mechanics, e.g. Scott (1980)) that equation (3.3) can be written as

$$F = \sigma_1(1 - \sin \phi_0) - \sigma_3(1 + \sin \phi_0) - 2C_0 \cos \phi_0 = 0 \tag{3.4}$$

It is noted from equation (3.4) that Mohr–Coulomb failure criterion does not depend on intermediate principal stress ($\sigma_2$).

Computing principal stress in three dimensions is also not trivial. A convenient way to handle equation (3.4) is to write it in terms of stress invariants—$\sigma_{\mathbf{m}}$, $\bar{\sigma}$ and $\theta$.

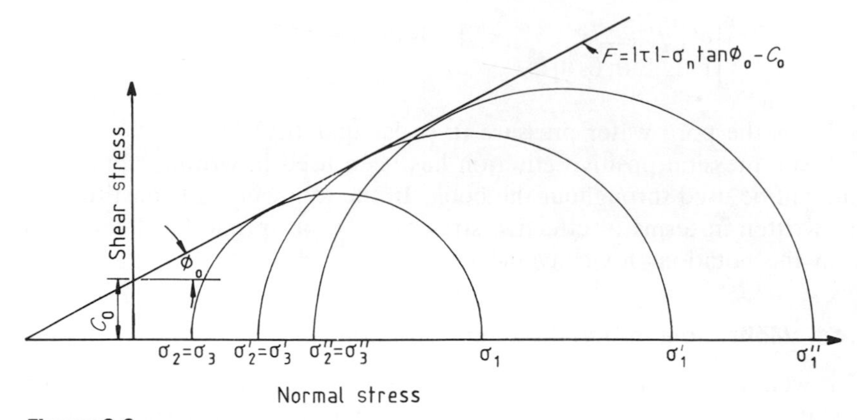


**Figure 3.3**
Mohr circles for stresses in triaxial tests at failure

In all numerical methods applied to rock engineering, stresses are worked out with reference to a fixed frame of axes. If different coordinate axes are chosen, the stress components will also be different. Stress invariants are quantities which are independent of the choice of reference axes. Readers not conversant with these should refer to Appendix I.

A convenient set of stress invariants, in terms $\boldsymbol{\sigma}$ are

$$\sigma_m = \frac{\sigma_x + \sigma_y + \sigma_z}{3} \tag{3.5}$$

$\sigma_m$ is referred to as 'mean stress'

$$\bar{\sigma}^2 = \tfrac{1}{2}[(\sigma_x - \sigma_y)^2 + (\sigma_y - \sigma_z)^2 + (\sigma_z - \sigma_x)^2 + 6(\tau_{xy}^2 + \tau_{yz}^2 + \tau_{zx}^2)] \tag{3.6}$$

$$\theta = \tfrac{1}{3}\sin^{-1}\left(\frac{3\sqrt{3}J_3}{2\bar{\sigma}^3}\right) \tag{3.7}$$

where $J_3$ is the third invariant of deviatoric stresses given by

$$J_3 = \begin{vmatrix} \sigma_x - \sigma_m & \tau_{xy} & \tau_{xz} \\ & \sigma_y - \sigma_m & \tau_{yz} \\ & & \sigma_z - \sigma_m \end{vmatrix}$$

and $|\,|$ denotes determinant of the matrix.

$\theta$ is referred to as Lode's angle after W. Lode (1926).

In terms of stress invariants, equation (3.4) can be written as

$$F = \bar{\sigma}\left(\cos\theta + \frac{\sin\theta\sin\phi_0}{\sqrt{3}}\right) - \sigma_m\sin\phi_0 - C_0\cos\phi_0 = 0 \tag{3.8}$$

Equation (3.8) is the form of Mohr–Coulomb criterion used in many rock mechanics programs. It involves two parameters, $C_0$ and $\phi_0$. These parameters can also be identified from uniaxial compressive strength ($f_{cu}$) and uniaxial tensile strength ($f_{ct}$).

Drawing Mohr circles for these two conditions, and taking an envelopping straight line as failure envelope, the parameters '$C_0'$' and '$\phi_0$' can be eliminated and $f_{cu}$ and $f_{ct}$ used instead. Figure 3.4 shows the Mohr circles, and using elementary geometry it can be shown that

$$\sin \phi_0 = \frac{1 - \alpha_z}{1 + \alpha_z} \tag{3.9}$$

and

$$C_0 = \frac{f_{ct}}{2\sqrt{\alpha_z}} = \frac{\sqrt{\alpha_z}}{2} f_{cu} \tag{3.10}$$

where

$$\alpha_z = \frac{f_{ct}}{f_{cu}}$$

Substituting equations (3.9) and (3.10) in equation (3.9) leads to

$$F = -\sigma_m \left(\frac{1 - \alpha_z}{1 + \alpha_z}\right) + \bar{\sigma}\left(\cos \theta \frac{\sin \theta}{\sqrt{3}} \frac{1 - \alpha_z}{1 + \alpha_z}\right) - \frac{\sqrt{\alpha_z}}{2} f_{cu} \frac{2\sqrt{\alpha_z}}{1 + \alpha_z} = 0 \tag{3.11}$$

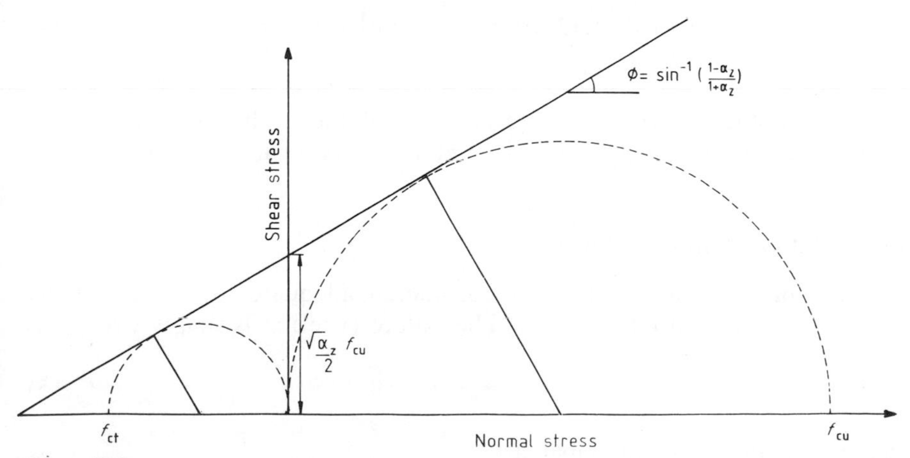


**Figure 3.4**
Relationship between $C_0$, $\phi_0$ and $f_{cu}$, $\alpha_z$

which can be rewritten as

$$F = a\sigma_m + \bar{\sigma}\left(b \cos \theta - \frac{a}{\sqrt{3}} \sin \theta\right) - f_{cu} = 0 \qquad (3.12)$$

where

$$a = -\frac{1 - \alpha_z}{a_z}$$

$$b = \frac{1 + \alpha_z}{\alpha_z}.$$

Equation (3.12) is an alternative representation of the Mohr–Coulomb failure condition using commonly available physical parameters like uniaxial tensile and compressive strengths.

In many practical situations, even two parameters ($f_{ct}$, $f_{cu}$) may not be available. Many rock classification systems use what is now well established 'point load' tests. International Society for Rock Mechanics (ISRM) has issued (1985) guidelines for determining point load strength. In these tests a piece of intact rock is crushed between two standard platens. The test is a version of Brazilian tensile strength test (Goodman, 1980). Using standard procedures set by ISRM, a value of strength corresponding to an intact rock sample having a diameter of 50 mm ($I_{S(50)}$) is calculated. $f_{ct}$ and $f_{cu}$ have been correlated with $I_{S(50)}$ for isotropic rocks and are given by

$$f_{ct} \simeq 1.25\, I_{S(50)}, \qquad f_{cu} \simeq 22\, I_{S(50)} \qquad (3.13)$$

Substitution of these equations in equations (3.11) or (3.12) leads to

$$F = -\sigma_m + \bar{\sigma}\left(1.12 \cos \theta + \frac{1}{\sqrt{3}} \sin \theta\right) - 1.33\, I_{S(50)} = 0 \qquad (3.14)$$

which is another alternative representation of the Mohr–Coulomb failure criterion using a single physical parameter, viz. point load strength.

### 3.4.2 Limited tension criterion

In many practical situations, it is considered adequate to describe intact rock as a limited tension material. The failure criterion is simply written as

$$F = -\sigma_3 - f_{ct} = 0 \qquad (3.15)$$

where $\sigma_3$ is the minor principal stress.

In terms of stress invariants used earlier, equation (3.15) can be written as

$$F = -\sigma_m + \bar{\sigma}\left(\sin\theta \cdot \frac{1}{\sqrt{3}} - \cos\theta\right) - f_{ct} = 0 \qquad (3.16)$$

### 3.4.3 Hoek–Brown criterion

Hoek and Brown (1980) have studied the published experimental results of a wide variety of rocks. On the basis of this study, they proposed the following condition for failure

$$\sigma_1 = \sigma_3 + (m f_{cu}\sigma_3 + s f_{cu}^2)^{1/2} \qquad (3.17)$$

where $m$ and $s$ are constants and depend on the properties of rock as well as the degree of fragmentation. Equation (3.17) can be written as a failure criterion in the following form:

$$F = \sigma_1 - \sigma_3 - (m f_{cu}\sigma_3 + s f_{cu}^2)^{1/2} = 0 \qquad (3.18)$$

A few points are worth noting with reference to this criterion:

(a) The failure, as with Mohr–Coulomb criterion, does not depend on the intermediate principal stress.
(b) The failure envelope is curved unlike Mohr–Coulomb which is a straight line. Figure 3.5 shows a plot of Hoek–Brown criterion for $f_{cu}$ = 100 Mpa, $s$ = 0.004 and $m$ = 1.7.
(c) The criterion, in a way relates to a rock mass as it takes into account the rock mass quality data.

In terms of stress invariants, equation (3.18) can be written as

$$F = 4\bar{\sigma}^2\cos\theta + m\bar{\sigma}f_{cu}\left(\cos\theta + \frac{\sin\theta}{\sqrt{3}}\right) - m\sigma_m f_{cu} - s f_{cu}^2 = 0 \qquad (3.19)$$

Experience with the use of this criterion in numerical methods is limited at present.

### 3.4.4 Other failure criteria

There are a number of other failure criteria proposed by various investigators. Notable amongst these are Griffith's criterion (1921) and its extension by Murrell (1963). In addition, there are a number of theories regarding

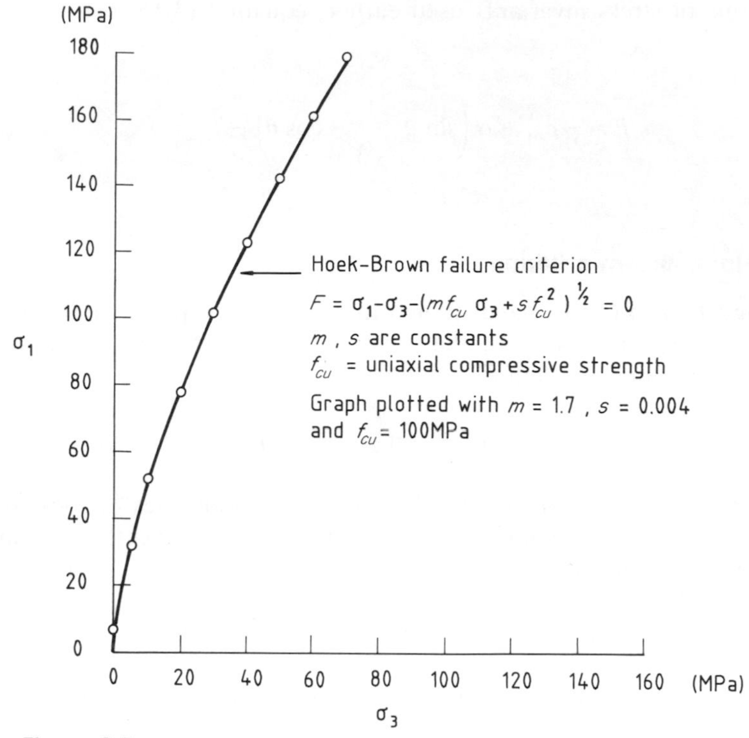


**Figure 3.5**
Relationship between $\sigma_1$ and $\sigma_3$ represented by Hoek–Brown failure criterion

anisotropic strength of intact rocks like shales, slates, etc. As discussed earlier, it is unlikely for intact rock in a jointed rock mass to fail except in very 'constrained' situations. In view of the overall complexity of the problems encountered in practice and likelihood of large variations in the properties of rock, it is hardly worthwhile to look for a 'perfect' failure criterion.

### 3.4.5 Flow rule

For describing the behaviour of intact rocks, we need to define a flow rule. In the absence of detailed information and also due to the fact that intact rock may never yield, precise knowledge of flow rule is unlikely to affect the final results of the numerical analysis. In view of this, it is almost universally assumed that the flow rule is 'associated', i.e. $Q \equiv F$.

### 3.4.6 Hardening/softening rules

If an ideally plastic behaviour of intact rock is assumed, then there is no need for a hardening rule. In this case plastic yield and failure have the same meaning and the same parameters, for example, $C_0$, $\phi_0$ define the yield and failure function.

Strain softening and post peak behaviour may have considerable influence in many problems. Simplest assumptions are those of brittle behaviour, i.e. sudden loss of strength, which are applicable for hard rocks. For soft rocks, residual strength parameters and a proper softening rule defining the post peak behaviour is required. A simplification in analysis can be achieved by assuming elastic, ideally plastic behaviour with residual strength parameters. This may, in many cases, lead to an uneconomic design. A full study of rock behaviour on stiff machines capable of capturing post peak behaviour accurately becomes important in such cases.

A word of warning here.

Numerical results obtained using a softening rule are not 'objective' meaning that they depend on the discretization of the problem into Finite/Boundary elements. Different Finite Element meshes may lead to different answers. Bazant and Oh (1983) have shown (in the context of concrete structures) that the slope of the strain-softening curve has to be related to the size of the mesh. The subject of strain softening is presently under active research.

## 3.5 DEFORMABILITY AND STRENGTH OF ROCK JOINTS

The behaviour of jointed rock masses is dominated by the behaviour of rock joints. A number of experimental studies have been conducted to understand the behaviour of natural as well as artificial joints. Artificial joints (mostly in plaster of Paris) have been studied mainly as they have the advantage in their reproducibility. Notable amongst these experimental studies are due to Patton (1966), Goodman (1970), Barton and Choubey (1977), Ladanyi and Archambault (1970), Bandis *et al.* (1981), Sun *et al.* (1985) and Yoshinaka and Yamabe (1986). Due to the large expense and time involved in experimental studies, coupled with the demands of highly accurate techniques for measurements, a number of investigators have recently attempted to study the behaviour of joints by analytical models.

Swan (1983, 1985) and Sun (1985) have used the concepts of mathematically defined shapes, sizes and density of asperites to predict the deformability and strength of rock joints and compared them with measured values on some real joints. Gerrard (1985) presents a very comprehensive study of the formulations for the mechanical properties of rock joints.

Yet another aspect of the studies has been the phenomenological models proposed by various investigators for describing deformability and strength of rock joints. More studies have been concerned with strength than with deformability. In the following paragraphs, we shall discuss some of these models of rock joint behaviour.

Joints in rock masses can vary widely in their physical state and mechanical behaviour. They can be fresh or weathered, asperites matching or mismatching, filled or unfilled with gouge material. Mechanical behaviour of filled joints is governed by the properties of the infilling gouge material, if the thickness of the joint is greater than twice the average height of asperites. This is so because the failure takes place through the infilling material and the characteristics of the joint walls play an insignificant role.

### 3.5.1 Nonlinear elastic model of rock joints

If a compressive normal stress ($\sigma_n$) is applied on a rock joint, it would cause its closure by a certain amount, say $\delta_n$. Figure 3.6(a) shows a typical relationship between $\sigma_n$ and $\delta_n$. The slope of the curve in Figure 3.5 gives

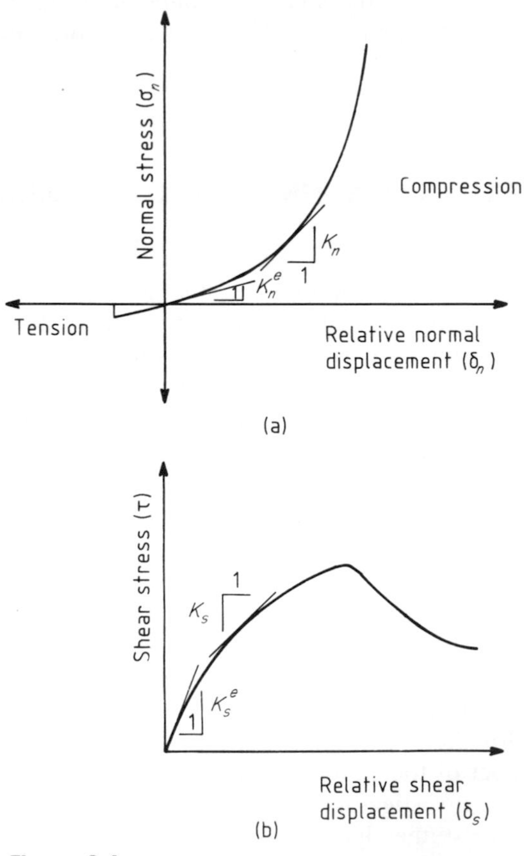


**Figure 3.6**
Typical stress–relative displacement relationships.
(a) $\sigma_n$ versus $\delta_n$; (b) $\tau$ versus $\delta_s$

the 'tangential normal stiffness' ($K_n$) of the joint and, at any stress level, is defined as

$$K_n = \frac{\Delta\sigma_n}{\Delta\delta_n} \tag{3.20}$$

where $\Delta$ denotes an increment.

It is noted that $K_n$ is small when $\sigma_n$ is small but rapidly builds up as the joint closes. There is actually a maximum limit of joint closure and $\sigma_n \rightarrow \infty$ as this limit ($\delta_{nc}$) is reached. Goodman *et al.* (1968) proposed a hyperbolic relationship given by

$$\sigma_n = \frac{\alpha}{\delta_{nc} - \delta_n} + \beta \tag{3.21}$$

where $\alpha$ and $\beta$ are constants defining the shape of the hyperbolic curve between $\sigma_n$ and $\delta_n$. Differentiating equation (3.21), we obtain the expression for $K_n$ as

$$K_n = \frac{\mathrm{d}\sigma_n}{\mathrm{d}\delta_n} = \frac{\alpha}{(\delta_{nc} - \delta_n)^2} \tag{3.22}$$

which can be rewritten as

$$K_n = \frac{(\sigma_n - \beta)^2}{\alpha} \tag{3.23}$$

It is noted that equation (3.23) is valid for compressive normal stresses only.

It is usual to assume that joints do not offer any reliable resistance to tensile normal stresses implying $K_n = 0$ if $\sigma_n$ is tensile.

If a shear stress ($\tau$) is applied on the joint, there will be a relative shear displacement ($\delta_s$) on the joint. Figure 3.6(b) shows a typical relationship between $\tau$ and $\delta_s$. It is now possible to define a 'tangential shear stiffness' ($K_s$) exactly in the same way as was done for the case of normal stress. Thus,

$$K_s = \frac{\Delta\tau}{\Delta\delta_s} \tag{3.24}$$

$K_s$ is roughly constant till a peak value of shear stress is reached. Nonlinear values can, however, be adopted if experimental results justify.

It is noted that the behaviour of a joint represented by this model is uncoupled. The shear stresses do not produce any relative displacement in the normal direction and vice versa normal stresses do not produce any relative displacement in the shear direction.

### 3.5.2 Elasto-plastic models of rock joints

In this class of models, the behaviour of joints is assumed to be elasto–plastic. The elastic behaviour is represented by the initial elastic tangential normal and shear stiffnesses ($K_n^e$, $K_s^e$). The peak strength and dilatancy of rock joints is represented by failure criterion and flow rule, respectively.

In the past, a number of empirical relationships have been proposed relating the strength and dilation of rock joints. Notable amongst these are those due to Patton (1966), Ladanyi and Archambault (1970), Barton and Choubey (1977). Any peak strength relationship can be treated as a failure criterion and a flow rule can be interpreted from the dilatancy relationships. Gerrard (1986) has critically examined many of the proposed strength and dilation models of rock joints and brought to attention the appropriate physical constraints which must apply to initially mated fresh joints.

Here, we shall restrict ourselves to some of the commonly used strength and 'flow' relationships keeping in mind that some of these may not be correct on theoretical considerations.

#### *3.5.2.1 Mohr–Coulomb model*

This is perhaps the crudest of rock joint models but has been extensively used in engineering analysis and design of rock structures. Here the failure strength of the rock joint is assumed to be given by:

$$F = |\sigma_s| - \sigma_n \tan \phi - c = 0 \tag{3.25}$$

where $|\sigma_s|$ is the absolute value of shear stress on the joint plane. $\sigma_n$ is the normal stress on the plane, $\phi$ and $c$ are 'friction angle' and 'cohesion', respectively, for the joint.

If an associated flow rule is adopted, rates of plastic normal strain ($\dot{\epsilon}_n^p$) and shear strain ($\dot{\gamma}^p$) are given by (using equation (2.9)).

$$\begin{Bmatrix} \dot{\epsilon}_n^p \\ \dot{\gamma}^p \end{Bmatrix} = \dot{\lambda} \begin{Bmatrix} \dfrac{\partial F}{\partial \sigma_n} \\ \dfrac{\partial F}{\partial \sigma_s} \end{Bmatrix} = \dot{\lambda} \begin{Bmatrix} -\tan \phi \\ \\ 1 \end{Bmatrix} \tag{3.26}$$

which implies

$$\frac{\dot{\epsilon}_n^p}{\dot{\gamma}^p} = -\tan \phi \tag{3.27}$$

The joints are, therefore, dilatant (note the negative sign on the right-hand side of the equation (3.27)), i.e. an increment of shear displacement ($\Delta\delta_s$) along the joint is accompanied by an increment in the normal displacement ($\Delta\delta_n$) given by

$$\Delta\delta_n = -\tan\phi \cdot \Delta\delta_s \tag{3.28}$$

The rate of dilation is constant and goes on unabated. This behaviour is quite unrealistic.

Dilatancy of rock joints is complex. The average inclination and height of asperities play a domineering role in determining the rate of dilation and the maximum dilation that can occur on a joint plane. Figure 3.7 shows a simple model proposed by Patton (1966). Two distinct types of behaviour can be identified from this model:

(a) At low normal stresses, there is a tendency for ride up action over the asperities leading to dilation of the joint. On a plane, coinciding with the inclination of asperities, there is no dilation.
(b) At high normal stresses, the asperities get sheared at the base inhibiting the tendency to dilate.

Roberts and Einstein (1978) present a very comprehensive model for rock discontinuities. From various studies it has been established that the flow rule for rock joints should be nonassociated. Following equation (2.8), a plastic potential function can be written by introducing a variable dilatancy angle ($\Psi$). Thus,

$$Q = |\sigma_s| - \sigma_n \tan\Psi = 0 \tag{3.29}$$

where $\Psi$ can be identified from the experimental results of rock joints. It is clear that when the normal displacement on the rock joint is equal to the average height of the asperities, dilation must cease, i.e. $\Psi \to 0$.

In certain situations, the strength of a rock mass is greatly influenced by the dilatancy of rock joints. If the rock mass is restrained, suppression of dilatancy due to boundary constraints leads to an increase in the normal stress on the joint. This increase in turn leads to an increase in the strength. The numerical analyst must, therefore, pay careful attention to the modelling of dilatancy of rock joints in such situations.

Sometimes, particularly in design situations, it is appropriate to incorporate a 'no tension' cut-off to the failure criterion (equation (3.25)) which is

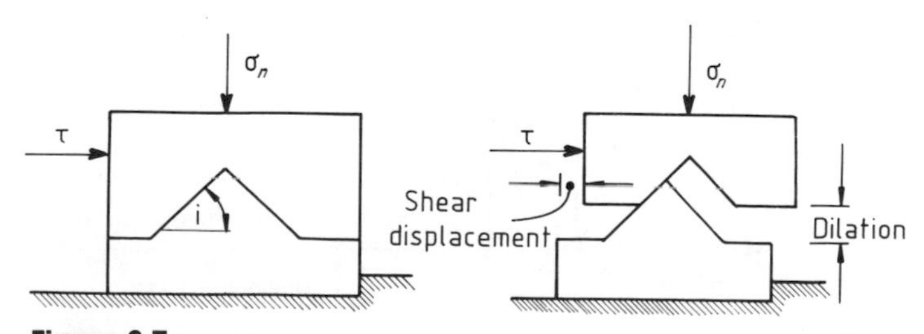


**Figure 3.7**
Conceptual mechanism of dilatancy in mated rock joints at low normal stresses

assumed to be valid only if $\sigma_n$ is compressive. Thus, an additional failure criterion

$$F = (-\sigma_n) = 0 \tag{3.30}$$

is adopted which is valid for $\sigma_n < 0$.

An associated flow rule has to be invoked in this case as otherwise it will not be possible to relax tensions across the rock joints which have to open. Our knowledge of joints subjected to a combined action of tension and shear is very limited since it is quite difficult to carry out experimental investigations in this stress regime. Conceptually, it appears that a joint will be able to withstand shear while it is opening under tensile normal stresses as the asperites will be 'interlocked'. However, when the joint opening is equal to or greater than the average height of asperities, interlocking should cease and the joint will be incapable of withstanding neither shear nor tensile stresses.

Figure 3.8 shows the Mohr–Coulomb criterion in $\sigma_s$, $\sigma_n$ space. Plastic potential functions in various zones are also shown. Assuming a pseudo-viscoplastic behaviour of joints, strategies for possible paths of stress relaxation of a few typical points are shown.

Mohr–Coulomb model has a major drawback. $c$ and $\phi$ in equation (3.25) are not truly constants. They depend on $\sigma_n$. The value of $\sigma_n$ on rock joints can vary by several orders of magnitude within the structure to be analysed. Choosing a single appropriate value of $c$ and $\phi$ for a joint set, therefore, becomes difficult, if not impossible.

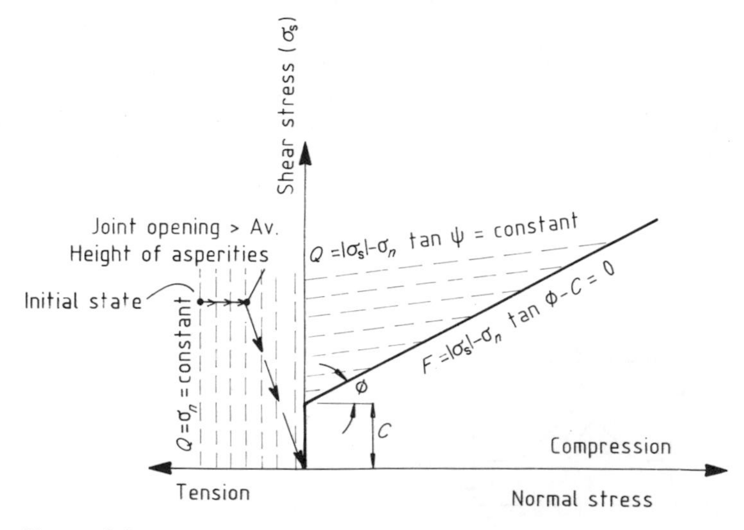


**Figure 3.8**
Mohr-Coulomb failure with tension cut-off

### 3.5.2.2 Barton and Chaubey model

Based on extensive experimental studies of artificial and natural joints, Barton and Chaubey (1977) proposed an empirical equation for the peak strength of rock joints which in the form of a failure criterion can be written as

$$F = |\sigma_s| - \sigma_n \tan\left(\mathrm{JRC} \log_{10}\left(\frac{\mathrm{JCS}}{\sigma_n}\right) + \phi_r\right) = 0 \tag{3.31}$$

where

JRC is joint roughness coefficient
JCS is joint wall compressive strength

and

$\phi_r$ is residual friction angle.

These three parameters are shown to be independent of normal stress on joints. A plot of equation (3.31) is shown in Figure 3.9. It is noted that the joint has no tensile strength and there is no need to adopt a 'no tension' cut-off as with Mohr–Coulomb yield function.

The parameters JRC, JCS and $\phi_r$ are easily determined by simple field tests. Methods of obtaining them and interpreting their values are contained in the guidelines prepared by the International Society for Rock Mechanics (1978).

If an associated flow rule is assumed, the dilation angle at peak strength can be readily computed by differentiating equation (3.31). However,

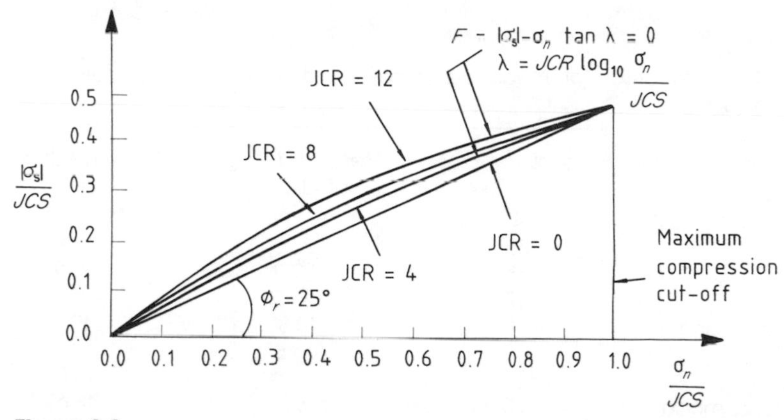


**Figure 3.9**
Normalized plot of failure criterion of Barton and Chaubey for rock joints (for $\phi_s = 25°$)

computed dilation angles ($\Psi$) based on an associated flow rule do not match with the experimentally observed values. Pande and Xiong (1982) proposed the following plastic potential function to match the experimental results of Barton and Chaubey.

$$Q = |\sigma_s| + \frac{\sigma_n \tan \lambda_1}{K_1} - \frac{\text{JRC}}{263.86} \frac{\sigma_n^2}{\text{JCS}} = \text{constant} \tag{3.32}$$

where

$$K_1 = (1 - \tan \lambda_1 \tan \phi_2) \quad \text{and} \quad \lambda_1 = \text{JRC} \log_{10} \left( \frac{\text{JCS}}{\sigma_n} \right)$$

Table 3.1 shows the comparison of experimental values with those computed using equation (3.32) as the plastic potential function. A close agreement is seen. The important point is that, if experimental data on dilatancy of joints are available, it is possible to fit a mathematical expression for the plastic potential function.

### 3.5.3 Post peak behaviour of joints

If a joint is tested at a constant normal stress $\sigma_n$, the peak shear strength is attained after a certain shear displacement. If a further displacement is applied, the shear stress actually reduces as shown in Figure 3.10. This reduction in strength (strain softening) can, in most cases, be assumed as linear.

For Mohr–Coulomb model, it is customary to assume residual cohesion as zero and residual angle of friction as 25–30°. Again, experimental results

**Table 3.1** Comparison of measured angle of dilation† with that predicted by equation (3.32)

| Rock type | No. of samples | Measured angle of dilation | Computed angle of dilation |
|---|---|---|---|
| Alpite | 36 | 25.5° | 23.C° |
| Granite | 38 | 20.9° | 20.2° |
| Hornfels | 17 | 26.5° | 26.2° |
| Calcareous shale | 11 | 14.8° | 19.1° |
| Slate | 7 | 6.8° | – |
| Gneiss | 17 | 17.3° | 15.5° |
| Soapstone model | 5 | 16.2° | 18.6° |
| Fractures | 130 | 13.2° | – |

†Initial data taken from Barton and Choubey (1977).

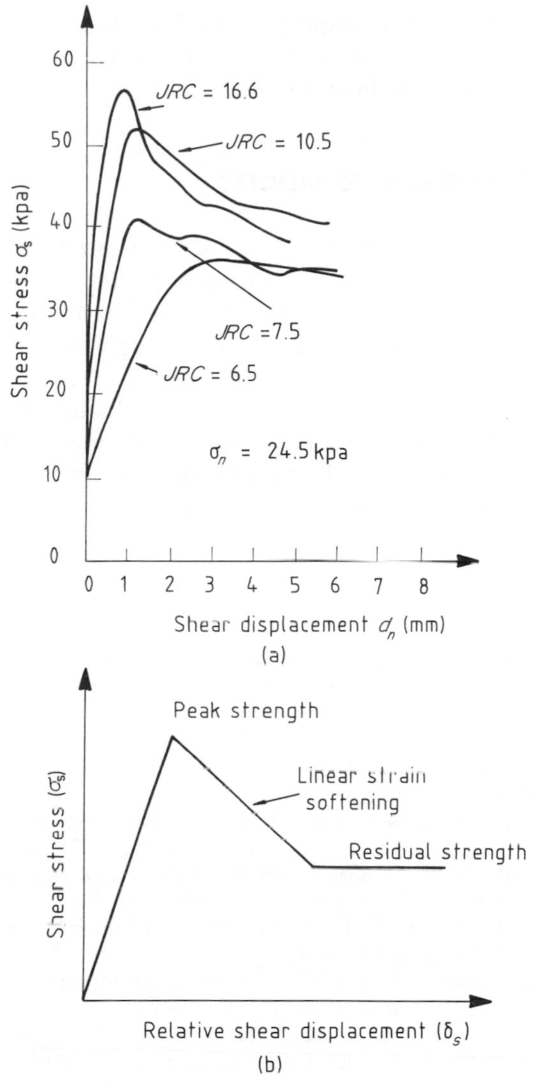


**Figure 3.10**
Post peak behaviour of rock joints. (a) $\tau$ versus $\delta_s$. (After Bandis *et al.*, 1981); (b) idealized relationship between $\tau$ and $\delta_s$

of tests of joints are required to be able to formulate an appropriate strain softening law.

For the model of Barton and Chaubey, it appears reasonable to assume that JRC varies from the peak value to zero as plastic shear strains accumulate on the joint. At the residual state, the strength is given by

$$\delta_s = \sigma_n \tan \phi_r \tag{3.33}$$

The question is what shear displacements (or strains) on the joint reduce the strength to residual value. This depends on the characteristics of joints such as average height, inclination and density of asperities.

## 3.6 MORE SOPHISTICATED MODELS

Conceptually, it is possible to build very sophisticated and complex models of the behaviour of rock joints. One such model has been proposed by Pande (1985). It is based on 'Bounding Surface' plasticity model of Dafalias and Popov (1975). Though it is capable of dealing with situations of loading, unloading and reloading of joints, model requires eight parameters. From the practical standpoint it is quite difficult to obtain reliable values of so many parameters without considerable expense of time and money. However, in critical structures such as nuclear waste depositories and problems relating to petroleum engineering, such an investment may be very worthwhile.

## REFERENCES

Bandis, S. C., Lurnsden, A. C., and Barton, N. R. (1981). 'Experimental studies of scale effects on the shear behaviour of rock joints.' *Intl. Jl. Rock Mech. Min. Sci.,* **18**, 1–21.

Barton, N. R., and Choubey, V. (1977). 'The shear strength of rock joints in theory and practice.' *Rock Mech.,* **10**, 1–54.

Barton, N. *et al.* (1978). 'Suggested methods for the quantitative description of discontinuities in rock masses, for ISRM Commission on Standardization of Lab and Field Tests.' *Int. J. Rock Mech. Min. Sci.,* **15**(6), 319–68.

Bazant, Z., and Oh, B. H. (1983). 'Crack band theory for fracture of concrete and geomaterials.' *A.S.C.E. Eng. Mech. Dn.,* **109**(3), 849–65.

Brady, B. H. G., and Brown, E. T. (1985). *Rock Mechanics for underground mining*, London: George Allen and Unwin.

Dafalias, Y. F., and Popov, E. P. (1975). 'A model of nonlinear hardening materials for complex loading.' *Acta Mechanica,* **21**, 173–92.

Farmer, I. W. (1983). *Engineering Behaviour of Rocks*, London: Chapman & Hall.

Goodman, R. E. (1970). 'The deformability of joints.' *ASTM Spec. Tech. Publ. 477.*

Goodman, R. E. (1980). *Introduction to Rock Mechanics*, New York: J. Wiley & Sons.

Goodman, R. E., Taylor, R., and Brekke, T. L. (1968). 'A model for the mechanics of jointed rocks' *Jl. A.S.C.E. Geotech. Dn.,* **94**SM(3), 637–59.

Gerrard, C. (1985). 'Formulations for the mechanical properties of rock joints.' *Proc. Intl. Symp. Fundamentals of Rock Joints*, Centek Press, Lulea, 405–22.

Gerrard, C. (1986). 'Shear failure of rock joints: appropriate constraints for empirical relations.' *Intl. Jl. Rock Mech. Min. Sci.,* **23**(20), (6), 421–29.

Griffith, A. A. (1921). 'The phenomenon of rupture and flow in solids.' *Phil. Trans. Roy. Soc. London A221,* 163–98.

ISRM—Commission on Testing Methods (1985). 'Suggested method for determining joint load strength.' *Intl. Jl. Rock Mech. Min. Sci.,* **22**(2), 51–60.

Hoek, E., and Brown, E. T. (1980). *Underground Excavation in Rock*. London: Institution of Mining and Metallurgy.
Jaeger, J., and Cook, N. G. W. (1976). *Fundamentals of Rock Mechanics*, 2nd edn, London: Chapman & Hall.
Ladanyi, B., and Archambault, G. (1970). 'Simulation of the shear behaviour of a jointed rock mass.' *Proc. 11th Symp. on Rock Mechanics* (AIME), 105–25.
Lode, W. (1926). *Zeits. Phys.*, **36**, 913.
Murrel, S. (1963). 'A criterion for brittle fracture of rocks and concrete under triaxial stress and the effect of pore pressure on the criterion.' *Proc. 5th Rock Mech. Symp.*, Pergamon Press.
Patton, F. D. (1966). 'Multiple modes of shear failure in rock.' *Proc. 1st Cong. ISRM* (Lisbon), **1**, 509–13.
Pande, G. N., and Xiong, W. (1982). 'An improved multilaminate model of jointed rock masses.' *Numerical Models in Geomech.* (eds R. Dungar, G. N. Pande and J. A. Studer), A. A. Balkema, Rotterdam, 218–26.
Pande, G. N. (1985). 'A constitutive model of rock joints.' *Proc. Intl. Symp. Fundamentals of Rock Joints* (ed. O. Stephansson), Centek Publ., Lulea, Sweden.
Roberts, W. J., and Einstein, H. A. (1978). 'Comprehensive model of rock discontinuities.' *Jl. A.S.C.E.*, **104**(GT5), 553–69.
Sun, Z. (1985). 'Asperity model for closure and shear.' *Proc. Intl. Symp. Fundamentals of Rock Joints*, Centek Press, Lulea, 173–83.
Sun, Z., Gerrard, C. M., and Stephansson (1985). 'Rock joint compliance tests for compression and shear loads.' *Intl. Jl. Rock Mech. Min. Sci.*, **22**(4), 197–213.
Swan, G. (1983). 'Determination of stiffness and other joint properties from roughness measurements.' *Rock Mech. & Rock Eng.*, **16**, 19–38.
Swan, G. (1985). 'Methods of roughness analysis for predicting rock joint behaviour.' *Proc. Intl. Symp. Fundamentals of Rock Joints*, Centek Press, Lulea, 153–61.
Yoshinaka, R., and Yambe, T. (1986). 'Joint stiffness and the deformation behaviour of discontinuous rock.' *Intl. Jl. Rock Mech. Min. Sci.*, **23**(1), 19–28.

# 4 Behaviour of Jointed Rock Masses

## 4.1 INTRODUCTION

Engineers are more concerned with the properties of jointed rock masses than intact rocks or rock joints. Their interest in intact rocks and rock joints is basically to derive the properties of the jointed rock mass from those of intact rock and rock joints for a known joint fabric. It is the deformability or strength of the jointed rock mass that is required to be assessed to predict the settlement or collapse of rock structures.

One of the obvious approaches is to determine the properties of the rock mass from *in situ* tests such as plate bearing, bore hole, radial or flat jack and dynamic tests. *In situ* tests are very expensive and to what extent they reflect the deformability of rock mass depends on the particular situation of the rock mass, size of the plate or jack used, spacing of rock joints, etc. The dynamic tests which are based on the velocity of a longitudinal or transverse wave in the rock mass are comparatively cheaper but the results of such tests can be highly variable in jointed rock masses. Readers requiring further information on *in situ* testing of jointed rock masses should refer to other texts, e.g. Jeager and Cook (1979), Obert and Duvall (1967), Goodman (1980).

In this chapter, we shall discuss a set of numerical models which can be used for the study of the behaviour of jointed rock masses. In Section 4.3 some qualitative one dimensional models are discussed for assessment of the deformability of rock mass. One of these models is generalized to a fully three-dimensional situation in Section 4.4. Strength of rock masses and factors influencing it are discussed in Section 4.6.

## 4.2 THE ROCK MASS FACTOR

If we consider a rock mass in a uniaxial situation, an elastic Young's modulus ($E_m$) for the direction of loading can be defined. The ratio of $E_m$ to the elastic Young's modulus of intact rock ($E_i$) is termed as 'rock mass factor' ($j$). Thus,

$$j = \frac{E_m}{E_i} \tag{4.1}$$

The rock mass factor reflects the decrease in the modulus due to the presence of joints in the rock mass, their characteristics and spacing. It has been used in the classification of rock masses (Deere, 1968; Coon and Merritt, 1970) and is of prime importance in settlement calculations.

$j$ values can be determined from *in situ* tests but depend on the method of measuring, size of test, etc. and considerable judgement has to be used in interpretation of test results. Figure 4.1 shows the variation of $j$ with joint frequency (number of joints/metre, $f$) plotted from *in situ* tests reported by various authors. $j$ values drop dramatically in the range of $f$ equal to 1–10. For higher values of $f$, the drop in $j$ value is gradual.

Unfortunately, most geotechnical problems arise when $1 < f < 10$. It is therefore useful to investigate theoretical models which can be used to predict $j$ values. Obviously various theoretical models are based on different assumptions and idealisations. The range of their applicability is restricted. However, they provide some insight into the problem of deformability.

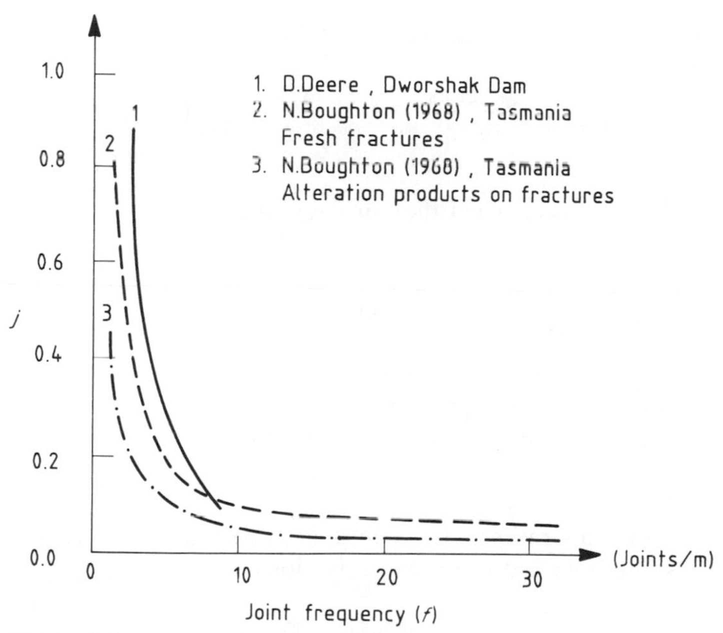


**Figure 4.1**
Variation of rock mass factor with joint frequency. (After Hobbs (1975))

## 4.3 SOME THEORETICAL MODELS FOR THE PREDICTION OF *j* VALUES

### 4.3.1 Walsh's model

Walsh and Brace (1966) have derived approximate mathematical expressions for apparent elastic moduli of a material which has dilute concentration of spherical cavities and elliptical cracks. He considered two conditions for elliptical cracks, viz. 'all cracks open' and 'all crack closed'. In the case of 'closed cracks', sliding can take place on the two faces of the crack and a coefficient of friction ($\mu$) defines the onset of sliding.

The expressions for the rock mass factor ($j$) based on his theory are as follows:

$$j = \left[1 + \frac{(1-\nu_i)(9+5\nu_i)}{7-5\nu_i} \cdot \frac{2\pi a^3}{v}\right]^{-1} \tag{4.2}$$

for spherical cavities

$$j = \left[1 + \frac{4\pi c^3}{3v}\right] \tag{4.3}$$

for planar (elliptic) cracks when they are open.

$$j = \left[1 + \frac{4\pi c^3}{15v}\left\{\frac{2+3\mu^2+2\mu^4}{(1+\mu^2)^{3/2}} - 2\mu\right\}\right]^{-1} \tag{4.4}$$

for planar (elliptic) cracks when they are closed.

In the above expressions:

$E_i$ = Young's modulus of intact rock
$\nu_i$ = Poisson's ratio of intact rock
$a$ = average diameter of spherical cavities
$c$ = average length of elliptical cracks
$v$ = volume of the rock mass associated with one cavity/crack.

The ratio of Poisson's ratio of rock mass ($\nu_m$) to that of intact rock ($i = \nu_m/\nu_i$) can also be obtained from his derivations

$$i = 1 + \frac{(1-\nu_i^2)(5\nu_i-1)}{7-5\nu_i} \cdot \frac{2\pi a^3}{3v} \tag{4.5}$$

for spherical cracks

$$i = \left[1 + \frac{4\pi c^3}{3v}\right] \tag{4.6}$$

for planar cracks when they are open

$$i = 1 + \frac{2\pi c^3(1 - 2\nu_i)(1 - \nu_i^2)}{15 v_i} \tag{4.7}$$

for planar cracks when they are closed.

Using equations (4.3) and (4.4), $j$ values plotted against fracture frequency ($f$) for all cracks open and all cracks closed case in Figure 4.2. A value of 0.65 has been adopted for $\mu$ and fracture frequency has been interpreted from average crack length per unit volume. These curves look strikingly similar to the ones obtained from the field tests (Figure 4.1).

Walsh and Brace (1966) have shown that elastic parameters of the rock mass are not constant when the assumptions of 'all cracks' closed is made. The nonlinearity arises due to the frictional behaviour of closed cracks and elastic parameters become stress path dependent.

The situation when some cracks are open and some closed is difficult to analyse but is undoubtedly more realistic. The equations (4.2) to (4.7) are not very useful for any particular rock mass but are of significant importance for making comparisons between rock masses having different characteristics.

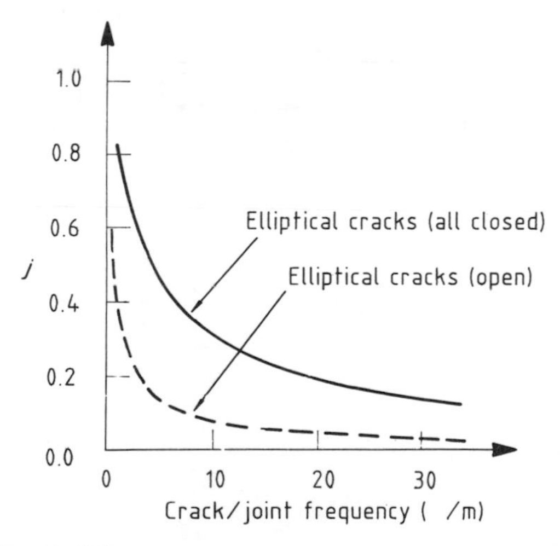


**Figure 4.2**
Variation of rock mass factor with crack/joint frequency. (After Walsh and Brace (1966))

### 4.3.2 Hobbs' model

Hobbs (1975) proposed that aspherities in rock joints (assumed normal to the direction of uniaxial loading) can be assumed as miniature circular loaded areas through which stress is transmitted (Figure 4.3). Let $b$ be the diameter of miniature circular areas and $n$ be the number of such idealised aspherities per unit area. In comparison to the diameter of loaded area, the joint spacing is such that the results of elastic settlements on semi-infinite elastic homogenous isotropic foundations are applicable. Settlement ($\delta$) of a uniformly loaded circular area is given by

$$\delta = \frac{\pi p(1 - \nu_i^2)b}{4E_i} \tag{4.8}$$

where $E_i$, $\nu_i$ are the Young's modulus and Poisson's ratio respectively of intact rock as defined earlier and $p$ is the intensity of load. Accounting for $n$ loaded areas, it can be shown that

$$j \simeq \left[1 + 2(1 - \nu_i^2)\frac{f}{bn} - 4bf\right]^{-1} \tag{4.9}$$

Assigning different values of $b$, $n$ and $f$, Hobbs (1975) has plotted graphs between $j$ and joint frequency $f$ and these are shown in Figure 4.4. $\nu^2$ has been assumed as 0.1. These graphs are again typically similar in nature to the graphs based on *in situ* tests shown in Figure 4.1.

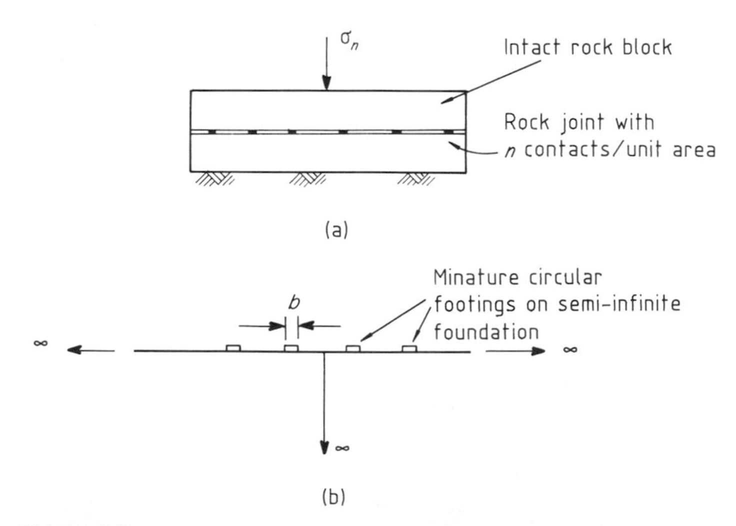


**Figure 4.3**
Conceptual model of a rock joint. (a) Transfer of normal stress ($\sigma_n$) to $n$ contacts; (b) each contact idealized as a miniature circular footing on semi-infinite foundation

In equation (4.9), $\nu_i$ and $f$ can be readily estimated. Techniques of photography can be used to digitize the profile of a joint (Tse and Cruden, 1979; Wu and Ali, 1978; Dight and Chia, 1981). Statistical analyses can then be performed to determine the parameters $b$ and $n$. Thus, a rough guidance on the $j$ value can be obtained.

### 4.3.3 A model based on joint stiffness

Consider a rock mass with horizontal joints have a frequency of $f$ subjected to uniaxial compressive loading as shown in Figure 4.5.

Let $k$ be the stiffness of the joint (see Section 3.5.1). This stiffness can be physically attributed to poor contact of intact rock blocks, resistance of the joints to closure, friction on the joint walls and compressibility of infilling gouge material. The closure of the joint under a uniaxial stress $\sigma$, is given by $\sigma f/k$. Deformation of the intact rock under the same stress $\sigma$ is $\sigma/E_i$.

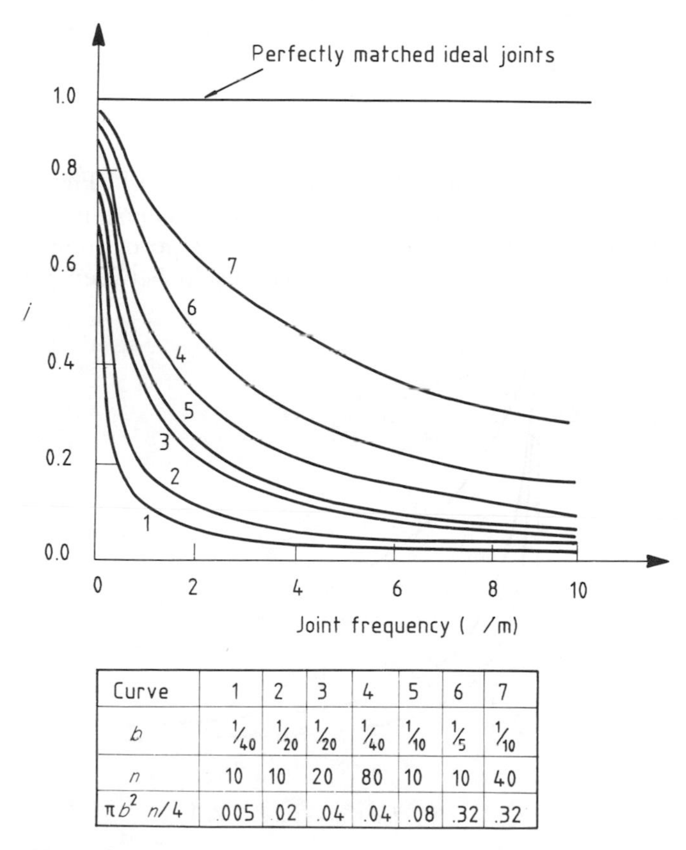


| Curve | 1 | 2 | 3 | 4 | 5 | 6 | 7 |
|---|---|---|---|---|---|---|---|
| $b$ | $1/40$ | $1/20$ | $1/20$ | $1/40$ | $1/10$ | $1/5$ | $1/10$ |
| $n$ | 10 | 10 | 20 | 80 | 10 | 10 | 40 |
| $\pi b^2 n/4$ | .005 | .02 | .04 | .04 | .08 | .32 | .32 |

**Figure 4.4**
Variation of rock mass factor with joint frequency. (After Hobbs (1975))

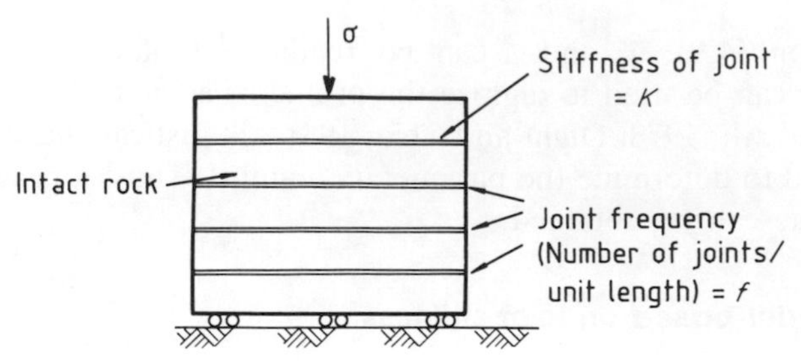


**Figure 4.5**
A simple model for rock mass

Equating the deformation of the rock mass to the sum of deformations in intact rock and rock joints leads to

$$\frac{\sigma}{E_m} = \frac{\sigma}{E_i} + \frac{\sigma f}{k} \tag{4.10}$$

which on rearranging gives

$$j = \frac{1}{1 + fE_i/k} \tag{4.11}$$

Equation (4.11) gives a graph between $j$ and $f$ as shown in Figure 4.6. It is noted that qualitatively the graph is again of the same type as obtained from experiments. For fixed values of $f/k$ (same type of joints and same joint frequency), Figure 4.7 shows the relationship between $j$ and $E_i$. It

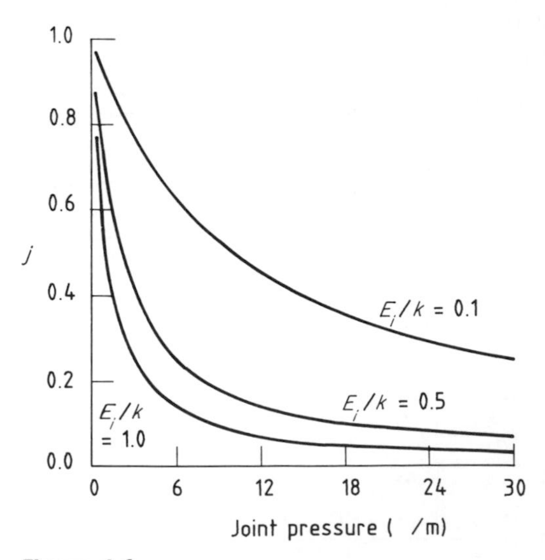


**Figure 4.6**
Variation of rock mass factor with joint frequency based on equation (4.11)

shows that joints affect the $j$ value most dramatically for hard rocks—a conclusion which can be logically supported.

## 4.4 ELASTICITY MATRIX OF A JOINTED ROCK MASS

The value of $j$ for a rock mass enables the designer to compute $E_m$ and thereby make use of various analytical formulae for computation of deformations/settlement. A value of Poisson's ratio is assumed and rock mass is assumed as isotropic homogenous and linear elastic material. It is obvious that it is a very crude approximation of the real situation. However, it would be unwise to ignore the usefulness of analytical solutions since the designers have considerable experience of using simple analytical methods, coupled with judgement for many practical situations.

The concept of a model based on joint stiffness discussed in Section 4.3.3 is a very useful one and can be extended to derive the complete elasticity matrix $[D_{RM}]$ of a jointed rock mass with multiple joint sets. Stiffnesses of the joints and their fabric is used to define a material that is 'equivalent' to the rock mass. The derivation is based on the assumption that the joints are continuous and parallel, and their spacing is uniform and much smaller than the dimensions of the structure to be analysed using the 'equivalent' material model. It is also assumed that the volumetric space occupied by joints is negligibly small.

Stephansson (1981) and Gerrard (1982a,b) have considered three orthogonal sets of joints of finite thickness in the rock mass and derived its stress–strain relationship. Gerrard (1982c) has generalized the model for

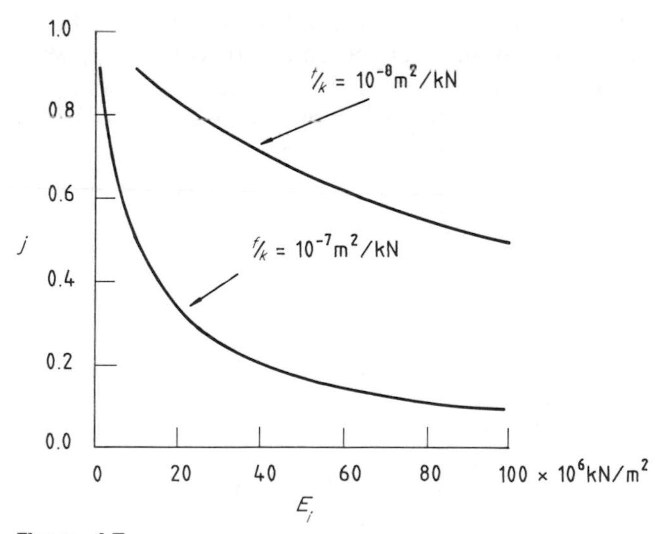


**Figure 4.7**
Variation of rock mass factor with Young's modulus of intact rock

multiple joints and presented a number of examples of special cases of rock mass properties. The derivation given here is largely based on his work.

### 4.4.1 Compliance matrix of a rock joint

Let us consider a continuous planar joint and, set up a system of local Cartesian coordinate ($x'$, $y'$, $z'$) axes of reference. $z'$ axis is chosen normal to the joint plane and the axes $x'$ and $y'$ lie on the joint plane. They are oriented arbitrarily but are mutually perpendicular (Figure 4.8). On the joint plane, the stresses are denoted by $\sigma_{z'}$, $\tau_{y'z'}$, $\tau_{z'x'}$ and correspond to the $x'$, $y'$ and $z'$ axes respectively. Movement of the upper block relative to the lower is denoted by $u$, $v$ and $w$ in the directions $x'$, $y'$ and $z'$, respectively. Application of a compressive normal stress ($\sigma_{z'}$) on the joint leads to a certain relative movement. Similarly, application of stresses $\tau_{x'z'}$ and $\tau_{y'z'}$ lead to relative movements along $x'$, $y'$ and $z'$ directions. One can, then, define a compliance matrix ($s$) of a typical joint which relates the stresses on the joint plane to the relative movements, i.e.

$$\begin{Bmatrix} w \\ v \\ u \end{Bmatrix} = \begin{bmatrix} s_{11} & s_{12} & s_{13} \\ s_{21} & s_{22} & s_{23} \\ s_{31} & s_{32} & s_{33} \end{bmatrix} \begin{Bmatrix} \sigma_{z'} \\ \tau_{y'z'} \\ \tau_{z'x'} \end{Bmatrix} \tag{4.12}$$

$s_{ij}$ is a $3 \times 3$ matrix and is fully populated. It may not be symmetrical. To understand the significance of the elements of $s_{ij}$, imagine a test on a joint in which $\sigma_z$, is applied and displacements $u$, $v$ and $w$ are measured. Graphs such as shown in Figure 4.9 can be drawn between $\sigma_{z'}$ and $v$, $u$ and $w$. The slopes of these curves represent $s_{11}$, $s_{21}$ and $s_{31}$. The relationship is nonlinear and $s_{11}$, etc. are not constant over the entire stress range. If we restrict our attention to elastic behaviour only, it is possible to identify $s_{11}$, $s_{12}$, etc. as the initial slopes. The nonlinear behaviour of rock mass can be introduced via the theory of plasticity and we shall see this in Chapter 7.

Now, repeating the test on the joint with $\tau_{y'z'}$, applied and $u$, $v$ and $w$'s measured, coefficients $s_{12}$, $s_{22}$, $s_{32}$ can be identified. Similarly, other coefficients can be identified. It is unlikely that test data will be available to quantify all the nine coefficients of $s$ matrix and approximations will have

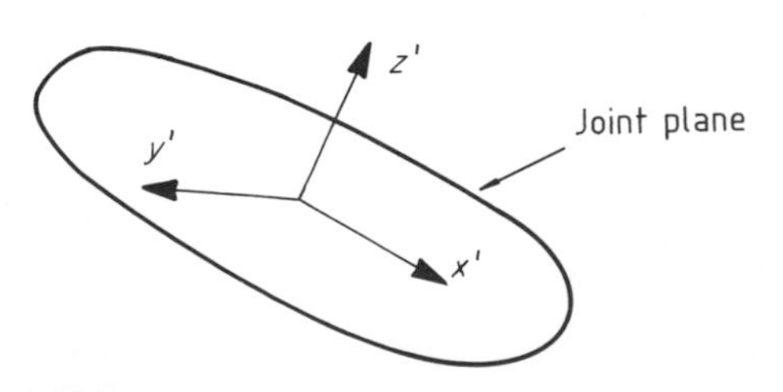


**Figure 4.8**
A local coordinate system

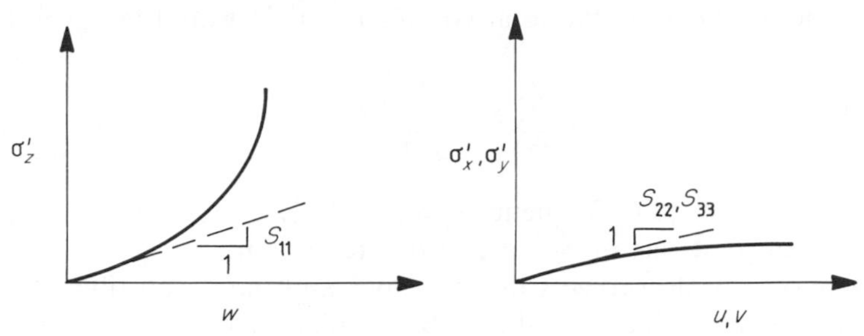


**Figure 4.9**
Schematic explanation of coefficients $S_{11}$, $S_{22}$ and $S_{33}$ in equation (4.12)

to be made. One of the assumptions which is particularly justified in the elastic range is that all of the non-diagonal terms are zero. Nonzero terms are $s_{11}$ which is the normal compliance ($s_n$) of the joint, and $s_{22}$ and $s_{33}$, which are shear compliances ($s_{y'z'}$, $s_{z'x'}$) along $y'$ and $x'$ axes, respectively. In many cases only resultant shear compliances ($s_s$) will be known and shear compliance component $s_{y'z'}$ and $s_{z'x'}$ may have to be assumed as $s_s/\sqrt{2}$. It may be noted that generally stiffnesses (which are inverse of compliance in special cases) are reported in the literature.

### 4.4.2 Compliance matrix of a set of continuous parallel joints

Consider once again rock mass with horizontal joints having an average frequency $f$ subjected to a uniaxial state of stress as shown in Figure 4.4. Adopting a procedure similar to the one used for deriving equation (4.10), it can be shown (Gerrard, 1982b) that compliance of a set of $f$ parallel joints ($S$) is given by

$$S = s_1 + s_2 + s_3 + \ldots s_f \tag{4.13}$$

where $s_1$, $s_2$ . . . represent the compliances of individual joints.

In the above example, the local axes of joint sets coincide with the global axes for the rock stress. In general, this is not the case and the compliance matrices of joints have to be transformed into global coordinate system before being summed up. This is achieved by premultiplying the compliance matrix by a transformation matrix ($T$) and postmultiplying it by transpose of $T(T^T)$. The elements of transformation matrix depend on the direction cosines of the normal to the joint plane and its exact form is given in Appendix II.

As generally, we would have the same compliance for all parallel joints in one set, equation (4.13) reduces to

$$S = fTsT^T \tag{4.14}$$

for an arbitrarily oriented joint set.

If there be $n$ joint sets, the total compliance ($S^*$) would be given by

$$S^* = \sum_{i=1}^{n} S_i = \sum_{i=1}^{n} f_i T_i s_i T_i^T$$

where $f_1$, $f_2$, etc. are joint frequencies in sets 1, 2, etc. and $S_1$, $S_2$ represent compliances of joint sets in global coordinate system.

The components of average strain ($\epsilon_J$) in a unit block contributed by the joint set can be identified from the global displacement of the block shown in Figure 4.4 and are given by

$$\boldsymbol{\epsilon}_J = S^* \boldsymbol{\sigma} \tag{4.15}$$

### 4.4.3 Final form of the elasticity matrix

To obtain the total strains, the strains in the intact rock should be calculated and added to the strains contributed by the joint sets. The strains contributed by intact rock are

$$\boldsymbol{\epsilon}_I = [D_I]^{-1} \boldsymbol{\sigma} \tag{4.16}$$

where $[D_I]$ represents the elasticity matrix of the intact rock. Thus, comparing this equation with

$$\boldsymbol{\sigma} = [\mathrm{D}_{RM}] \boldsymbol{\epsilon} \tag{4.17}$$

where $D_{RM}$ is the complete elasticity matrix of the jointed rock mass, we have

$$[D_{RM}] = [[D_I]^{-1} + S^*]^{-1} \tag{4.18}$$

which gives the final complete form of the elasticity matrix of an 'equivalent material' of jointed rock mass.

The above equation takes into account the characteristics of joints (through compliance), the spacing of joints, the fabric of joint sets and the characteristics of intact rock in a systematic manner.

## 4.5 COMMENTS ON COMPLIANCE MATRIX OF A JOINT

The nine coefficients of compliance matrix relate the relative displacement of the joint to the applied stresses. This matrix may not be symmetric. The deformational process of rock mass is generally nonconservative. The origin of this nonconservativeness lies in the aspherities which have a certain shape, inclination, amplitude and are not always perfectly matching.

Compliance matrix may depend on the sign of stress. It is obvious that, if tensile stresses are applied on the joint or the direction of shearing stresses is reversed, the displacements may not be the same. Thus, conceivably a completely different compliance matrix may be applicable with signs of stress reversed.

A dual valued nonsymmetric compliance matrix is almost impossible to deal with in practical computations. The model has a useful role in studying the influence of various joint fabrics on elastic parameters. It is apt to recall here that description of a general elastic material involves 21 constants and there is no material for which these constants have been measured or determined.

From the practical stand point, the compliance matrix of joints should have only leading diagonal elements nonzero. Assuming this, the form of complete elasticity matrices can be worked out for any joint fabric using the formulation given in Section 4.4 above.

## 4.6 STRENGTH OF JOINTED ROCK MASSES

In Chapter 3 strength of intact rock and rock joint was discussed. Here, we shall take into account the strengths of intact rock and rock joints to deduce the strength of jointed rock mass which would be strongly influenced by the orientation of joint sets. A rock mass with a single set of joints is considered first. The case of multiple joints is obtained by simple extension.

There are many engineering problems in which rock structures are subjected to simple state of stress, e.g. uniaxial compression in mine pillars, tension due to bearing in roofs. It is therefore useful to discuss strength of jointed rock masses under various simple states of stress.

### 4.6.1 Rock mass with a single set of joints

Consider a block of rock mass having a single joint† as shown in Figure 4.10. This block can be viewed to be mechanically equivalent to another block of the same dimensions with the influence of the single joint smeared over the cubic block. The general aim of considering equivalent material is that one can incorporate its constitutive relations in numerical models and then solve complex boundary value problems. As discussed in Section 4.4 validity of 'equivalent' material approach depends on the scale of the problems under consideration.

† The joint has been assumed in the plane of paper. For a joint orientated in space, or graphical methods based on vectors (suitable for computing) are more convenient. The basic equations follow the same pattern.

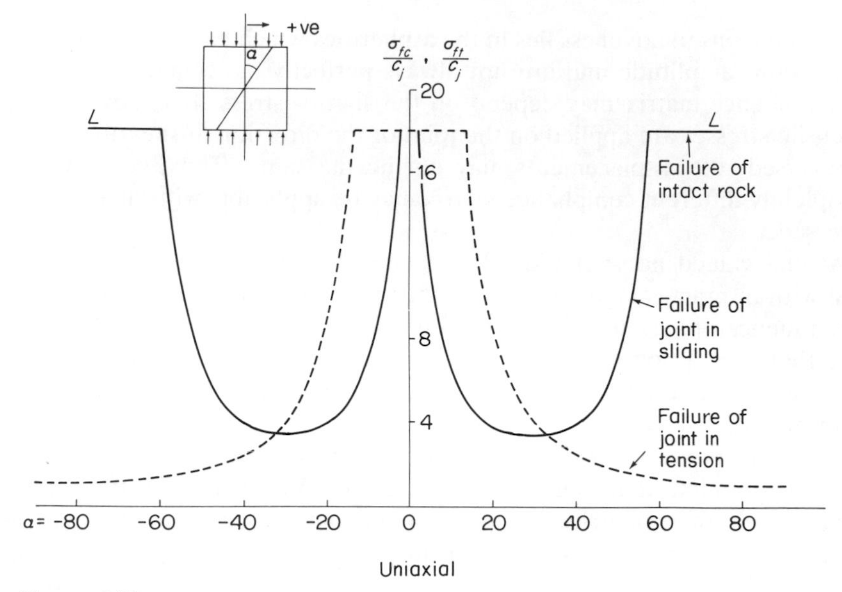


**Figure 4.10**
Variation of uniaxial strength of jointed rock mass with $\alpha$ ($\phi_j = 30°$)

*Uniaxial compression*

The joint set makes an angle $\alpha$ with the vertical. The uniaxial stress $\sigma$ can be resolved to give normal ($\sigma_n$) and shear stresses ($\tau$) on the joints plane. Depending on the orientation of the joint, failure would be induced in the intact rock or joint set. Here any of the failure criteria for intact rock and rock joint discussed in Chapter 3 can be used to determine the uniaxial failure load.

Using Mohr–Coulomb failure for rock joints, failure in rock joint takes place when

$$\tau - \sigma_n \tan \phi_j - C_j \geqslant 0 \qquad (4.19)$$

where $\phi_j$ is the peak angle of friction and $C_j$ is the cohesion intercept. In the uniaxial situation

$$\tau = \tfrac{1}{2}\sigma \sin 2\alpha \qquad (4.20)$$

and

$$\sigma_n = \sigma \sin^2 \alpha \qquad (4.21)$$

substituting equations (4.20) and (4.21) in equation (4.19) gives

$$\frac{\sigma}{C_j} \geqslant \frac{1}{\sin \alpha(\cos \alpha - \sin \alpha \tan \phi_j)}$$

Denoting the failure stress by $\sigma_{fc}$ and varying $\alpha$, a graph between $(\sigma_{fc}/C_j)$ and $\alpha$ can be plotted and is shown in Figure 4.10.

It is noted that when

$$\alpha = 0 \frac{\sigma_{fc}}{C_j} = \infty. \quad \frac{\sigma}{C_j} \text{ again becomes } \infty$$

when

$$\cot \alpha = \tan \phi_j$$

or

$$\alpha = \cot^{-1} (\tan \phi_j) = (90^\circ - \phi_j)$$

The joint failure can take place when $-(90^\circ - \phi_j) < \alpha < (90^\circ - \phi_j)$. For angles for which failure of joint does not take place, the uniaxial strength of the rock mass is only limited by the uniaxial failure strength of intact rock. Thus, adopting Mohr–Coulomb failure for intact rock, the uniaxial strength is given by

$$2C_i \tan \left(45^\circ + \frac{\phi_i}{2}\right)$$

A limiting horizontal line $(L - L)$ is therefore drawn in Figure 4.10. It is obvious that any other failure criterion could be used for rock joints and intact rock to draw uniaxial strength curves for jointed rock mass.

### *Uniaxial tension*

Tensile strength of the rock mass is obviously dependent on the orientation of joint sets. For certain values of $\alpha$, joint failure can not take place and failure strength of intact rock determines the strength. Assuming that the tensile strength of joint is limited to $C_j$, graphs can be drawn between normalised failure stress in uniaxial tension $(\sigma_{ft}/C_j)$ against $\alpha$ and is shown in Figure 4.10.

### *Rock mass in pure shear*

Pure shear gives rise to diagonal tension and compression of the same magnitude as the applied pure shear (Figure 4.11). This can be seen by finding out principal stresses due to pure shear. The strength of the rock mass in pure shear is thus influenced by (a) failure strength of joint in compression, (b) failure strength of joint in tension, (c) failure strength of

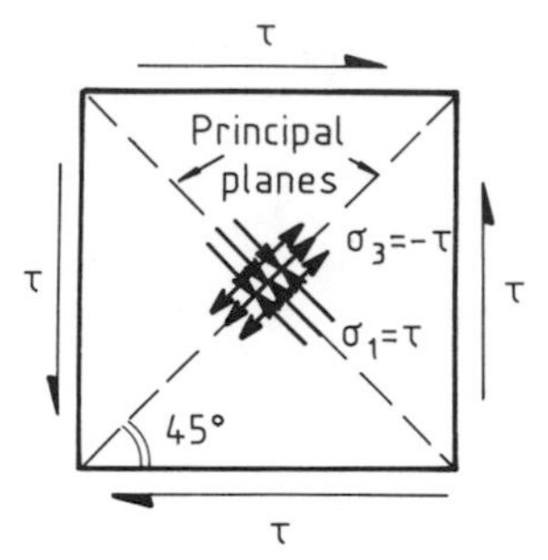


**Figure 4.11**
Equivalence of pure shear to a diagonal system of principal stress

intact rock in compression, (d) failure strength of intact rock in tension, (e) orientation of joint plane. Figure 4.12 shows the variation of failure strength in shear ($\sigma_{fs}/C_j$) with $\alpha$ varying from $-90°$ to $+90°$.

### 4.6.2 Rock mass with multiple sets of joints

The strength of a rock mass with multiple sets of joints can be determined by examining the strength corresponding to each set of joints in turn and adopting the least value. This is based on the assumption that presence of a set of joints does not affect the failure characteristics of the other.

The presence of joints makes the rock mass highly anisotropic, i.e. strength depends on the orientation of joints with respect to the directions of applied stress. However, as the number of joint sets increases the degree of anisotropy decreases. In the limiting case when a number of joint sets approaches infinity and failure characteristics of all joint sets are the same, the rock mass (virtually a soil mass) becomes isotropic again. This fact can be used with advantage in modelling of soil masses (Pande and Sharma, 1981, 1983).

## 4.7 FACTORS INFLUENCING STRENGTH OF ROCK MASSES

In Section 4.6 strength of jointed rock masses was derived from that of its constituents, e.g. intact rock, rock joints and their fabric. There are a number of other factors which influence the strength of rock mass and these will be discussed here.

### 4.7.1 Influence of water in rock joints

Rock masses in many practical situations, may be saturated. Rock joints as well as voids in intact rock would have water pressure. This water pressure would affect the strength of rock mass. Strength calculations can be carried out in the manner similar to that described in Section 4.6, except that 'effective' stresses should be considered everywhere. The presence of water in joints does not influence the shear stress on the joint but normal stress

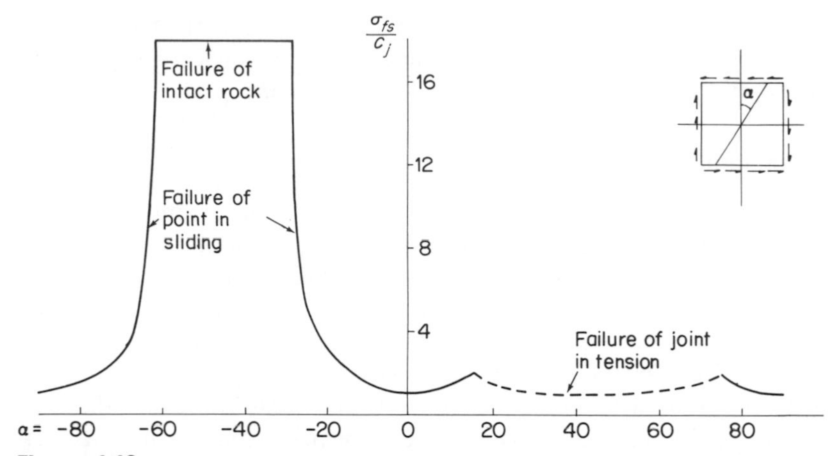


**Figure 4.12**
Variation of strength of rock mass in pure shear with $\alpha$ ($\phi_j = 30°$)

in all strength equations should be taken as 'effective' normal stress ($\sigma_n'$) which are given by

$$\sigma_n' = \sigma_n - \mathrm{p}$$

where $\sigma_n$ is the total normal stress and $p$ is the water pressure in joint. In general, the effective stresses ($\boldsymbol{\sigma}'$) are given by

$$\boldsymbol{\sigma}' = \boldsymbol{\sigma} - \boldsymbol{m}p$$

where $\boldsymbol{\sigma}$ is the vector of total stress components and $\boldsymbol{m} = [111\ 000]^T$.

While applying failure criteria for intact rock, components of *effective* stress should be used.

How water pressure in joints affects the deformability and strength of rock mass depends on the permeability of the rock mass and rate of loading. If permeability is high and rate of loading slow, no excess pore pressure will take place and the rock mass will practically behave as 'drained' implying that there will be no excess pore pressure over and above the initial ones (before loading was imposed). On the other hand, if permeability is low and rate of loading is rapid, the rock mass behaves as undrained. In this situation excess pore pressure will take place which reduces the effective stresses. Reduction in effective stresses leads to reduction in strength of the rock mass.

In the Finite Element context it is easy to perform a 'drained' as well as 'undrained' analysis. Further discussion on this topic is given in Chapter 7.

Some rocks and rock joints are weakened by the presence of water due to the chemical action of the rock constituents and/or washing out of the joints. Some sandstones may lose as much as 15 percent of their uniaxial strength on saturation. Such effects have to be taken care of in analysis based on experimental data and practical judgement.

### 4.7.2 Influence of size on strength

All rocks exhibit the so-called 'scale' effect to some extent, e.g. a larger sample has lower strength than a smaller sample of the same rock. The explanation for this behaviour lies in the discontinuous nature of rock structure which includes cracks, fissures, etc. In a small specimen, fewer discontinuities are present and many new ones are to be created to cause the rock to fail. On the other hand, in a larger many more discontinuities pre-exist and only a few new ones have to be created to cause the failure.

Weibull (1939) gave a statistical theory which explains the 'size' effect semiquantitatively. It is an interesting theory and has not received much attention from researchers. The present-day computational capabilities should be able to enhance this theory for a better understanding of 'scale' effects.

Few studies on scale effects through physical tests have been made in the past. Pratt *et al.* (1972) and Bieniawski (1968) have studied Diorite and coal respectively in uniaxial compression. Bandis *et al.* (1981) have studied scale effects on the strength of artificial simulated joints. Much more research remains to be done for a better quantitative understanding of this topic.

## REFERENCES


Bandis, S. C., Lumsden, A. C., and Barton, N. R. (1981). 'Experimental studies of scale effects on the shear behaviour of rock joints.' *Intl. Jl. Rock Mech. Min. Sci.,* **18**, 1–21.

Bieniawski, Z. T. (1968). 'The effect of specimen size on compressive strength of coal.' *Intl. Jl. Rock Mech. Min. Sci.,* **5**, 325–35.

Coon, R. F., and Merritt, A. H. (1970). 'Predicting *in situ* modulus of deformation using rock quality indexes.' *A.S.T.M.,* STP 477, 154–73.

Deere, D. U. (1968). 'Geological considerations', Ch. 1 in *Rock Mechanics in Engineering Practice*, (eds K. G. Stagg, and O. C. Zienkiewicz), J. Wiley & Sons.

Dight, P. M., and Chia, H. K. (1981). 'Prediction of shear behaviour of joints using profiles.' *Intl. Jl. Rock Mech. Min. Sci.,* **18**, 369–86.

Gerrard, C. M. (1982a). 'Equivalent elastic moduli of a rock mass consisting of orthorhombic layers.' *Intl. Jl. Rock Mech. Min. Sci. & Geomech. Abstr.,* **19**, 9–14.

Gerrard, C. M. (1982b). 'Elastic models of rock masses having one, two and three sets of joints.' *Intl. Jl. Rock Mech. Min. Sci.,* **19**, 15–23.

Gerrard, C. M. (1982c). 'Joint compliances as a basis for rock mass properties and the design of supports.' *Intl. Jl. Rock Mech. Min. Sci.,* **19**, 285–305.

Goodman, R. E. (1980). *An Introduction to Rock Mechanics*, J. Wiley & Sons, Inc.

Hobbs, N. B. (1975). 'Factors affecting the prediction of settlement of structures on rock: with particular reference to the Chalk and Triass in settlement of structures,' London: Pentech Press. pp. 579–610.

Jeager, J. C., and Cook, N. G. W. (1979). *Fundamentals of Rock Mechanics*, London: Chapman & Hall.
Obert, L., and Duvall, W. I. (1967). *Rock Mechanics and the Design of Structures in Rock*, J. Wiley & Sons, Inc.
Pande, G. N., and Sharma, K. G. (1981). 'A multilaminate model of clays—a numerical study of the influence of rotation of principal stress axes.' *Proc. Implementation of Computer Procedures and Stress–strain Laws in Geotechnical Engineering*, Vol. II, Durham, N.C.: Acorn Press.
Pande, G. N., and Sharma, K. G. (1983). 'Multilaminate model of clays—a numerical evaluation of the influence of rotation of the principal stress axes.' *Int. Jl. Num. Anal. Meth. Geomech.*, **7**, 397–418.
Pratt, H. R., Black, A. D., Brown, W. S. and Brace, W. F. (1972). 'The effect of specimen size on the mechanical properties of unjointed Diorite'. *Intl. Jl. Rock Mech. Min. Sci.*, **9**, 513–29.
Stephansson, O. (1981). 'The Naslider project—rock mass investigations.' London: *Applications of Rock Mechanics to Cut and Fill Mining, I.M.M.* 161.
Tse, R. and Cruden, R. M. (1979). 'Estimating joint roughness coefficients.' *Int. Jl. Rock Mech. Sci. & Geomech. Abstr.*, **16**, 303–7.
Walsh, J. B., and Brace, W. F. (1966). 'Elasticity of rock: a review of some recent theoretical studies.' *Rock Mech. and Engng Geol.*, **4**(4).
Weibull, W. (1939). 'A statistical theory of strength of materials.' *Proc. R. Swed. Acad. Eng. Sci.*
Wu, T. H., and Ali, E. M. (1978). 'Statistical representation of joint roughness.' **15**, 259–62.

# 5 The Finite Element Method

## 5.1 INTRODUCTION

In recent years, the finite element method has become the most popular numerical method in many branches of engineering. It has now applications in a wide variety of fields such as solid mechanics, fluid mechanics, biomechanics, electricity and magnetism, heat transfer, semiconductor devices, etc. It was first applied to aircraft structures in the 1940s. The term 'Finite Element' was first used by Turner *et al.* in 1956. Since then, there has been an exponential growth in the number of publications relating to the method. A number of text books have been written, a few of which are listed at the end of this chapter. In addition, there are a few books dealing primarily with the programming aspects of the finite element method (Hinton and Owen, 1977; Smith, 1981).

The main objective of this chapter is to explain the method in a simple form from the point of view of the practising engineers dealing with problems of rock mechanics in the fields of civil and mining engineering. An attempt has been made to avoid mathematical complexities as far as possible, since there are many texts dealing with the method rigorously.

We restrict ourselves here to load-displacement relationships of rock masses (structures) and a 'direct' derivation of the finite element equations. Generally, any physical phenomenon governed by a differential equation can be modelled by the finite element method formulated via the principles of variational calculus. The 'direct' derivation of load displacement relations has close similarity with the 'stiffness method' of analysis of the framed structures and is easier to follow for people with engineering background. The derivation via a differential equation and variational calculus is required for seepage problems.

A structural frame is a problem with finite degrees of freedom. In the stiffness method of analysis a relationship between the forces and the displacements and rotations at the joints or nodes is formulated. In a space frame, for example, every joint has six degrees of freedom: three displacements and three rotations. If the frame has $n$ free joints, the number of degrees of freedom are $6n$ and $6n$ stiffness equations are required to be set up to solve the problem.

In contrast to this situation, a two- or three-dimensional body (Gower potato of Chapter 1) has an infinite number of points having two or three degrees of freedom: displacements along two or three axes of reference. The number of stiffness equations in this case will be infinite and obviously the problem can not be solved. However, if we choose a finite but sufficiently large number of 'nodes', the problem can be solved and an approximate deformed shape can be obtained. This is, though rather superficially, the basis of the finite element method.

In the finite element method, the structure under consideration is divided into smaller zones, known as elements. The elements are assumed to be connected to each other at certain points (usually at the corners) called nodes. It is at the nodes that we compute the displacements. Thus the body with infinite number of degrees of freedom is approximated by a body having degrees of freedom equal to two or three times the number of nodes. It is obvious, though there are rigorous mathematical proofs available, that as the number of nodes is increased, a better (closer to the exact) solution is obtained.

The displacements at any point within an element are related to the displacements at the nodes by making certain assumptions. Displacements are fundamental variables. From the displacement field within the element, strains can be calculated. From the strains, using the stress–strain relationship, stresses can be calculated.

We shall now discuss details of two dimensional analysis. Once the concepts of two-dimensional analysis are understood, extension to three dimensional analysis would be straight forward.

## 5.2 TWO-DIMENSIONAL ANALYSIS

In two-dimensional problems, the body is divided into elements which may be triangular, rectangular or four-sided regions with straight or curved sides. Figure 5.1 shows these types of elements with the position of nodes.

The lower set of elements having nodes on the sides as well as corners are called 'parabolic' elements. This is because a parabolic interpolation rule is used for the displacements at any point within an element. The line elements are used for modelling of anchors and ties, props and flexibile lining. They have no resistance or stiffness in bending mode. Plate-bending elements can be used for representing stiff liners but are not very common.

Many commercially available packages have a 'library' of elements and any of these can be used. These include elements with more than one node on the sides of elements, called 'higher order' elements. Plate-bending elements are useful for modelling tunnel liners, but are available on relatively few programs. Usually two dimensional thin elements can be used instead of plate bending elements.

Parabolic elements appear to be the most popular and perhaps represent best value in computation time per node. Higher order elements are seldom used in geotechnical analysis. There has been a recent shift in popularity towards 'linear' elements (the upper set in Figure 5.1). This is due to microcomputers becoming more popular. The 'stiffness' of linear elements can be exactly integrated and programmed, while that of parabolic and higher-order elements requires numerical integration which can become time consuming on not too fast microcomputers.

### 5.2.1 Finite element mesh

The first step in the Finite Element analysis is to prepare a mesh of elements. The choice of mesh is arbitrary. The total number of elements (and nodes) to be used in a particular problem will depend on the desired accuracy of results, cost of computing and effort in data preparation. Generally, it is possible to obtain quite good results with only a few parabolic elements. The mesh does not have to be uniform, i.e. elements need not be of the same size. The general rule is to have smaller elements where sharp changes in stress are expected. Thus, smaller elements are desirable, for example, at the base of a footing, around the periphery of an underground cavity, and the toe of an excavation. Figure 5.2 shows typical finite element meshes for two problems.

Each of the elements, in theory, can have different material properties. In practice, however, one hardly ever needs to model more than a few materials. The finite element mesh will have to be organized in such a way that any element is in one material zone only, i.e. it does not saddle over two material zones.

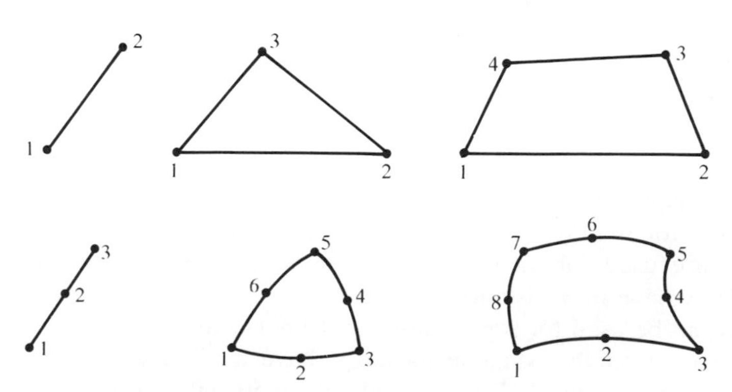

**Figure 5.1**
Shape and node position of some elements used in two-dimensional analysis

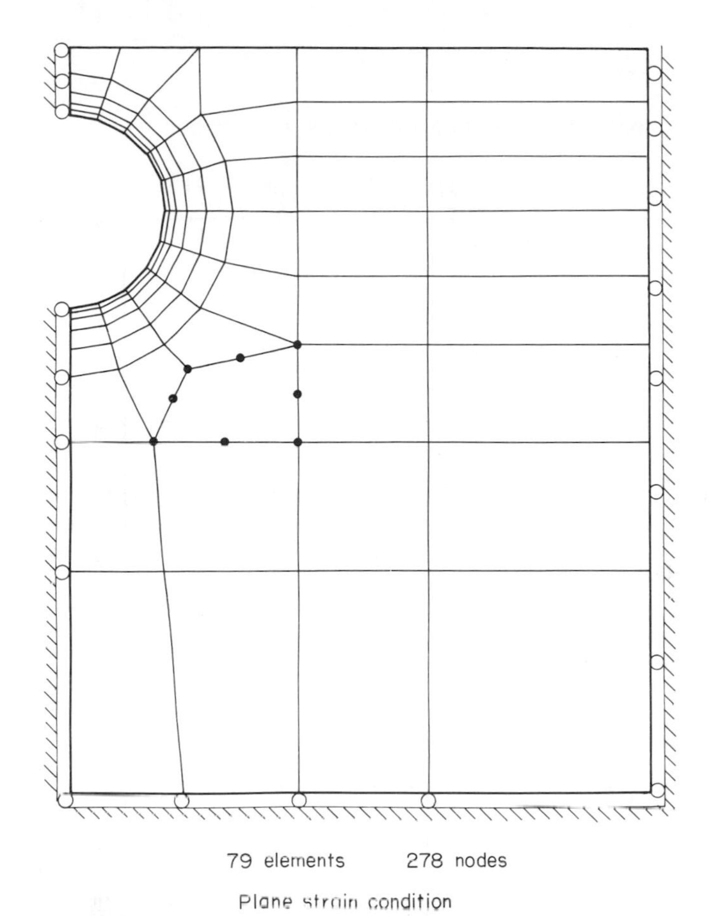


**Figure 5.2**
(a) Finite element mesh for a shallow model tunnel

## 5.2.2 Shape functions

As discussed earlier, the displacements are calculated at the nodes. Displacements at any point within the element are related to the displacements of the nodes ($\boldsymbol{\delta}$) through shape functions. Let us consider a square element with four nodes as shown in Figure 5.3. The sides of the square are two units. The coordinate axes are $\xi$, $\eta$, with the origin at the centre of the square. The reader may at this stage find the choice of shape and size of the element and coordinate axes a bit too restrictive or peculiar. This choice has been deliberately made. The reason is that any quadrilateral element of whatever shape or size, referred to any arbitrary set of axes, can be 'mapped' into an element of Figure 5.3 by transformation of coordinate system. How this is done is discussed in detail in Section 5.2.3.

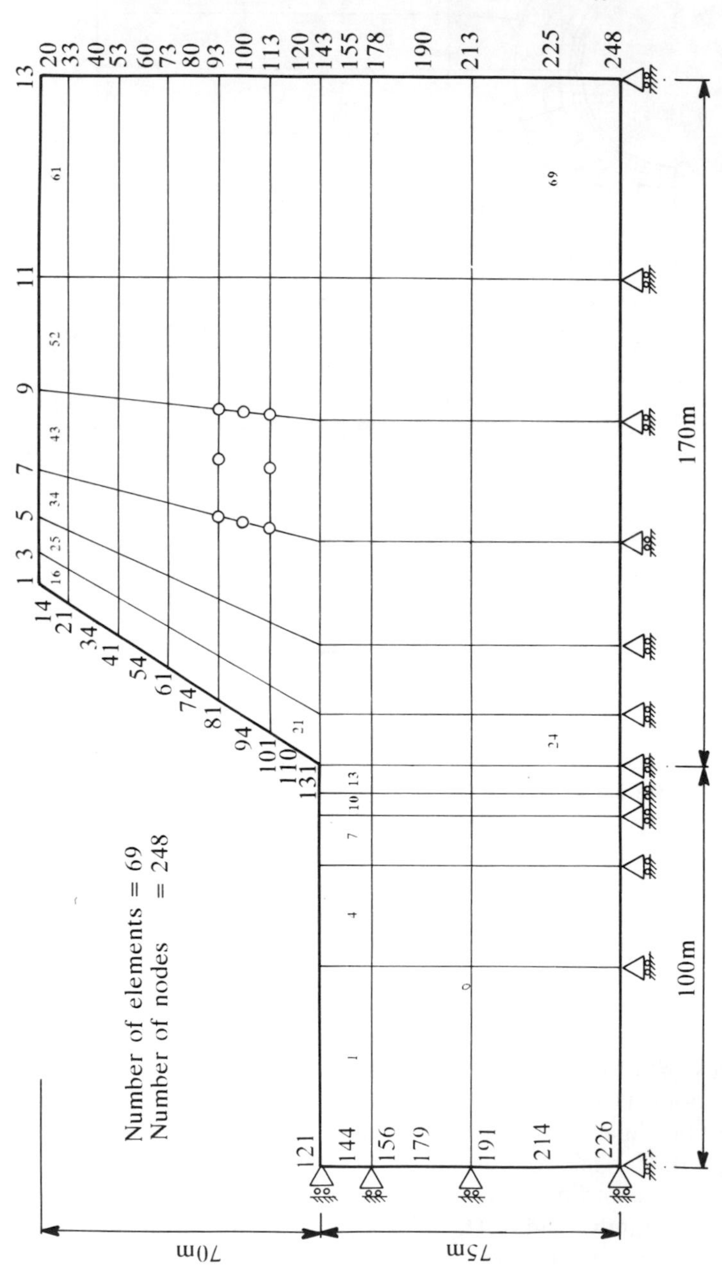


**Figure 5.2**
(b) Finite element mesh for a rock slope

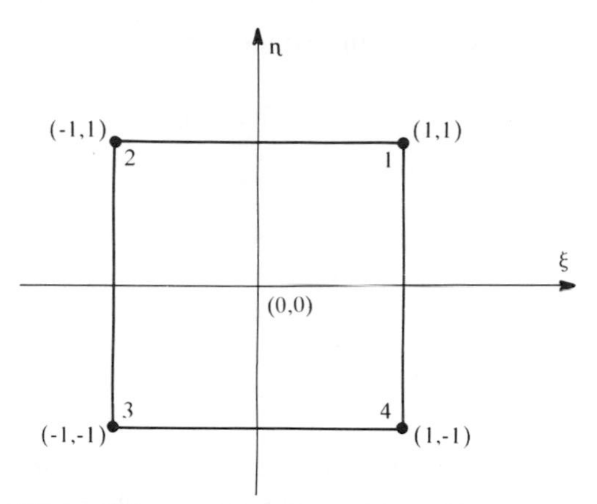


**Figure 5.3**
A quadrilateral element mapped into a 2 × 2 unit square element

The displacements at any point within the element ($u$, $v$) are represented by the equation:

$$\left\{\begin{matrix} u \\ \overline{v} \end{matrix}\right\} = \begin{bmatrix} N_1 & 0 & N_2 & 0 & N_3 & 0 & N_4 & 0 \\ 0 & N_1 & 0 & N_2 & 0 & N_3 & 0 & N_4 \end{bmatrix} \left\{\begin{matrix} u_1 \\ v_1 \\ u_2 \\ v_2 \\ u_3 \\ v_3 \\ u_4 \\ v_4 \end{matrix}\right\} \tag{5.1}$$

where $(u_1, v_1, \ldots u_4, v_4\} = \boldsymbol{\delta}^T$ are nodal displacements and $N_1$, $N_2$, $N_3$, $N_4$ are shape functions associated to nodes 1, 2, 3 and 4 respectively. The shape functions $N_i$ ($i = 1, 2, \ldots 4$) are functions of the coordinates of the points whose displacements are required. Thus

$$N_i = N_i\{\xi, \eta)$$

They must have the property that when coordinates of the $i$th node are substituted, then $N_i$ should equal 1 and $N_{j \neq i}$ should be equal to zero. The reason is obvious from equation (5.1). If we substitute the coordinates of, say, node 1, $N_1 = 1$, $N_2 = N_3 = N_4 = 0$, then we have $u = u_1$ and $v = v_1$, which is the desired result.

Now, let us inspect the following shape functions

$$
\begin{aligned}
N_1 &= \frac{(1+\xi)(1+\eta)}{4} \\
N_2 &= \frac{(1-\xi)(1+\eta)}{4} \\
N_3 &= \frac{(1-\xi)(1-\eta)}{4} \\
N_4 &= \frac{(1+\xi)(1-\eta)}{4}
\end{aligned}
\tag{5.2}
$$

Noting that the coordinates of nodes 1, 2, 3 and 4 are (1, 1), (−1, 1), (−1, −1) and (1, −1) respectively, the shape functions satisfy the criterion laid down above.

Thus, equation (5.2) represents the shape functions of a four-noded quadrilateral element. The shape functions of most commonly used elements were found out simply by intuition and inspection and such shape functions are said to belong to serendipity† family. For higher-order elements, one has to use far more formal mathematical procedures to derive expressions for shape functions.

Table 5.1 gives most commonly used elements in two-dimensional analysis, their mapping in the local coordinates and the shape functions.

### 5.2.3 Coordinate transformation

The main aim of coordinate transformation is to facilitate integration of certain quantities required for calculation of stiffness matrices of the elements. Referring to Figure 5.3, if a function, $f$, is to be integrated over the area of the square, we could write

$$
F = \int_{-1}^{+1}\int_{-1}^{+1} f(\xi, \eta)\,d\xi\,d\eta \tag{5.3}
$$

evaluation of which is much easier than integration over the area of irregular shaped parent rectangle.

Let us now consider a four-noded element in global system of coordinates. This element is to be mapped on to a square of sides equal to two units.

† This word was coined by Horace Walpole, 1754, after princes of Serendip, Sri Lanka who had amazing powers of discovering things by chance.

We can use the shape functions to map as well. Let the coordinates of a point within the element be given by

$$x = N_1x_1 + N_2x_2 + N_3x_3 + N_4x_4$$
$$y = N_1y_1 + N_2y_2 + N_3y_3 + N_4y_4 \tag{5.4}$$

This can be written in a more compact form

$$\begin{Bmatrix} x \\ y \end{Bmatrix} = \begin{bmatrix} N_1 & 0 & N_2 & 0 & N_3 & 0 & N_4 & 0 \\ 0 & N_1 & 0 & N_2 & 0 & N_3 & 0 & N_4 \end{bmatrix} \begin{Bmatrix} x_1 \\ y_1 \\ x_2 \\ y_2 \\ x_3 \\ y_3 \\ x_4 \\ y_4 \end{Bmatrix} \tag{5.5}$$

$$x = \sum_{i=1}^{4} N_i x_i \tag{5.6}$$

$$y = \sum_{i-1}^{4} N_i y_i$$

where $(x_1, y_1)$ $(x_2, y_2)$ etc. are the coordinates of the nodes in a $x$, $y$ coordinate system and $N_1$, $N_2$, $N_3$ and $N_4$ are shape functions discussed in Section 5.2.2 and explicitly given by equation (5.2). These shape functions are functions of $\xi$, $\eta$. Readers may note that if $\xi$, $\eta$ values of any of the corner nodes are substituted in equation (5.5), coordinates of that corner node are recovered. If $\xi = 0$ and $\eta = 0$ is substituted, the coordinates of the centre of the quadrilateral are obtained. It is easy to check that the equations of the sides of the quadrilateral are $\xi = +1$, $\eta = +1$, $\xi = -1$ and $\eta = -1$.

From mathematics, it follows that an infinitesimal area dA is given by

$$\mathrm{d}A = \mathrm{d}x \cdot \mathrm{d}y = |J|\, \mathrm{d}\xi\, \mathrm{d}\eta \tag{5.7}$$

where $|J|$ represents the determinant of a matrix called Jacobian matrix, given by

$$J = \begin{bmatrix} \partial x/\partial \xi & \partial y/\partial \xi \\ \partial x/\partial \eta & \partial y/\partial \eta \end{bmatrix} \tag{5.8}$$

**Table 5.1**
Shape functions of various one- and two-dimensional finite elements

| Element | Shape | Mapped shape | Shape functions |
|---|---|---|---|
| 1. 6-noded curved triangle | 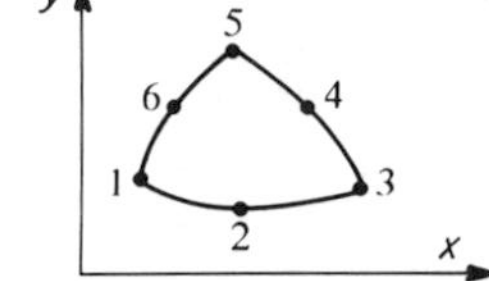 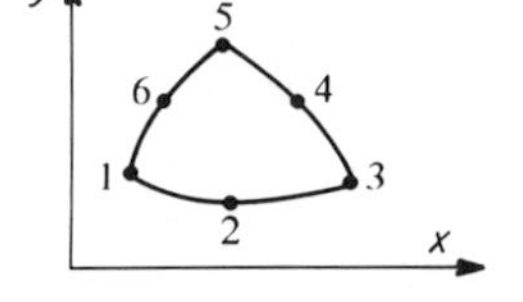 | | $N_1 = 1 - 3(\xi + \eta) + 2(\xi + \eta)^2$<br>$N_2 = 4\xi(1 - \xi - \eta)$<br>$N_3 = \xi(2\xi - 1)$<br>$N_4 = 4\xi\eta$<br>$N_5 = \eta(2\eta - 1)$<br>$N_6 = 4\eta(1 - \xi - \eta)$ |
| 2. 4-noded quadrilateral | | 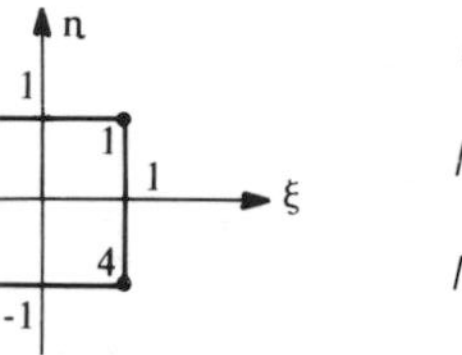 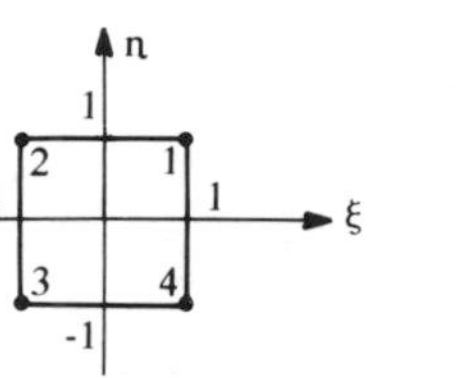 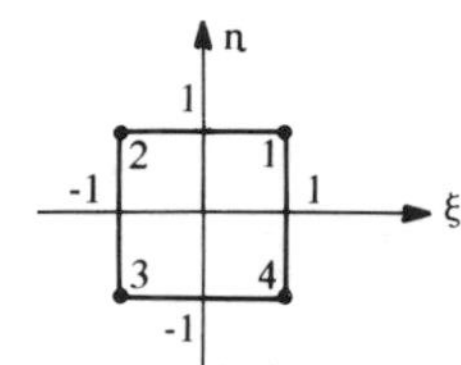 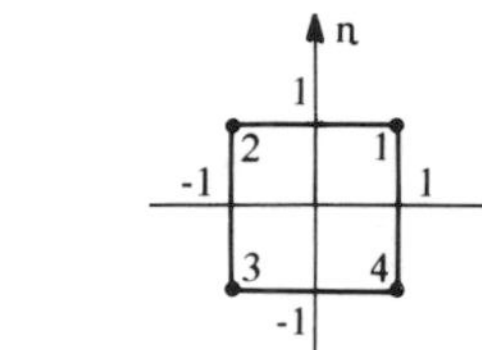 | $N_1 = \frac{(1 + \xi)(1 + \eta)}{4}$<br>$N_2 = \frac{(1 - \xi)(1 + \eta)}{4}$<br>$N_3 = \frac{(1 - \xi)(1 - \eta)}{4}$<br>$N_4 = \frac{(1 + \xi)(1 - \eta)}{4}$ |

3. 8-noded parabolic

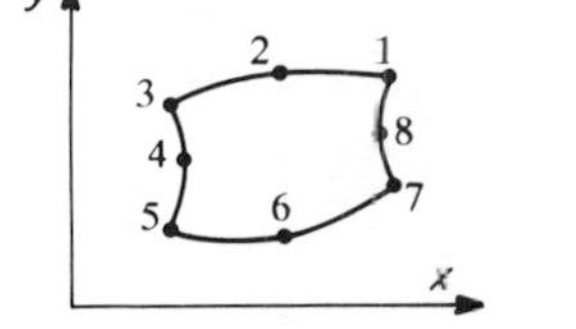
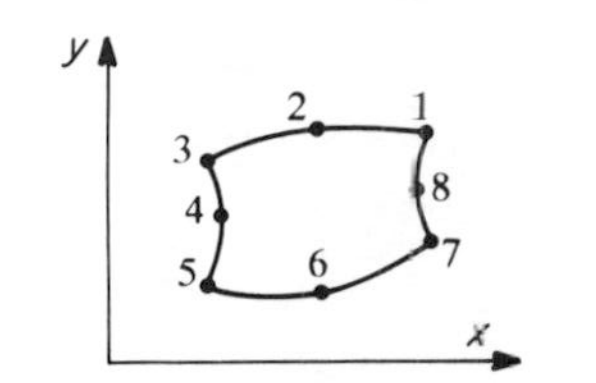

$N_1 = -\frac{1}{4}(1+\xi)(1+\eta)(1-\xi-\eta)$
$N_2 = \frac{1}{2}(1+\eta)(1-\xi^2)$
$N_3 = -\frac{1}{4}(1+\xi)(1-\eta)(1+\xi-\eta)$
$N_4 = \frac{1}{2}(1-\xi)(1-\eta^2)$
$N_5 = -\frac{1}{4}(1-\xi)(1-\eta)(1+\xi+\eta)$
$N_6 = \frac{1}{2}(1-\eta)(1-\xi^2)$
$N_7 = -\frac{1}{4}(1+\xi)(1-\eta)(1-\xi+\eta)$
$N_8 = \frac{1}{2}(1+\xi)(1-\eta^2)$

4. 2-noded line element

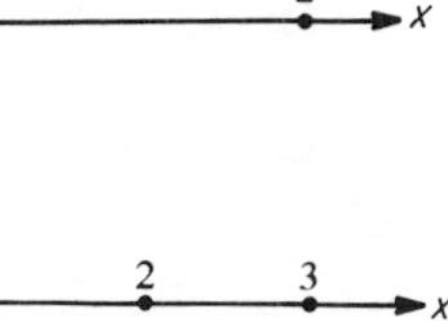

$$N_1 = \frac{1-\xi}{2}$$
$$N_2 = \frac{1+\xi}{2}$$

5. 3-noded line element

$N_1 = -\frac{1}{2}\xi(1-\xi)$
$N_2 = 1-\xi^2$
$N_3 = \frac{1}{2}\xi(1+\xi)$

The elements of the Jacobian matrix can be readily found by differentiating equation (5.6)

$$\begin{aligned}\partial x/\partial \xi &= \sum_{i=1}^{4} (\partial N_i/\partial \xi)x_i \\ \partial x/\partial \eta &= \sum_{i=1}^{4} (\partial N_i/\partial \eta)x_i \\ \partial y/\partial \xi &= \sum_{i=1}^{4} (\partial N_i/\partial \xi)y_i \\ \partial y/\partial \eta &= \sum_{i=1}^{4} (\partial N_i/\partial \eta)y_i\end{aligned} \tag{5.9}$$

$\partial N_i/\partial_\xi$ and $\partial N_i/\partial_\eta$ are in turn found out by differentiating the expressions for shape functions, $N_i$, given in Table 5.1.

In the formulation of stiffness matrix, we shall also require derivatives of the shape functions with respect to Cartesian coordinates, i.e.

$$\frac{\partial N_i}{\partial_x}, \qquad \frac{\partial N_i}{\partial_y}$$

Applying the chain rule, we can write

$$\mathrm{d}N_i = \frac{\partial N_i}{\partial x} \cdot \mathrm{d}x + \frac{\partial N_i}{\partial y} \cdot \mathrm{d}y \tag{5.10}$$

Partially differentiating the above equation leads to

$$\begin{Bmatrix} \partial N_i/\partial \xi \\ \partial N_i/\partial \eta \end{Bmatrix} = [J] \begin{Bmatrix} \partial N_i/\partial x \\ \partial N_i/\partial y \end{Bmatrix} \tag{5.11}$$

which can be rewritten as

$$\begin{Bmatrix} \partial N_i/\partial x \\ \partial N_i/\partial y \end{Bmatrix} = [J]^{-1} \begin{Bmatrix} \partial N_i/\partial \xi \\ \partial N_i/\partial \eta \end{Bmatrix} \tag{5.12}$$

### 5.2.4 Strain displacement relations

The components of strain ($\boldsymbol{\epsilon}^T = \epsilon_x, \epsilon_y, \epsilon_z, \gamma_{xy}$) at any point are related to displacements by the following relations of mechanics:

(a) Plane stress or plane strain

$$\begin{aligned} \epsilon_x &= \partial u/\partial x \\ \epsilon_y &= \partial v/\partial y \\ \gamma_{xy} &= (\partial u/\partial y + \partial v/\partial x) \\ \epsilon_z &= 0 \quad \text{for plane strain} \end{aligned} \tag{5.13}$$

where $u$, $v$ are displacements along Cartesian coordinate axes $x$ and $y$.

(b) Axi-symmetric: Here, assuming $y$ as the axis of symmetry and $x$ as the radial direction

$$\begin{aligned} \epsilon_x &= \partial u/\partial x = \epsilon_r \quad && \text{(radial strain)} \\ \epsilon_y &= \partial v/\partial y = \epsilon_a && \text{(axial strain)} \\ \epsilon_z &= \quad u/x = \epsilon_t && \text{(hoop or tangential strain)} \end{aligned} \tag{5.14}$$

We can write the strains as a function of nodal displacements by making use of equation (5.1). Thus, for a plane strain case,

$$\boldsymbol{\epsilon} = \begin{Bmatrix} \epsilon_x \\ \epsilon_y \\ \gamma_{xy} \\ \epsilon_z \end{Bmatrix} = B \begin{Bmatrix} u_1 \\ v_1 \\ u_2 \\ v_2 \\ \vdots \\ u_n \\ v_n \end{Bmatrix} = B\,\boldsymbol{\delta} \tag{5.15}$$

It is seen that the size of $B$ matrix is $(4 \times 2n)$ where $n$ is the number of nodes in the element. $B$ matrix consists of $n$ submatrices, $B_i(i = 1, \ldots n)$, of size $(4 \times 2)$ whose elements are

$$B_i = \begin{Bmatrix} \partial N_i/\partial x & 0 \\ 0 & \partial N_i/\partial y \\ \partial N_i/\partial y & \partial N_i/\partial x \\ 0 & 0 \end{Bmatrix} \tag{5.16}$$

The need for deriving expressions for $\partial N_i/\partial x$ and $\partial N_i/\partial y$ in the previous section is now apparent.

### 5.2.5 Stress–strain relations

These relations are the crux of the whole analysis. We shall be discussing these in detail in Chapter 7. In the most general form, these can be expressed as

$$\Delta\boldsymbol{\sigma} = D_T \Delta\boldsymbol{\epsilon} \tag{5.17}$$

where $\boldsymbol{\sigma} = (\sigma_x, \sigma_y, \tau_{xy}, \sigma_z)^T$ is the vector of stress components and $\boldsymbol{\epsilon}$ represents corresponding components of strains. $D_T$ is a square matrix which is constant in elastic case but in the case of a general nonlinear behaviour, is stress or strain path dependent. Equation (5.17) is incremental, i.e. increments of stress and strain are related and therefore $D_T$ represents a tangent material modulus matrix.

### 5.2.6 Stiffness equations of an element

Similar to the procedures in structural analysis, we first derive the equations of stiffness for an element and then 'assemble' them to obtain the global stiffness of the structure.

We consider a typical element in isolation. The element is subjected to forces $\mathbf{F}^e$ acting at the nodes given by

$$\mathbf{F}^e = \begin{Bmatrix} F_{x1} \\ F_{y1} \\ F_{x2} \\ F_{y2} \\ \vdots \\ F_{x_n} \\ F_{y_n} \end{Bmatrix} \tag{5.18}$$

and deforms with associated nodal displacements $\boldsymbol{\delta}^e$ given by

$$\boldsymbol{\delta}^e = \begin{Bmatrix} u_1 \\ v_1 \\ u_2 \\ v_2 \\ \vdots \\ u_n \\ v_n \end{Bmatrix} \tag{5.19}$$

and is in equilibrium.

The 'principle of virtual work' states that if a structure is in equilibrium, the work done by the external forces through a set of virtual displacements ($\boldsymbol{\delta}^*$) must equal the work done by internal stresses through the virtual strains ($\boldsymbol{\epsilon}^*$) caused by the applied virtual displacements.

Applying this principle, we can write

$$(\boldsymbol{\delta}^*)^T \mathrm{F}^e = \int_V (\boldsymbol{\epsilon}^*)^T \boldsymbol{\sigma}\, \mathrm{d}V = \int_V (B\boldsymbol{\delta}^*)^T \boldsymbol{\sigma}\, \mathrm{d}V \tag{5.20}$$

Integration of the right-hand side of equation (5.19) is performed over the volume of the element ($V$). Substituting equation (5.15) in (5.20) leads to

$$\mathbf{F}^e = \int_V B^T \boldsymbol{\sigma}\, \mathrm{dV} \tag{5.21}$$

Note that $\boldsymbol{\delta}^*$ are arbitrary virtual displacements which cancel out to obtain equation (5.21). Equation (5.21) is an important equation and is used to find equivalent nodal forces due to any stress field in equilibrium.

Equation (5.21) can also be written in an incremental form, i.e.

$$\Delta\mathbf{F}^e = \int_V B^T \Delta\boldsymbol{\sigma}\, \mathrm{d}V \tag{5.22}$$

where $\Delta$ denotes an increment of the variable. Substituting $\Delta\boldsymbol{\sigma}$ in equation (5.22) from equation (5.17) and using equation (5.15) leads to

$$\Delta\mathbf{F}^e = \int_V B^T D_T\, \Delta\boldsymbol{\epsilon}\, \mathrm{d}V \tag{5.23}$$

$$\Delta\mathbf{F}^e = \int_V B^T D_T\, B\, \Delta\boldsymbol{\delta}^e\, \mathrm{d}V \tag{5.24}$$

$$\Delta\mathbf{F}^e = K^e\, \Delta\boldsymbol{\delta}^e \tag{5.25}$$

where

$$K^e = \int_V B^T D_T\, B\, \mathrm{d}V \tag{5.26}$$

$K^e$ is the stiffness matrix of the element. If $D_T$ is constant, $K^e$ is also constant. Otherwise, $K^e$ represents a variable tangent stiffness matrix and applies only to the appropriate $\Delta\boldsymbol{\sigma}$ and $\Delta\boldsymbol{\epsilon}$.

In problems of rock mechanics, the rock structure has 'initial' stresses. For example, before the tunnel is excavated, there are stresses in the rock mass which already exist and must be accounted for. The initial strains are usually taken as zero. This simply implies that we look for deformations and strains subsequent to the application of loads or excavation. In many situations where we try to simulate excavation, rock bolting, grouting, etc., in stages the strains which occurred up to the previous stage can be regarded as 'initial' strains. Moreover, initial strains may arise due to causes other than external loads, e.g. due to changes in temperature, creep, saturation, etc. The constitutive relation of equation (5.17) applies only to the structural stresses and strains and the general form of equation (5.16) is

$$\Delta(\boldsymbol{\sigma} - \boldsymbol{\sigma}_0) = D_T \Delta(\boldsymbol{\epsilon} - \boldsymbol{\epsilon}_0) \tag{5.27a}$$

where $\boldsymbol{\sigma}_0$ and $\boldsymbol{\epsilon}_0$ are initial stresses and strains, respectively. Equation (5.27a) can be rearranged as

$$\Delta\boldsymbol{\sigma} = D_T \Delta\boldsymbol{\epsilon} - D_T \Delta\epsilon_0 + \Delta\sigma_0 \tag{5.27b}$$

which, when substituted in equation (5.22) gives

$$\Delta\mathbf{F}^e = \int_v B^T D_T \Delta\boldsymbol{\epsilon}\, \mathrm{d}v - \int_v B^T D_T \Delta\boldsymbol{\epsilon}_0\, \mathrm{d}v + \int_v B^T \Delta\boldsymbol{\sigma}_0\, \mathrm{d}v \tag{5.28}$$

Equation (5.28) can be rearranged to give

$$K^e \Delta\boldsymbol{\delta}^e = \Delta\mathbf{F}^e + \int_v B^T D_T \Delta\boldsymbol{\epsilon}_0\, \mathrm{d}v - \int_v B^T \Delta\boldsymbol{\sigma}_0\, \mathrm{d}v \tag{5.29}$$

$$K^e \Delta\boldsymbol{\delta}^e = \Delta R^e \tag{5.30}$$

where

$$\Delta\mathbf{R}^e = \Delta\mathbf{F}^e + \int_v B^T D_T \Delta\boldsymbol{\epsilon}_0\, \mathrm{d}v - \int_v B^T \Delta\boldsymbol{\sigma}_0\, \mathrm{d}v \tag{5.31}$$

represents equivalent nodal forces acting on the element; the second and third terms on the right-hand side of equation (5.31) arise due to 'initial strains' and 'initial stresses' respectively.

$\Delta\mathbf{R}^e$ is generally referred to as the 'right-hand side' and includes equivalent forces due to all loads. Distributed pressures and body forces also are converted into equivalent nodal forces and added on the right-hand side before calculating for displacements. How equivalent nodal forces are found for body forces and surface pressures and other actions is discussed in Sections 5.2.8 and 5.2.9 respectively.

Before we discuss the next step, i.e. assembly of element stiffnesses to formulate 'global' stiffness, a few remarks on the element stiffness, $K^e$, are appropriate here. The stiffness is given by

$$K^e = \int_v B^T D_T B \, \mathrm{d}v \qquad (5.26) \text{ bis}$$

The size of $B$ is $4 \times 2n$, where $n$ is the number of nodes in the element. $B^T$ will be of size $2n \times 4$ and the final matrix will have size $2n \times 2n$. If the problem was three dimensional, the size of $K^e$ will be $3n \times 3n$.

Stiffness matrix $K^e$ is symmetrical if, and only if, $D_T$ is symmetric. There are many situations of nonlinear problems where $D_T$ may not be symmetric and in these situations, element stiffness and consequently global stiffness, of the structure will not be symmetric.

### 5.2.7 Global stiffness

In the last section, we have derived the expressions for stiffness of an element. Using equation (5.26), the stiffness of each of the elements is computed (the integration involved can be performed analytically for simple elements, e.g. constant strain triangle, but techniques of numerical integration are required for higher order elements). These stiffnesses have to be 'added' to obtain a stiffness matrix for the whole structure which will have a size corresponding to the total number of degrees of freedom in the whole mesh, e.g. twice the number of nodes for 2D problems and three times the number of nodes for 3D problems. A node which is common to a number of elements, will have contribution to stiffness from all those elements. The procedure of assembly of the overall stiffness is exactly the same as for framed structures and the reader not familiar with this should refer to standard text books on structural mechanics such as Ghali and Neville (1972).

The right-hand side of the global stiffness equations is also assembled from the right hand side vector $\mathbf{R}^e$, again following the well established principles of structural mechanics. Finally, the global stiffness equations have the form

$$K \, \Delta \boldsymbol{\delta} = \Delta \mathbf{R} \qquad (5.32)$$

where $\Delta \boldsymbol{\delta}$ *is the unknown vector of increments of nodal displacements due to an increment of force* $\Delta \mathbf{R}$.

Equation (5.32) has been written in the 'incremental' form assuming $K$ is changeable for each load (or right-hand side) increment. If the material behaviour is linear elastic, $K$ is constant and equation (5.32) can be written in the total form, i.e.

$$K \boldsymbol{\delta} = \mathbf{R} \qquad (5.33)$$

### 5.2.8 Computation of nodal forces

In the finite element method all forces, either due to gravity, temperature, initial stress or strain, pore pressures, centrifugal, inertia, etc. are converted into equivalent forces acting at the nodes. In Section (5.2.6) we have already seen that initial strains ($\Delta\boldsymbol{\epsilon}_0$) give rise to nodal forces ($\Delta\mathbf{R}_{\epsilon 0}$) given by

$$\Delta\mathbf{R}^e_{\epsilon 0} = \int_v B^T D_T \,\Delta\boldsymbol{\epsilon}_0 \,\mathrm{d}v \tag{5.34}$$

and initial stresses ($\Delta\boldsymbol{\sigma}_0$) give rise to nodal forces ($\Delta\mathbf{R}_{\sigma_0}$) given by

$$\Delta\mathbf{R}^e_{\sigma_0} = \int_v B^T \,\Delta\boldsymbol{\sigma}_0 \,\mathrm{d}v \tag{5.35}$$

Note that equations (5.34) and (5.35) are in 'incremental' form.

The expressions for equivalent nodal forces for other actions likely to be encountered in rock engineering problems are given below:

(1) *Body forces:* Equivalent forces due to body forces for an element ($\mathbf{R}^e_b = \{R_{bx}, R_{by}\}^T$) are given by

$$\mathbf{R}^e_{bx} = \int_v \rho_x N_i \,\mathrm{d}v \tag{5.36}$$

$$\mathbf{R}^e_{by} = \int_v \rho_y N_i \,\mathrm{d}v$$

where $\rho_x$, $\rho_y$, are the body forces (forces per unit volume) in the $x$, $y$ directions, respectively and $N_i$ are shape functions at a point whose coordinates are $x$, $y$.

Note that a coordinate transformation will be required for integration of the terms over the volume of element on the right-hand side of equation (5.36).

(2) *Surface tractions:* Equivalent nodal forces due to surface tractions ($\mathbf{p} = [\tau, \sigma_n]^T$) acting on an edge or face of an element $\Delta\mathbf{R}^e_p$ are given by

$$\Delta\mathbf{R}^e_p = \int \mathbf{p}\, T N_i \,\mathrm{d}A \tag{5.37}$$

where $T$ is a transformation matrix given by

$$T = \begin{bmatrix} \cos\theta & \sin\theta \\ -\sin\theta & \cos\theta \end{bmatrix}$$

Here, $\theta$ is the angle that the loaded edge of the element makes with the global $x$-axis and the integration is carried out over the area of loaded edge ($A$).

(3) *Pore pressures:* The pore pressures can be dealt with in exactly the same manner as initial stresses. Thus, the equivalent nodal forces ($\Delta\mathbf{R}_u^e$), due to a change of pore pressure ($\Delta u$) for an element are given by

$$\Delta\mathbf{R}_u^e = -\int_v B^T \Delta u \begin{Bmatrix} 1 \\ 1 \\ 0 \\ 1 \end{Bmatrix} \mathrm{d}v \tag{5.38}$$

Note that initial pore pressures should be included as initial stresses in $\sigma_0$ term when applying the general constitutive equation (5.27a, b).

(4) *Temperature changes:* In a rock mass which is restrained, a change in temperature would induce strains and stresses. If the coefficient of thermal expansion of the rock is $\alpha$, a change in temperature of $\Delta\theta$ produces linear strain of $\alpha\,\Delta\theta$. The equivalent nodal forces due to a change of temperature for an element ($\Delta\mathbf{R}_T$) are dealt with in a manner similar to initial strains. Thus,

$$\Delta\mathbf{R}_T^e = -\int_v B^T D_{T\Delta}\,\boldsymbol{\epsilon}_T\,\mathrm{d}v \tag{5.39}$$

where $\Delta\boldsymbol{\epsilon}_T$ are thermal strains given by

$$\Delta\boldsymbol{\epsilon}_T = \begin{Bmatrix} \alpha \\ \alpha \\ 0 \\ \alpha \end{Bmatrix} \Delta\theta \tag{5.40}$$

and must be included as initial strains in $\epsilon_0$ term when applying the general constitutive equation (5.27a, b).

## 5.3 THREE-DIMENSIONAL ANALYSIS

Once the concepts of finite element method in two dimensions are understood, extension to three dimensions is straight forward. In this section, we shall go over the various aspects of the formulation in three dimensions following closely the pattern of two-dimensional analysis discussed in Section 5.2.

### 5.3.1 Element shapes

Here the body is divided into smaller three-dimensional regions. The equivalent of a triangular element is a tetrahedron in 3D and that of a four-noded rectangle is the eight-noded brick element. Figure 5.4 shows some commonly used elements in three-dimensional analysis.

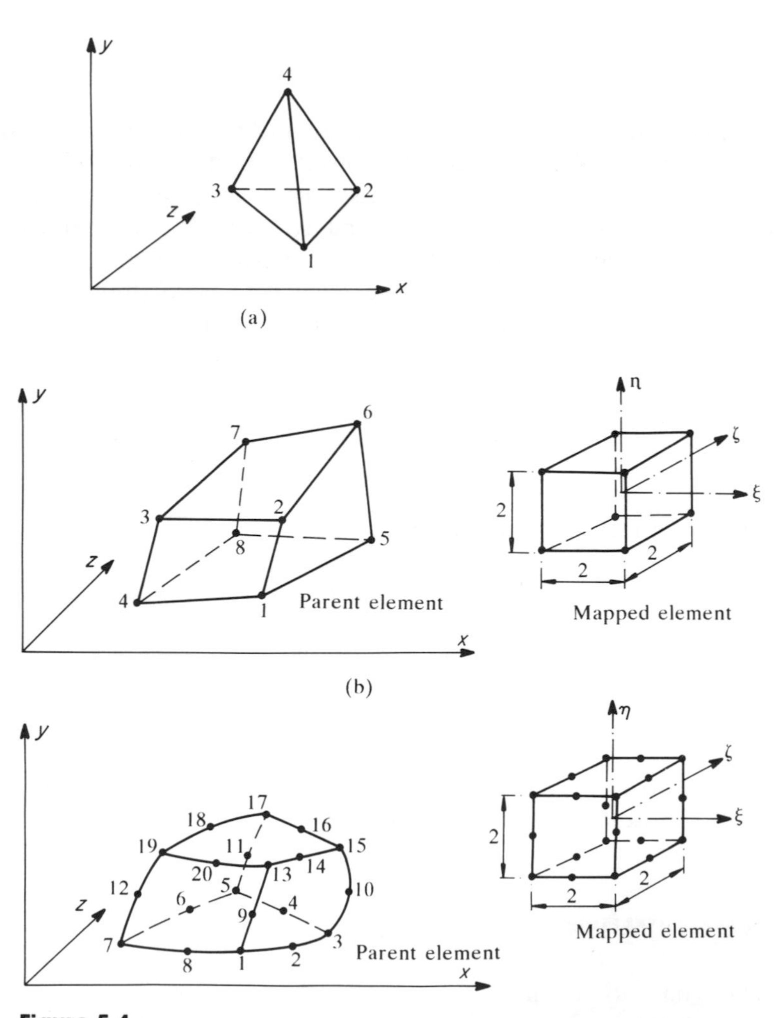


**Figure 5.4**
Some commonly used 3D elements and their mapping in local co-ordinate system. (a) Tetrahedron; (b) 8-noded brick; (c) 20-noded brick

### 5.3.2 Finite element mesh

The task of generating Finite Element meshes in three dimensions is not trivial. Here computer programs which generate the mesh, number the nodes and elements are essential and indeed a number of such programs are commercially available. The size of these programs in terms of numbers of lines of coding is many times more than the entire finite element program.

The size and grading of elements in the mesh is again chosen based on the gradient of stresses expected in the problem. More (smaller) elements in the region where concentrations of stresses are expected and less (larger) elements in the regions where stresses are likely to be uniform.

It is apparent that considerable attention is to be paid in plotting and displaying the mesh. Generally a number of plots of the mesh from various positions of the eye are required to ensure that the mesh is correctly set up.

### 5.3.3 Shape functions of 3D elements

Here again the displacement at any point $(u, v, w)$ inside the element is related to the displacements at the nodes $(u_1, v_1, w_1$, etc.) through shape functions $(N_i)$. The shape functions are generally defined in local system of coordinates $(\xi, \eta, \zeta)$. The relationship, similar to the two-dimensional case, takes the form

$$\begin{Bmatrix} u \\ v \\ w \end{Bmatrix} = \begin{bmatrix} N_1 & 0 & N_2 & 0 & \cdots \\ 0 & N_1 & 0 & N_2 & \cdots \\ 0 & 0 & N_1 & 0 & \cdots \end{bmatrix} \begin{Bmatrix} u_1 \\ v_1 \\ w_1 \\ u_2 \\ v_2 \\ w_2 \\ \vdots \end{Bmatrix} \tag{5.41}$$

The shape functions for tetrahedron, 8-noded and 20-noded brick elements are given below.

Readers requiring further information on other types of elements should refer to texts on the finite element method.

*Tetrahedron*: In the case of tetrahedron, it is quite convenient to write the shape functions in the global coordinate system. Referring to the Figure 5.4(a), the shape functions are written as

$$N_i = (a_i + b_i x + c_i y + d_i z)/6V \tag{5.42}$$

where $i = 1, 2, 3, 4$ and $V$ is the volume of the tetrahedron given by

$$6V = \det \begin{vmatrix} 1 & x_2 & y_2 & z_2 \\ 1 & x_3 & y_3 & z_3 \\ 1 & x_4 & y_4 & z_4 \\ 1 & x_1 & y_1 & z_1 \end{vmatrix} \tag{5.43}$$

and $a_i$, $b_i$, $c_i$, $d_i$ are constants. These are obtained from the expressions for $6 \times V$ similar to equation (5.43). The way it is done will be explained using (5.43) as an example. If we wish to write the shape function for any node, look at the opposite face of the tetrahedron with the eye placed at that node. Write the expression for $6 \times V$ similar to equation (5.43) placing the coordinate of the nodes in the opposite face in an anticlockwise order in the second third and fourth rows of the determinant. Complete the determinant by placing the coordinates of the node for which shape functions are required in the first row. Equation (5.43) has been written for node 2. $a_2$, $b_2$, $c_2$ and $d_2$ are cofactors when the determinant is expanded using the first row. Thus,

$$a_2 = \det \begin{vmatrix} x_3 & y_3 & z_3 \\ x_4 & y_2 & z_4 \\ x_1 & y_1 & z_1 \end{vmatrix} \tag{5.44}$$

$$b_2 = -\det \begin{vmatrix} 1 & y_3 & z_3 \\ 1 & y_4 & z_4 \\ 1 & y_1 & z_1 \end{vmatrix} \tag{5.45}$$

and so on.

Substituting the values of $a_i$ in equation (5.42) gives explicit expressions for $N_1$, $N_2$, $N_3$ and $N_4$. The displacements at any point $(u, v, w)$ within the tetrahedron are given by

$$\begin{Bmatrix} u \\ v \\ w \end{Bmatrix} = [IN_1 \quad IN_2 \quad IN_3 \quad IN_4] \begin{Bmatrix} u_1 \\ v_1 \\ w_1 \\ \vdots \\ u_4 \\ v_4 \\ w_4 \end{Bmatrix} \tag{5.46}$$

where $I$ is a $3 \times 3$ identity matrix and $u_1$, $v_1$, $w_1$, etc. are displacements of nodes 1, 2, 3 and 4. It is noted that shape functions of the tetrahedron element are written in the global system of coordinates.

*Eight-noded brick element:* The shape functions for eight-noded brick elements can be conveniently written in the local coordinate system ($\xi$, $\eta$, $\zeta$), Figure 5.4(b). They are obtained by induction and extrapolation of shape functions given in equation (5.2) and are given in Table 5.2.

**Table 5.2**
Shape functions of various three-dimensional elements

| Element | Mapped shape | Shape functions |
|---|---|---|
| Linear tetrahedron | 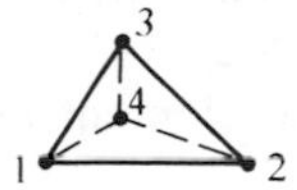 | $N_i = L_i$ where $L_i = (a_i + b_i x + c_i y + d_i z)/6V$ and $a_i$, $b_i$ etc are given by equations 5.44 and 5.45 as explained in Section 5.3.3 |
| Quadratic tetrahedron | 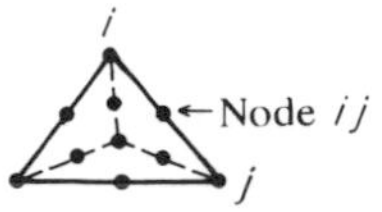 | $N_i = L_i\,(2L_i - 1)$ for nodes at the vertices of tetrahedron for<br>$N_{ij} = 4L_iL_j$ mid-side nodes |
| 8-noded brick | 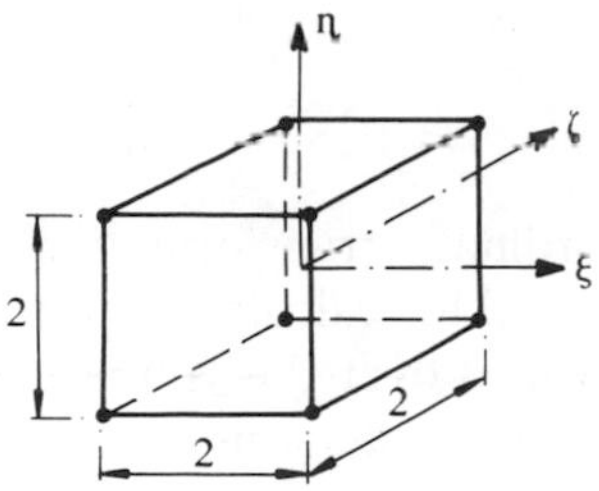 | $N_i = \frac{1}{8}(1 + \xi_0)(1 + \eta_0)(1 + \zeta_0)$ where $\xi_0 = \xi\xi_i$, $\eta_0 = \eta\eta_i$ and $\zeta_0 = \zeta\zeta_i$ |
| 20-noded brick | 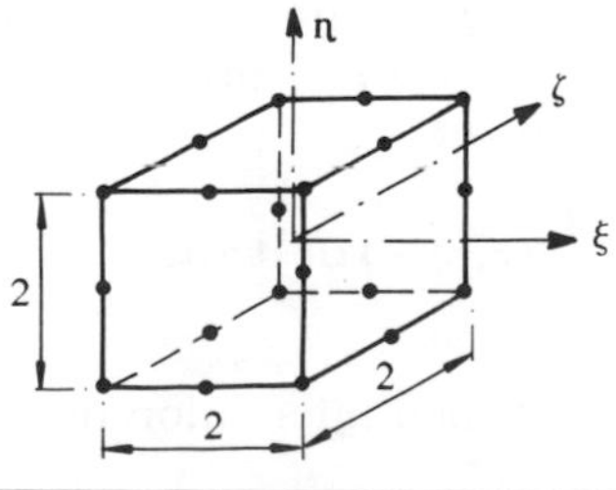 | $N_i = \frac{1}{8}(1 + \xi_0)(1 + \eta_0)(1 + \zeta_0)(\xi_0 + \eta_0 + \zeta_0 - 2)$ for corner nodes<br>$N_i = \frac{1}{4}(1 - \xi_0^2)(1 + \eta_0)(1 + \zeta_0)$ for typical mid-side nodes on $\xi_i = 0$, $\eta_i = \pm 1$, $\zeta_i = \pm 1$ |

The displacements at any point within the element are now given by

$$\begin{Bmatrix} u \\ v \\ w \end{Bmatrix} = [IN_1 \quad IN_2 \quad \ldots \quad IN_8] \begin{Bmatrix} u_1 \\ v_1 \\ w_1 \\ \vdots \\ u_8 \\ v_8 \\ w_8 \end{Bmatrix} \tag{5.47}$$

*20-Noded brick elements:* The shape functions for 20-noded brick elements are also written in local coordinate system ($\xi$, $\eta$, $\zeta$), Figure 5.4(c). This element is an extension of two-dimensional eight-noded parabolic element. Its shape functions are also listed in Table 5.2.

The displacements at any point within the element are now given by

$$\begin{Bmatrix} u \\ v \\ w \end{Bmatrix} = [IN_1 \quad IN_2 \quad \ldots \quad IN_{20}] \begin{Bmatrix} u_1 \\ v_1 \\ w_1 \\ \vdots \\ u_{20} \\ v_{20} \\ w_{20} \end{Bmatrix} \tag{5.48}$$

### 5.3.4 Coordinate transformation

Here again, the main aim of coordinate transformation is to facilitate integration of certain quantities required for calculation of stiffness matrices of the elements. The function now varies over the volume of the elements.

For tetrahedron elements, the shape functions are written in global coordinates. Hence, no coordinate transformation is required.

For 8-noded and 20-noded brick elements, the parent irregular brick (with trapeziodal or curved sides) is mapped into a regular brick with length, width and height equal to two units. The integration of a function, say $f(\xi, \eta, \zeta)$, is carried out over the limits $-1$ to $+1$. Thus,

$$F = \int_{-1}^{+1} \int_{-1}^{+1} \int_{-1}^{+1} f(\xi, \eta, \zeta)\, \mathrm{d}\xi\, \mathrm{d}\eta\, \mathrm{d}\zeta \tag{5.49}$$

evaluation of which is more convenient than integration over the volume of the parent element.

The shape functions of elements are again used for coordinate transformation exactly in the same way as discussed in Section 5.2.3. The coordinates of a point within the element $(x, y, z)$ are given by

$$x = \sum_{i=1}^{n} N_i x_i$$
$$y = \sum_{i=1}^{n} N_i y_i \tag{5.50}$$
$$z = \sum_{i=1}^{n} N_i z_i$$

where $x_i$, $y_i$, $z_i$ are the coordinates of the nodes of the elements and $n$ is the number of nodes. It is noted that the above equation is an extension of equation (5.6) for the two-dimensional case.

An infinitesimal volume of the element (d$V$) is given by

$$\mathrm{d}V = \mathrm{d}x\,\mathrm{d}y\,\mathrm{d}z = |J|\,\mathrm{d}\xi\,\mathrm{d}\eta\,\mathrm{d}\zeta \tag{5.51}$$

where the Jacobian matrix is $3 \times 3$ and is given by

$$[J] = \begin{bmatrix} \dfrac{\partial x}{\partial \xi} & \dfrac{\partial y}{\partial \xi} & \dfrac{\partial z}{\partial \xi} \\ \dfrac{\partial x}{\partial \eta} & \dfrac{\partial y}{\partial \eta} & \dfrac{\partial z}{\partial \eta} \\ \dfrac{\partial x}{\partial \zeta} & \dfrac{\partial y}{\partial \zeta} & \dfrac{\partial z}{\partial \zeta} \end{bmatrix} \tag{5.52}$$

The derivatives of the shape functions with respect to global Cartesian coordinate system are written as

$$\begin{Bmatrix} \dfrac{\partial N_i}{\partial x} \\ \dfrac{\partial N_i}{\partial y} \\ \dfrac{\partial N_i}{\partial z} \end{Bmatrix} = [J]^{-1} \begin{Bmatrix} \dfrac{\partial N_i}{\partial \xi} \\ \dfrac{\partial N_i}{\partial \eta} \\ \dfrac{\partial N_i}{\partial \zeta} \end{Bmatrix} \tag{5.53}$$

### 5.3.5 Strain displacement relations

In three-dimensional analysis, strains ($\boldsymbol{\epsilon}^T = \epsilon_x, \epsilon_y, \epsilon_z, \gamma_{xy}, \gamma_{yz}, \gamma_{zx}$) at any point in the continuum are related to the displacement at the same point by

$$
\begin{aligned}
\epsilon_x &= \frac{\partial u}{\partial x}, \qquad \epsilon_y = \frac{\partial v}{\partial y}, \qquad \epsilon_z = \frac{\partial w}{\partial z} \\
\gamma_{xy} &= \left(\frac{\partial u}{\partial y} + \frac{\partial v}{\partial x}\right), \qquad \gamma_{yz} = \left(\frac{\partial v}{\partial z} + \frac{\partial w}{\partial y}\right) \\
\gamma_{zx} &= \left(\frac{\partial w}{\partial x} + \frac{\partial u}{\partial z}\right)
\end{aligned}
\tag{5.54}
$$

We can now write strains at any point as a function of nodal displacement by using equation (5.46) or (5.47) or (5.48), depending on the type of element to be used. In general,

$$
\boldsymbol{\epsilon} = \begin{Bmatrix} \epsilon_x \\ \epsilon_y \\ \epsilon_z \\ \gamma_{xy} \\ \gamma_{yz} \\ \gamma_{zx} \end{Bmatrix} = B \begin{Bmatrix} u_1 \\ v_1 \\ w_1 \\ \vdots \\ u_n \\ v_n \\ w_n \end{Bmatrix}
\tag{5.55}
$$

where $B$ is a matrix of size $(6 \times 3n)$, where $n$ is the number of nodes in the elements. $B$ consists of $n$ submatrices $B_i (i = 1, \ldots n)$, of size $(6 \times 3)$ whose elements are:

$$
B_i = \begin{Bmatrix}
\frac{\partial N_i}{\partial x} & 0 & 0 \\
0 & \frac{\partial N_i}{\partial y} & 0 \\
0 & 0 & \frac{\partial N_i}{\partial z} \\
\frac{\partial N_i}{\partial y} & \frac{\partial N_i}{\partial x} & 0 \\
0 & \frac{\partial N_i}{\partial z} & \frac{\partial N_i}{\partial y} \\
\frac{\partial N_i}{\partial z} & 0 & \frac{\partial N_i}{\partial x}
\end{Bmatrix}
\tag{5.56}
$$

### 5.3.6 Stress–strain relations

Nothing more need be added at this stage except that for three dimensional analysis, the $D_T$ matrix discussed in Section 5.2.5 is a $6 \times 6$ matrix.

### 5.3.7 Stiffness equations of an element

The stiffness equations are derived in exactly the same manner as discussed in Section 5.2.6. The final result is again

$$K^e \, \Delta\boldsymbol{\delta}^e = \Delta R^e \tag{5.57}$$

where $R^e = \begin{Bmatrix} F_{x_1} \\ F_{y_1} \\ F_{z_1} \\ F_{xn} \\ F_{yn} \\ F_{zn} \end{Bmatrix}$ are nodal forces along $x$, $y$, $z$ axes and $\boldsymbol{\delta}^e$ are associated nodal displacements given by

$$\boldsymbol{\delta}^e = \begin{Bmatrix} u_1 \\ v_1 \\ w_1 \\ \vdots \\ u_n \\ v_n \\ w_n \end{Bmatrix} \tag{5.58}$$

The element stiffness matrix $K^e$ is given by

$$K^e = \int_v B^T D_T B \, \mathrm{d}v \tag{5.59}$$

where $B$ and $D_T$ matrices have been already discussed. The sizes of various matrices should be noted. $B^T$ is $3n \times 6$, $D_T$ is $6 \times 6$ and $B$ is $6 \times 3n$, leading to the size of $K^e$ as $3n \times 3n$.

### 5.3.8 Global stiffness and computation of nodal forces

The procedure of assembling global stiffness of three-dimensional bodies is the same as for two-dimensional bodies and follows the procedure adopted in structural mechanics.

Equations for nodal force vectors with various types of forces such as gravity, temperature, initial stress and strains, etc. have been derived in

Section 5.2.8. They apply for the three-dimensional case as well. Appropriate sizes of various matrices should, however, be substituted and it should be noted that the forces now have $x$, $y$ and $z$ components.

## 5.4 QUASI-THREE-DIMENSIONAL ANALYSIS

In many problems of tunnels and mines, situations arise that make the assumptions of a two-dimensional plane strain analysis invalid. These situations are:

(a) The axes of material orthotropy of rock mass (see Chapter 7) do not coincide with the plane of analysis of two-dimensional problem.
(b) The axes of *in-situ* principal stresses do not coincide with the plane of two-dimensional analysis.
(c) The fabric of rock joints is such that the normals to the joint planes do not lie in the plane of two-dimensional analysis.

In Figure 5.5(a), a tunnel is excavated in a rock mass where the orientation of rock joints is such that its normal does not lie in the plane of analysis. In Figure 5.5(b), the axes of *in-situ* stresses do not coincide with the axis of the tunnel. In such situations a fully three-dimensional analysis will have to be carried out. However, it is possible to reduce the analysis to a thick slice by the observation that full three-dimensional analysis has implications of repeatability.

For the case shown in Figure 5.5(a), although a fully three-dimensional analysis is required it can be confined to a slice of thickness $H$ which is repeatable. Two points $A$ and $B$ separated by a distance $H$ are homologous and have the same displacement, strains and stresses. This leads to a special case of three-dimensional analysis which we shall name as 'quasi-three-dimensional'.

If the 'quasi-three-dimensional' analysis is to be carried out for the reasons of material anisotropy of for *in-situ* stresses, the repeatability distance is quite arbitrary and can as well be 'unity' as in the plane strain case. For the case, when joints are spaced at a spacing $H$, analysis of a slice of thickness $H$ is to be carried out. In many cases, the joints are smeared (see Chapter 7). Here again, analysis with unit thickness is sufficient.

For the quasi-three-dimensional problem, fully three-dimensional displacement, strain and stress components are considered. The constraint imposed is that the displacements ($u$, $v$, $w$) at any point are functions of $x$ and $y$ only and independent of $z$-coordinate (perpendicular to the plane of analysis). Thus the strain-displacement relationship (equation (5.54), for quasi-three-dimensional case gets revised to

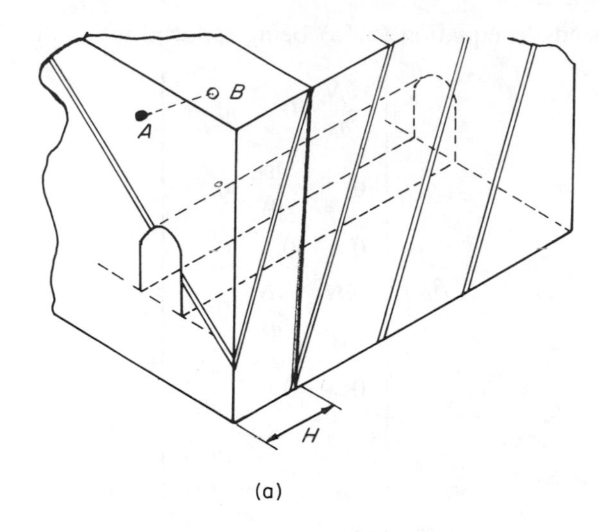


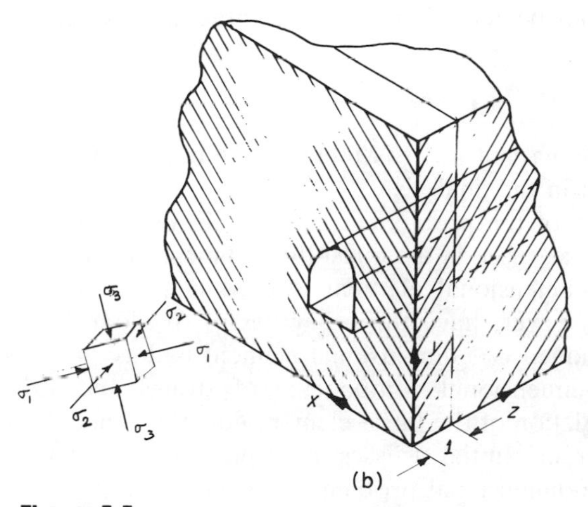


**Figure 5.5**
Situations where a quasi-three-dimensional analysis is applicable. (a) repeated set of joints; (b) *in-situ* stress principal axes not coincident with the plane of analysis

$$\epsilon = \frac{\partial u}{\partial x}, \qquad \epsilon_y = \frac{\partial v}{\partial y}, \qquad \epsilon_z = 0 \tag{5.60}$$

$$\gamma_{xy} = \left(\frac{\partial u}{\partial y} + \frac{\partial v}{\partial x}\right), \qquad \gamma_{yz} = \frac{\partial w}{\partial y}, \qquad \gamma_{zx} = \frac{\partial w}{\partial x}$$

This in turn leads to equation (5.56) being reduced to

$$B_i = \begin{bmatrix} \frac{\partial N_i}{\partial x} & 0 & 0 \\ 0 & \frac{\partial N_i}{\partial y} & 0 \\ 0 & 0 & 0 \\ \frac{\partial N_i}{\partial y} & \frac{\partial N_i}{\partial x} & 0 \\ 0 & 0 & \frac{\partial N_i}{\partial y} \\ 0 & 0 & \frac{\partial N_i}{\partial x} \end{bmatrix} \tag{5.61}$$

It is noted that in equation (5.61) terms corresponding $\epsilon_z$ have been retained though they are zero. The reason is that in nonlinear analysis, the plastic part of $\epsilon_z$ ($\epsilon_z^p$) may not be zero. For the quasi-three-dimensional as well as plane strain case,

$$\boldsymbol{\epsilon}_z = \boldsymbol{\epsilon}_z^e + \boldsymbol{\epsilon}_z^p = 0$$

where $\boldsymbol{\epsilon}_z^e$ is the elastic part of $\epsilon_z$. It is therefore convenient to retain all six components of strain in the analysis.

Since the geometry of the problem is specified in two dimensions, two-dimensional elements are used in the quasi-three-dimensional analysis. Thus, an eight-noded two-dimensional element is used in place of a 20-noded brick. The nodes, however, have three degrees of freedom. The stiffness matrix of the eight-noded two-dimensional element is thus $24 \times 24$.

Since fully three-dimensional components of strains and stresses are employed, the formulation of the finite element equations, nodal forces due to pressure, initial strains, initial stresses, pore pressures etc. is the same as discussed for a three-dimensional problem in Section 5.3.

## REFERENCES

Ghali, A. and Neville (1972). *Structural Analysis*, Intext Educational Publishers.

Hinton, E., and Owen, D. R. J. (1977). *Finite Element Programming*, Academic Press.

Reddy, J. N. (1985). An Introduction to the Finite Element Method. McGraw Hill.

Smith, I. M. and Griffiths, D. V. (1988). *Programming the Finite Element Method*, J. Wiley & Sons.

Turner, M. J., Clough, R. W., Martin, H. C., and Topp, L. J. (1956). 'Stiffness and deflection analysis of complex structures,' *J. Aero. Sci.*, **23**, 805–23.

Zienkiewicz, O. C. (1977). Texts on the finite element method. The Finite Element Method, 3rd edn., McGraw Hill.

Zienkiewicz, O. C. and Taylor, R. L. (1989). The Finite Element Method, 4th edn., Vol. I. Basic Formulation and Linear Problems, McGraw Hill.

# 6 Joint Elements and Infinite Elements

## 6.1 INTRODUCTION

We have seen in Chapter 3 that rock joints play a domineering influence on the behaviour of rock masses. Their fabric and strength greatly influences the properties of the rock mass. It is observed that the strength of joints which are filled with gouge material is primarily governed by the strength of the gouge material if the thickness of infilling is greater than twice the mean height of asperities on the faces of the rock joint. Major discontinuities in rock mass like faults, shear zones and joints, etc. can be modelled by special elements called 'joint' elements. The thickness of the fault or shear zones is usually small in comparison to the size of the rock structure. If one attempts to model these zones with ordinary elements, problems of numerical ill-conditioning may arise. In this chapter, we shall discuss some useful joint elements.

Another type of element which is discussed in this chapter is 'infinite' element. In geotechnical problems, we usually have a semi-infinite domain. Though we can discretize the area away from the zone of interest with large elements, the discretization boundary should truly be at infinite distance away. The infinite elements simulate the decay of displacements at the far end of the element.

If the number of joints in the rock mass to be considered is very large, it is impossible to model them by discrete joint elements. In such situations appropriate stress–strain laws which account for the fabric of joints are required. These are discussed in Chapter 7.

## 6.2 JOINT ELEMENTS

Physical characteristics of various types of joints were discussed in Chapter 3. Infilled joints having a finite measurable thickness are easier to deal with in finite element analysis. Contact joints with no thickness are difficult. In this case we have a multivalued problem. Figure 6.1(a) shows a block of

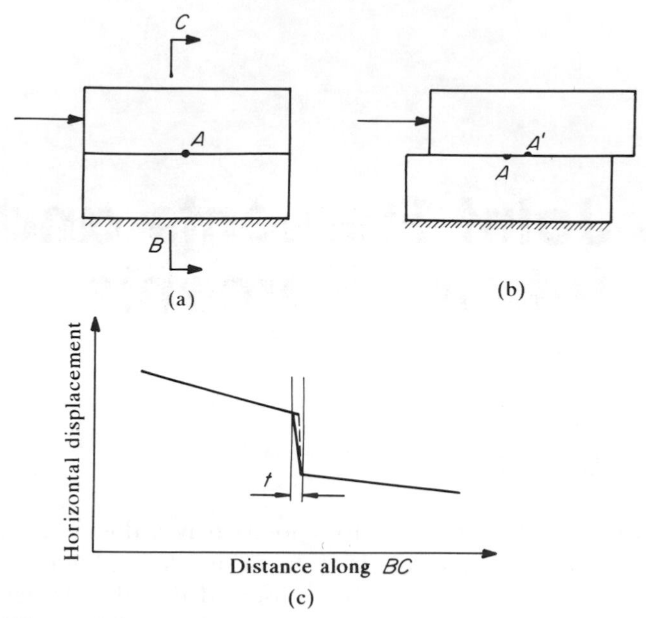


**Figure 6.1**
Interpretation of joint element. (a) Block of rock before sliding; (b) discontinuous displacement field; (c) discontinuity smoothed over the thickness of a joint element

rock resting on another. Point $A$ is on the joint between the two blocks. If a force is applied on the top block, the deformed position of the blocks may be as shown in Figure 6.1(b). The horizontal displacement of the point $A$ has two values. Since the finite element analysis always seeks to obtain a single unique value of displacement of very point in the continuum, this problem can not be solved by this method. However, if we introduce a certain arbitrary small thickness '$t$' of the joint (taking away $t/2$ from both the blocks) no anomaly arises and every point has a single unique displacement. The joint elements are therefore a sort of smooth interpolating functions to make an otherwise discontinuous function, continuous (see Figure 6.1(c)).

For three-dimensional analysis, the joint element is a two-dimensional element, with compatible number of nodes, while for two-dimensional analysis, joint element is one dimensional, again with compatible number of nodes.

### 6.2.1 Goodman's joint element

This element proposed by Goodman, Taylor and Brekke in 1968 is a linear line element suitable for two-dimensional analysis with four-noded elements. Figure 6.2 shows the four-noded element of length $L$. The thickness is initially zero. The pair of nodes 1, 4 and 2, 3 are coincident before

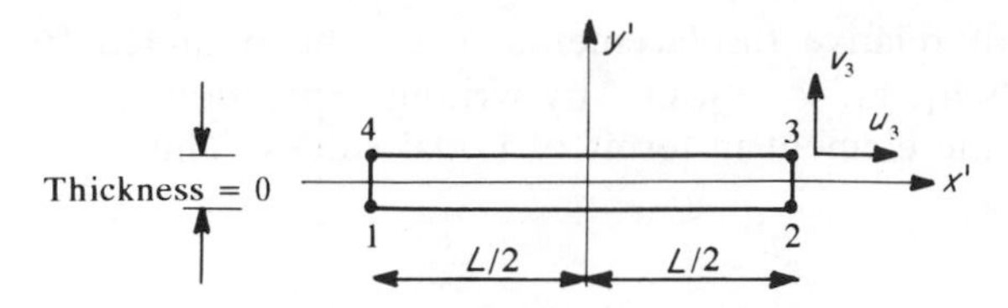


**Figure 6.2**
Joint element after Goodman *et al.*

deformation. Adopting a system of local axes ($x'$, $y'$), the vector of relative displacements ($w$) is written as:

$$\mathbf{w} = \begin{Bmatrix} w_s \\ w_n \end{Bmatrix} = \begin{Bmatrix} u_T - u_B \\ v_T - v_B \end{Bmatrix} \tag{6.1}$$

where $w_s$ and $w_n$ represent relative displacement of the top line of the element in the shear direction (along $x'$-axis) and normal direction (along $y'$-axis) respectively. $u$ and $v$ are displacements along $x'$ and $y'$ axes, respectively. Subscripts $T$ and $B$ are used to signify 'top' and 'bottom' blocks, respectively. If

$$\mathbf{P} = \begin{Bmatrix} P_s \\ P_n \end{Bmatrix}$$

is a vector of forces/unit length of the joint element, then the strain energy stored ($\phi$) in such an element is given by

$$\phi = \int_{-L/2}^{+L/2} \mathbf{w}^T \mathbf{P} \, dx' \tag{6.2}$$

where $L$ is the length of the element.

The force vector $P$ is related to the vector of relative displacements by stiffness of the joint, i.e.

$$\mathbf{P} = \begin{Bmatrix} P_s \\ P_n \end{Bmatrix} = \begin{bmatrix} K_s & 0 \\ 0 & K_n \end{bmatrix} \begin{Bmatrix} w_s \\ w_n \end{Bmatrix} \tag{6.3}$$

where $K_s$, $K_n$ are the shear and normal stiffnesses of the joint. These values may not be constant and may vary with stresses. Substitution of equations (6.3) in (6.2) leads to

$$\phi = \int_{-L/2}^{+L/2} \mathbf{w}^T \begin{bmatrix} K_s & 0 \\ 0 & K_n \end{bmatrix} \mathbf{w} \, dx' \tag{6.4}$$

The vector of relative displacements $w$ can be obtained from the nodal displacements $u_1$, $v_1$, $u_2$, $v_2$, etc. by writing expressions for $u$ and $v$ at any point within the element in terms of nodal values. Thus,

$$\begin{Bmatrix} u \\ v \end{Bmatrix}_T = \frac{1}{2}\begin{bmatrix} 1 + 2x/L & 0 & 1 - 2x'/L & 0 \\ 0 & 1 + 2x'/L & 0 & 1 - 2x'/L \end{bmatrix} \begin{Bmatrix} u_1 \\ v_1 \\ u_2 \\ v_2 \end{Bmatrix} \tag{6.5}$$

where subscript $T$ indicates that $u$, $v$ are the ones for the top face of the element. Writing similar expressions for the bottom face

$$\begin{Bmatrix} u \\ v \end{Bmatrix}_B = \frac{1}{2}\begin{bmatrix} 1 - 2x'/L & 0 & 1 + 2x'/L & 0 \\ 0 & 1 - 2x'/L & 0 & 1 + 2x'/L \end{bmatrix} \begin{Bmatrix} u_3 \\ v_3 \\ u_4 \\ v_4 \end{Bmatrix} \tag{6.6}$$

subtracting equation (6.6) from equation (6.5) leads to

$$\mathbf{w} = \frac{1}{2}\begin{bmatrix} -\alpha & 0 & -\beta & 0 & \beta & 0 & \alpha & 0 \\ 0 & -\alpha & 0 & -\beta & 0 & \beta & 0 & \alpha \end{bmatrix} \begin{Bmatrix} u_1 \\ v_1 \\ u_2 \\ v_2 \\ u_3 \\ v_3 \\ u_4 \\ v_4 \end{Bmatrix} \tag{6.7}$$

where $\alpha = 1 - 2x'/L$ and $\beta = 1 + 2x'/L$. Equation (6.7) can also be written as

$$\mathbf{w} = \tfrac{1}{2}B\mathbf{u} \tag{6.8}$$

where $B$ is defined by the $2 \times 8$ matrix in equation (6.7) and $u$ represents the nodal displacement vector. Substituting (6.8) in equation (6.4) leads to

$$\phi = \int_{-L/2}^{+L/2} \tfrac{1}{4}\mathbf{u}^T B^T \begin{bmatrix} K_s & 0 \\ 0 & K_n \end{bmatrix} B\mathbf{u}\, \mathrm{d}x' \tag{6.9}$$

For equilibrium, the displacements $\mathbf{u}$ are such that the stored energy $\phi$ is a minimum. Differentiating $\phi$ with respect to $\mathbf{u}$, performing the integration (note that $B$ is a function of $x'$) and substituting the limits of integration we obtain

$$\mathbf{Ku} = \mathbf{P} \tag{6.10}$$

where K is the stiffness matrix of the element and is given by

$$K = \frac{1}{6}\begin{bmatrix} 2K_s & & & & & & & \\ 0 & 2K_n & & & \text{Symmetric} & & & \\ K_s & 0 & 2K_s & & & & & \\ 0 & K_n & 0 & 2K_n & & & & \\ -K_s & 0 & -2K_s & 0 & 2K_s & & & \\ 0 & -K_n & 0 & -2K_n & 0 & 2K_n & & \\ -2K_s & 0 & -K_s & 0 & K_s & 0 & 2K_s & \\ 0 & -2K_n & 0 & -K_n & 0 & K_n & 0 & 2K_n \end{bmatrix} \tag{6.11}$$

It is noted that the stiffness matrix given by equation (6.11) above is in the local system of coordinates ($x'$, $y'$). It will have to be transformed to the global system of coordinates before assembling of the stiffness for the whole rock mass being analysed. If $\delta$ is the angle $x'$ axis makes with respect to the global $x$ axis, then the global element stiffness ($K_g$) is given by

$$K_g = T^T K T \tag{6.12}$$

where $T$ is a transformation matrix given by

$$T = \begin{bmatrix} \cos\delta & \sin\delta \\ -\sin\delta & \cos\delta \end{bmatrix} \tag{6.13}$$

### *Some remarks on Goodman's joint element*

(1) In the derivation above, the properties of joint are assumed to be represented by stiffness of joints $K_s$ and $K_n$. The stiffness matrix of the joint

$$\begin{bmatrix} K_s & 0 \\ 0 & K_n \end{bmatrix}$$

has no off-diagonal terms. This implies that a shear and normal loads applied on the joint produce shear and normal displacements, respectively. It also implies that there is no dilatancy of joints which is an important aspect of the behaviour of real joints. To model this, off-diagonal terms $K_{sn}$ and $K_{ns}$ must be introduced. Physical interpretation of these terms has been discussed in Chapter 3. One can follow a very similar derivation as given in Section 6.2.1 above for the case when $K_{sn} = K_{ns} \neq 0$. It should, however, be noted that when $K_{sn} \neq K_{ns}$, the stiffness of the joint element will not be symmetric.

(2) $K_s$, $K_n$, $K_{sn}$ and $K_{ns}$ are generally not constant and depend on the relative displacements. A constant value used in the analysis may lead to anamolous results, i.e. elements penetrating or overlapping each other. To avoid this, nonlinear values are used; specially large values of $K_n$ as the joint closes.

(3) It is possible to formulate higher-order joint elements on the basis of Goodman's joint element. A procedure of numerical integration will have to be adopted as direct explicit integration is quite cumbersome.

(4) Mehtab and Goodman (1970) have extended the formulation of the joint element suitable for three-dimensional analysis. The joint element is a two-dimensional eight-noded quadrilateral with the nodes in the thickness direction being coincident.

### 6.2.2 Transversely isotropic paralinear element

One of the ways of modelling discrete joints is by introducing joint elements of an arbitrarily small thickness. Zienkiewicz *et al.* (1970) proposed a 6-noded element as shown in Figure 6.3. In theory, the joint element can be curved-parabolic in shape, but in practice one hardly needs such shapes. The thickness of the joint element is assigned arbitrarily. The material properties for the element can be orthotropic (designated by elastic moduli and Poisson's ratios in the two directions and a shear modulus) or isotropic (elastic modulus, $E_j$ and Poisson's ratio, $\mu_j$).

If the thickness of the joint element ($t_j$) is chosen relatively small, numerical ill-conditioning can arise. The main drawback of this element is that one has to choose thickness as well as elastic properties of the joint material almost arbitrarily. This approach is useful for parametric studies to judge the influence of the joint on the behaviour of the rock mass, but quantitative results may not be reliable.

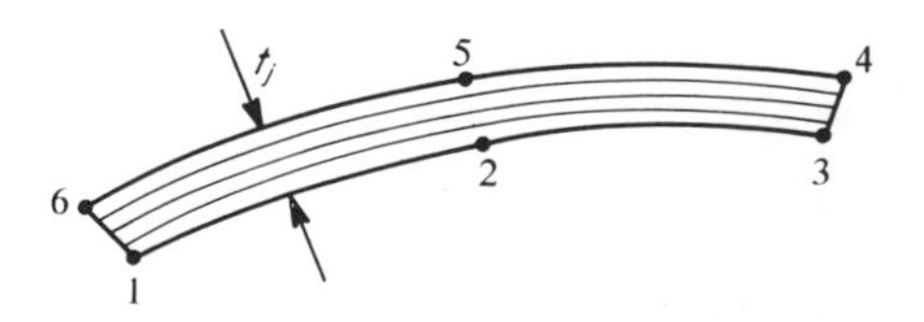


**Figure 6.3**
Para-linear element

If one assumes the joint material to be linear elastic and isotropic, it is possible to obtain equivalent values of $K_s$ and $K_n$ involved in Goodman's joint element. Simple algebra shows that

$$K_n = \frac{E_j}{t_j}$$

and

$$K_s = \frac{E_j}{2(1 + \mu_j)t_j} \tag{6.14}$$

or conversely

$$E_j = K_n \cdot t_j$$

$$\mu_j = \left(\frac{K_n}{2K_s} - 1\right) \tag{6.15}$$

It is noted that for any values of $E_j$, $\mu_j$ and $t_j$, it is possible to find $K_s$ and $K_n$ ($K_{ns} = K_{sn} = 0$ by implication of elasticity). However, a meaningful value of $\mu_j$ will be obtained only if

$$2 \geqslant \frac{K_n}{K_s} \geqslant 3$$

### 6.2.3 Ghaboussi–Wilson–Isenberg element

Ghaboussi, Wilson and Isenberg (1973) proposed an element for rock joints and interfaces based on the concepts of the theory of plasticity. They also pointed out the possibility of numerical ill-conditioning of stiffness matrix which might occur in joint elements due to very large off-diagonal terms or very small diagonal terms.

Their joint element, which connects two continuum elements, has a finite thickness '$t$' and uses relative displacement between the two continuum elements as an independent degree of freedom. Figure 6.4 shows 'top' and 'bottom' continuum elements with joint elements between them. The degrees of freedom for the bottom elements are the nodal displacements in the global coordinate system as in normal finite element procedure. The degrees of freedom for the top element for the nodes on the side of the joint element are relative displacements (displacements relative to the nodes of the bottom element). The joint element has relative displacement as the degree of freedom. Using superscripts $T$ and $B$ for 'top' and 'bottom' elements,

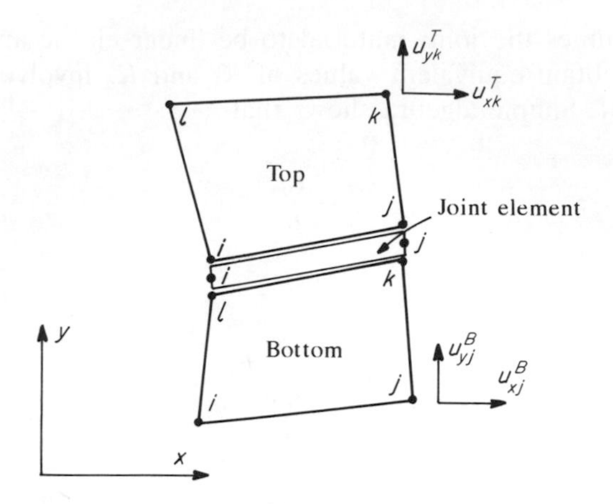


**Figure 6.4**
Definitions of symbols used in the joint element

respectively, and denoting the nodes in the elements as $i$, $j$, $k$ and $l$, the displacement ($u$) of the nodes $i$ and $j$ for the top element can be written as

$$
\begin{aligned}
u_{xi}^T &= u_{xl}^B + \Delta u_{xi} \\
u_{yi}^T &= u_{yl}^B + \Delta u_{yi} \\
u_{xj}^T &= u_{xk}^B + \Delta u_{xj} \\
u_{yj}^T &= u_{yk}^B + \Delta u_{yj}
\end{aligned}
\tag{6.16}
$$

We adopt here a joint element with two nodes to be compatible with the 'top' and 'bottom' continuum elements. It is possible to have higher order joint elements based on the approach outlined here. The stiffness matrix of

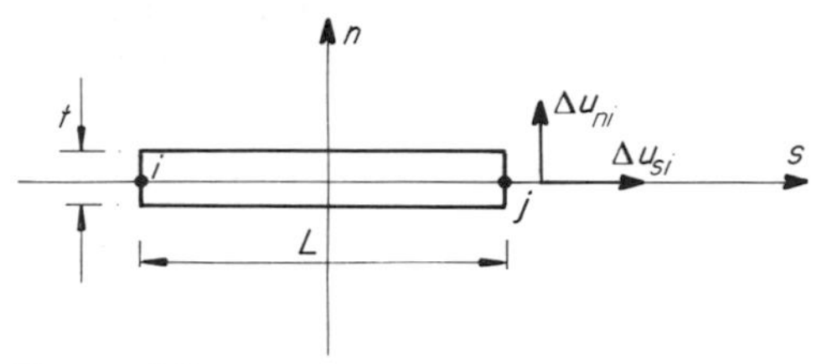


**Figure 6.5**
Joint element in local system of co-ordinates

the joint element is developed in a local coordinate system (see Figure 6.5). The nodes $i$ and $j$ of the joint element have normal and tangential displacements ($\Delta u_n$ and $\Delta u_s$) as degrees of freedom. They are assumed to vary linearly along the length of the element. Denoting the local axes by $s$ and $n$, and measuring the distance in normalized coordinates $\xi_0$ ($=2s/L$, where $L$ is the length of the jont) along the joint

$$\Delta u_n = \frac{1-\xi_0}{2}\Delta u_{ni} + \frac{1+\xi_0}{2}\Delta u_{nj}$$
$$\Delta u_s = \frac{1-\xi_0}{2}\Delta u_{si} + \frac{1+\xi_0}{2}\Delta u_{sj} \tag{6.17}$$

where $\Delta u_{ni}$ and $\Delta u_{si}$ represent relative normal and tangential displacements, respectively at the node $i$ and $\Delta u_{nj}$ and $\Delta u_{sj}$ represent the same quantities at the node $j$.

Since the thickness of the joint element is small, it is assumed that the joint has only two components of strain: normal strain ($\epsilon_n$) and shear strain ($\epsilon_s$) which are given by

$$\epsilon_n = \frac{\Delta u_n}{t}$$
$$\epsilon_s = \frac{\Delta u_s}{t} \tag{6.18}$$

Substituting equation (6.17) in (6.18) and writing in the matrix form, leads to

$$\begin{Bmatrix} \epsilon_n \\ \epsilon_s \end{Bmatrix} = \frac{1}{2t} \begin{bmatrix} (1-\xi_0) & 0 & (1+\xi_0) & 0 \\ 0 & (1-\xi_0) & 0 & (1+\xi_0) \end{bmatrix} \begin{Bmatrix} \Delta u_{ni} \\ \Delta u_{si} \\ \Delta u_{nj} \\ \Delta u_{sj} \end{Bmatrix} \tag{6.19}$$

which in a more compact notation can be written as

$$\boldsymbol{\epsilon} = B\,\Delta\mathbf{u} \tag{6.20}$$

The stresses and strains are related through material property matrix ($D$) which in this case is $2 \times 2$

$$\begin{Bmatrix} \sigma_n \\ \tau \end{Bmatrix} = \begin{bmatrix} D_{nn} & D_{ns} \\ D_{sn} & D_{ss} \end{bmatrix} \begin{Bmatrix} \epsilon_n \\ \epsilon_s \end{Bmatrix} \tag{6.21}$$

where $\sigma_n$ and $\tau$ are normal and shear stresses respectively. The coefficients of $D$ matrix will depend on the material model chosen to represent joint behaviour. If purely linear elastic behaviour is required, $D_{ns} = D_{sn} = 0$, $D_{nn}$ = elastic modulus and $D_{ss}$ = shear modulus. There is no dilatancy and the normal and shear behaviours are uncoupled. If plasticity based models are adopted, the coefficients of the $D$ matrix will be stress dependent.

The stiffness matrix for the joint element in the local coordinate system is given by (see Chapter 5 for detailed derivation)

$$K_{\xi\eta} = \frac{Lt}{2} \int_{-1}^{+1} B^T D B \, \mathrm{d}\xi_0 \tag{6.22}$$

Transformed into the global $x - y$ coordinate system, the global stiffness ($K_g$) is given by

$$K_g = T^T K_{\xi\eta} T \tag{6.23}$$

where $T$ matrix is given by equation (6.13). Equation (6.22) can be explicitly integrated and substituting in equation (6.23), the global stiffness ($K_g$) is given by

$$K_g = \frac{L}{6t} \begin{bmatrix} 2(A_1 - 2B_1) & & & \textit{Symmetric} \\ 2(A_3 + B_2) & 2(A_2 + 2B_1) & & \\ (A_1 - 2B_1) & (A_3 + B_2) & 2(A_1 - 2B_1) & \\ (A_3 + B_2) & (A_2 + 2B_1) & 2(A_3 + B_2) & 2(A_2 + 2B_2) \end{bmatrix} \tag{6.24}$$

where

$$A_1 = D_{ss}a^2 + D_{nn}b^2$$

$$B_1 = D_{ns}ab$$

$$A_2 = D_{ss}b^2 + D_{nn}a^2$$

$$B_2 = D_{ns}(a^2 - b^2)$$

$$A_3 = (D_{nn} - D_{ns})ab$$

and $$a = \frac{1}{L}(x_j - x_i); \quad b = \frac{1}{L}(y_j - y_i).$$

In the above ($x_i$, $y_i$) and ($x_j$, $y_j$) are the coordinates of the nodes $i$ and $j$ and $L$ is the length of the joint element.

### 6.2.4 Other joint element formulations

The joint elements also have applications in a wider area of soil/rock structure interaction. For example, the contact between the foundation and the structure, pipe culvert and the surrounding fill and the contact between tunnel lining and surrounding rock can be modelled by joint elements. In fact, a number of joint elements have been developed with this application in view. They can as well be used for rock joints. Katona (1981) developed a contact-friction interface element in which the constraint equations along the interface were incorporated in a virtual work statement. The mating nodes are assumed at the same location before loading. For a two-noded line interface element, explicit constraint equations can be written for three states—(a) Fixed, (b) Slip and (c) Free. Katona (1981) has applied this element to the problems of buried culverts.

Desai (1984) has developed a 'thin layer element' which basically is a solid element with a constitutive law for contact, sliding, separation and rebonding applied to it.

Multilaminate model of rocks described in the next chapter can also be applied to joint elements to simulate the contact characteristics. Here, only one set of joints along the length of the element is specified on which an appropriate strength criterion together with a 'joint monitor' enables modelling of joint characteristics. Pande *et al.* (1979) have applied it to the problem of strengthening of a concrete dam by buttresses.

### 6.2.5 Some remarks on numerical ill-conditioning

When joint elements are used, problems of numerical ill-conditioning may arise. If the joint elements have a large aspect ratio (ratio of length to thickness), small values of coefficients at the diagonal of the stiffness matrices can create problems in solution routine with loss in accuracy. Whether ill-conditioning takes place or not depends on the problem, aspect ratio and the accuracy of computer used. Formulations which use relative displacement as an independent degree of freedom (such as Ghaboussi *et al.*, Section 6.2.3) are believed to be more robust. The problem of ill-conditioning is not so critical on accurate main frame machines (Pande and Sharma (1979)) but with increasingly popular powerful microcomputers, one has to be careful. Parametric studies with varying thickness must be made to ascertain the influence of ill-conditioning.

## 6.3 INFINITE ELEMENTS

Almost all geotechnical problems involve an infinite domain, i.e. a domain which is much larger than the region to be analysed. For example, in Figure 5.2, the finite element meshes cover only a finite zone around the rock structures. At the boundary of these meshes it is assumed that appropriate components of displacements are zero. This is only an approximation and a more accurate result will be obtained as the boundaries of prescribed displacement are moved farther and farther away from the structure. The question arises as to how far the finite element mesh should extend to obtain reasonably accurate results. The answer is problem dependent and is not unequivocal.

In the last decade or so, special elements have devised which extend in one or two directions to infinity. These are called 'infinite elements'. They can be used at the boundary of the finite element meshes to model the infinite domain. The shape functions for these elements are so chosen that the displacement at the node located at infinity are zero (or a prescribed values).

The idea of infinite elements was proposed by Anderson and Ungless (1977) and Bettess (1977) who applied it to the problems of fluid mechanics. Beer and Meck (1981) proposed 'infinite domain elements' and Beer has pioneered their application in geotechnical and mining problems (Beer, 1983), Beer and Swoboda (1985). Zienkiewicz *et al.* (1983) have proposed an infinite element which is extremely simple to incorporate in a finite element program. Which infinite element is the best is a matter of controversy and it is hoped that further research and application to engineering problems would answer to this question.

### 6.3.1 Mapped infinite element (Zienkiewicz *et al.* (1983))

It is convenient to discuss this element in one dimension first. Figure 6.6 shows a three-noded element. The node 3 has coordinate $x_3 = \infty$ and $x_2 - x_1 = a$. This element can be mapped into an element of length equal to two units by transformation to $\xi$ coordinate system in which it extends from $\xi = -1$ to $\xi = +1$ (see Figure 6.6). Let us consider the transformation given by

$$x = \left(\frac{-\xi}{1-\xi}\right)x_0 + \left(\frac{1}{1-\xi}\right)x_2 \tag{6.25}$$

where $x_0$ is the coordinate of a 'pole' and is equal to $(x_1 - a)$.

When $\xi = -1$,

$$\begin{aligned} x &= \tfrac{1}{2}(x_1 - a) + \tfrac{1}{2}x_2 \\ &= \tfrac{1}{2}(x_1 - a) + \tfrac{1}{2}(x_1 + a) = x_1, \end{aligned}$$

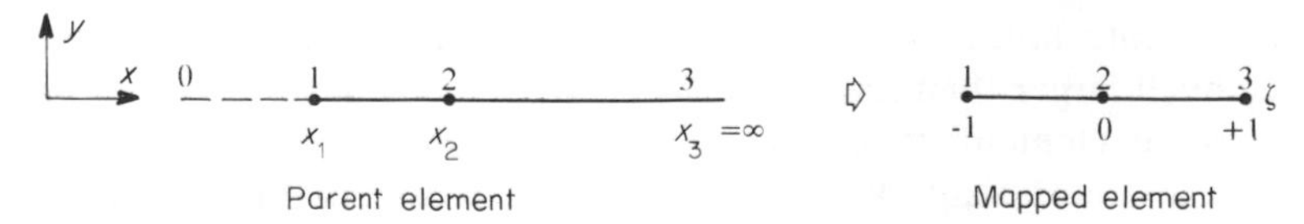


**Figure 6.6**
One-dimensional infinite element

when

$$\xi = 0$$
$$x = x_2$$

and when

$$\xi = +1$$
$$x = x_3 = \infty$$

Thus, the infinite element can be mapped into a finite element by the proposed transformation of equation (6.25) which can be rewritten in the form

$$x = N_0(\xi)x_0 + N_2(\xi)x_2 \tag{6.26}$$

which is the form similar to equation (5.5) used for finite elements and where

$$N_0(\xi) = \frac{-\xi}{1-\xi}$$

and

$$N_2(\xi) = \frac{1}{1-\xi} \tag{6.27}$$

We can now use parabolic shape functions (see Table 5.1) to describe the variation of displacements ($u$) within the element, i.e.

$$u = \{\tfrac{1}{2}\xi(\xi-1), 1-\xi^2, \tfrac{1}{2}\xi(\xi+1)\}\begin{Bmatrix} u_1 \\ u_2 \\ u_3 \end{Bmatrix} \tag{6.28}$$

where $u_1$, $u_2$, and $u_3$ are the nodal displacements. It is noted that as $\xi \rightarrow 1, x \rightarrow \infty$ and displacement approaches $u_3$ which is usually zero in geotechnical applications. The decay in displacements from the value at node 2 to zero at node 3 is parabolic. This may not be a good choice and other forms of shape functions such as polynomial form of

$$u = \alpha_0 + \alpha_1\xi + \alpha_2^2\xi\ldots \tag{6.29}$$

can be adopted. Here, $\alpha$'s are constants.

Equation (6.25) can be rearranged in the form

$$\xi = 1 - \frac{2a}{x - x_0} = 1 - \frac{2a}{r} \tag{6.30}$$

where $r = x - x_0$ is the distance of a point in the element from the pole. Substituting equation (6.30) in equation (6.29) leads to

$$u = \beta_0 + \frac{\beta_1}{r} + \frac{\beta_2}{r^2} + \ldots \tag{6.31}$$

where $\beta_0$, $\beta_1$, $\beta_2$ etc are coefficients which can be evaluated in terms of $\alpha_0$, $\alpha_1$, $\alpha_2$ etc. From equation (6.31) it is clear that as $r$ increases, $u \rightarrow \beta_0$ which may be set to zero to simulate zero displacements at infinity.

*Two-dimensional infinite elements:* Two-dimensional elements will extend to infinity in one direction which may be chosen as $\xi$ direction. A two-dimensional element can then be mapped into a square element as shown in Figure 6.7 with limits of integration $-1 \leqslant \xi \leqslant 1$ and $-1 \leqslant \eta \leqslant 1$. The shape functions for mapping in the $\eta$ direction are the usual shape functions and have to be parabolic (as there are three nodes in $\eta$ direction) whilst shape function for $\xi$ direction are $N_0$ and $N_2$ as defined by equation (6.27). They are explicitly given by

$$x = [A_1, A_1', A_2, A_2', A_3, A_3'] \begin{Bmatrix} x_1 \\ x_2 \\ x_3 \\ x_4 \\ x_5 \\ x_6 \end{Bmatrix} \tag{6.32}$$

where

$$A_i = 2N_i(\eta)N_0$$

$$A_i' = -N_i(\eta)N_0 + N_2$$

$$N_1(\eta) = \tfrac{1}{2}\eta(\eta - 1)$$

$$N_2(\eta) = 1 - \eta^2$$

$$N_3(\eta) = \tfrac{1}{2}\eta(\eta + 1)$$

$$N_0 = -\frac{\xi}{1 - \xi}$$

$$N_2 = \frac{1}{1 - \xi}$$

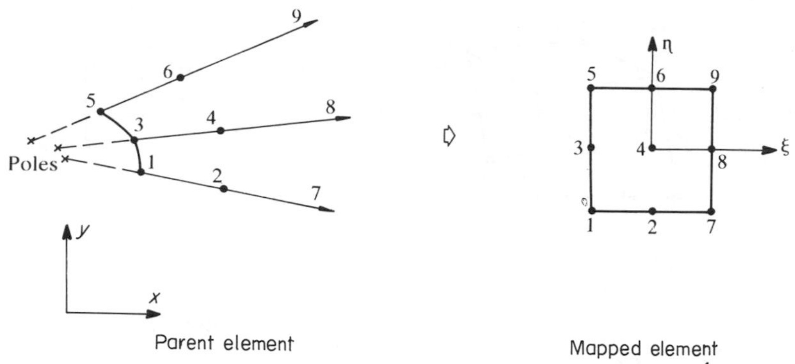


**Figure 6.7**
Two-dimensional infinite element

A similar expression is written for the $y$-coordinate, i.e.

$$y = [A_1, A_1', A_2, A_2', A_3, A_3'] \begin{Bmatrix} y_1 \\ y_2 \\ y_3 \\ y_4 \\ y_5 \\ y_6 \end{Bmatrix} \tag{6.33}$$

The reader can verify that the nodes 7, 8 and 9 lie at infinity when $\xi = +1$. Equations (6.32) and (6.33) can be differentiated to obtain elements of the Jacobian matrix (see Chapter 5) viz.

$$\frac{\partial x}{\partial \xi}, \quad \frac{\partial y}{\partial \xi}, \quad \frac{\partial x}{\partial \eta} \quad \text{and} \quad \frac{\partial y}{\partial \eta}.$$

The shape functions given in equations (6.32) and (6.33) can also be used for variation of displacements within the element. Thus implementation of infinite elements of the type discussed in this section in finite element program is quite straight forward.

*Three-dimensional infinite elements:* Here again, the direction extending to infinity is chosen as $\xi$ direction and a linear element is mapped into a cube in the local coordinate system (Figure 6.8). The shape functions in $\eta$ and $\xi$ direction are normal shape functions while shape functions in the $\xi$ direction are given by equation (6.27). They are explicitly given by

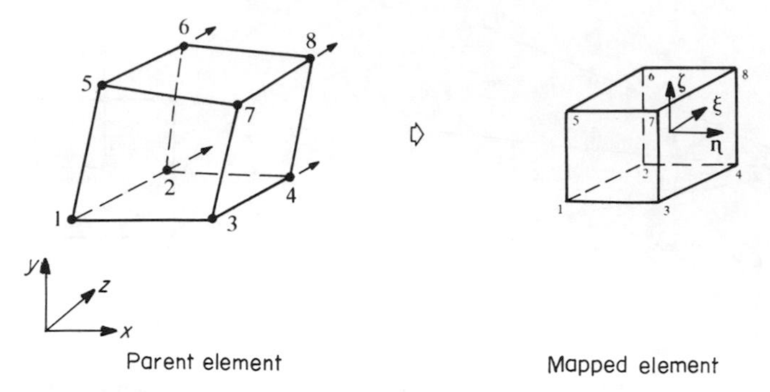


**Figure 6.8**
Three-dimensional infinite element

$$x = N_1(\zeta)N_1(\eta)[(2x_1 - x_2)N_0 + x_2N_2] + N_2(\eta)N_1(\zeta)[(2x_3 - x_4)N_0 + x_4N_2] + N_1(\eta)N_2(\zeta)[(2x_5 - x_6)N_0 + x_6N_2] + N_2(\eta)N_2(\zeta)[(2x_7 - x_8)N_0 + x_8N_2] \quad (6.34)$$

with similar expressions for $y$ and $z$.

Integration of stiffness is usually carried out in local coordinates with limits for $\xi$, $\eta$, $\zeta$ as $-1$ to $+1$.

### 6.3.2 Infinite domain elements (Beer and Meek 1981)

Infinite domain elements proposed by Beer and Meek (1981) differ from the mapped infinite elements discussed in Section 6.4.1 in two respects:

(1) The shape functions of these elements have a singularity at $\xi = +1$ (the direction in which the domain extends to infinity).
(2) In the infinite direction the displacements are assumed to decay from the value at the boundary according to a specified function of the distance to a 'decay origin'.

The advantage of this approach is that no additional nodes are required to be located at infinity. For example, in Figure 6.7, there is no need for nodes 7, 8 and 9 at infinity. The mapping of two- and three-dimensional infinity domain elements is shown in Figures 6.9 and 6.10. The shape functions used for mapping in the infinite direction are:

$$x = N_1x_1 + N_2x_2 + \ldots \quad (6.35)$$

with

$$N_j = 1 + \xi_0 + \xi_j$$

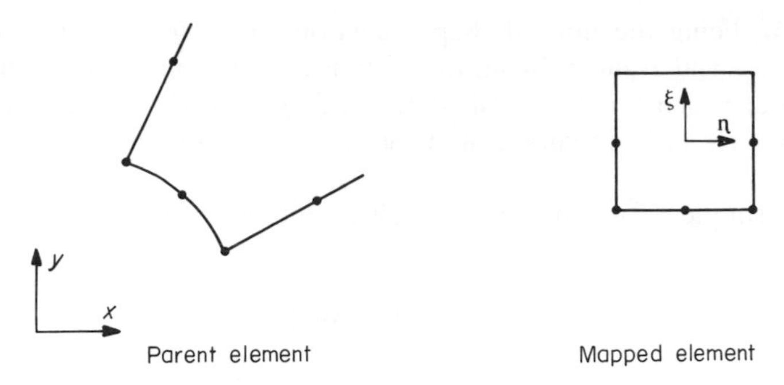


**Figure 6.9**
Two-dimensional infinite domain elements

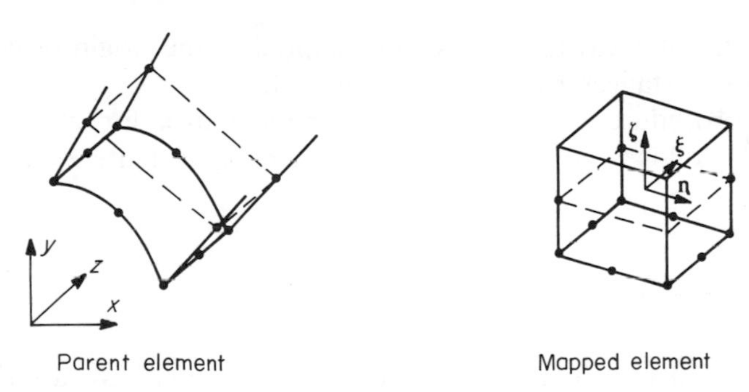


**Figure 6.10**
Three-dimensional infinite domain elements

where

$$\xi_0 = 2(\xi + \tfrac{1}{2})\xi \quad \text{for } \xi < 0$$

and

$$\xi_0 = 2(\xi + \tfrac{1}{2})\xi/(1 - \xi) \quad \text{for } \xi > 0$$

It is noted that when

$$\xi = +1, \; x = \infty.$$

The variation of displacement within the element is given by

$$u = \bar{M}_1 u_1 + \bar{M}_2 u_2 + \ldots \tag{6.36}$$

where $u_1$, $u_2$ are nodal displacements and $\bar{M}_1$, $\bar{M}_2$ are shape functions of the form

$$\bar{M}_j = M_j f\left(\frac{r_j}{r}\right) \tag{6.37}$$

with $M_j$ being the normal shape function which interpolates along the boundary and $r$ and $r_j$ being the distances from the decay origin to the points at which $u$ is to be calculated and nodal point $j$ respectively. The function $f(r_j/r)$ is so chosen as to give a unit value at $j$ and zero value at $r = \infty$.

The simplest form of $f(r_j/r)$ which is suitable is given by:

$$f\left(\frac{r_j}{r}\right) = \frac{r_j}{r} \quad \text{from plane strain problems} \tag{6.38}$$

$$f\left(\frac{r_j}{r}\right) = \left(\frac{r_j}{r}\right)^2 \quad \text{for three-dimensional problems} \tag{6.39}$$

The centre of excavation is usually adopted as the origin of decay from which the distances $r$ and $r_j$ are measured.

The procedure for developing stiffness matrix for infinite domain elements follows the standard pattern of mapped finite elements.

## 6.4 SOME COMMENTS

The infinite and particularly infinite domain elements have a considerable similarity with the Boundary Element Method discussed in Chapters 9 and 10. This has led to coupled finite element/boundary element methods in which 'near' region is modelled by the finite elements and the 'far' region by the boundary elements, thus retaining the advantages of both the methods.

## REFERENCES


Anderson, D. L., and Ungless, R. L. (1977). 'Infinite finite elements.' Int. Symp. Innovative Num. Anal. Appl. Eng. Science, France.

Beer, G., and Meek, J. L., (1981). 'Infinite domain elements.' *Intl. J. Num. Meth. Eng.*, **17**(1), 43–52.

Beer, G. (1983). 'Infinite domain elements in finite element analysis of underground excavations.' *Int. J. Num. Anal. Meth. Geomech.*, **7**, 1–7.

Beer, G., and Swoboda, G. (1985). 'Efficient analysis of shallow tunnels.' *Comp. Geotech.* 15–31.

Bettess, P. (1977). 'Infinite elements.' *Int. J. Num. Meth. Engin.*, **11**, 53–64.

Desai, C. S., Zaman, M. M., Lightner, J. G., and Siriwardane, H. J. (1984). 'Thin-layer element for interfaces and joints.' *Intl. J. Num. Anal. Meth. Geomech.*, **8**, 19–43.

Ghaboussi, J., Wilson, E. L., and Isenberg, J. (1973). 'Finite elements for rock joints and interfaces.' *Jl. Soil Mech. Dn. A.S.C.E.*

Goodman, R. E., Taylor, R. L., and Brekke, T. L. (1968). 'A model for the mechanics of jointed rock.' *J. Soil Mech. Dn. A.S.C.E.*, **94**, SM3.

Katona, M. G. (1983). 'Efficient analysis of shallow tunnels.' *Comp. Geotech.* 15–31.

Mehtab, M. A., and Goodman, R. E. (1970). 'Three-dimensional finite element analysis of jointed rock slopes.' *Proc. 2nd Congress of the Intl. Soc. of Rock Mech.*, **3**, Belgrade.

Pande, G. N., and Sharma, K. G. (1979). 'On joint/interface elements and associated problems of numerical ill-conditioning.' Short Comm., *Intl. J. Num. Anal. Meth. Geomech.*, **3**, 293–300.

Pande, G N., Bicanic, N., and Zienkiewicz, O. C. (1979). 'Influence of joint/interface nonlinearity on strengthening of dams.' Paper No. R36, Q.48, 13th Intl. Cong. Large Dams, New Delhi.

Zienkiewicz, O. C., Best, B., Dullage, C., and Stagg, K. G. (1970). 'Analysis of non-linear problems in rock mechanics with particular reference to jointed rock systems.' *Proc. 2nd Intl. Congress on Rock Mechanics*, Belgrade.

Zienkiewicz, O. C., Emson, C., and Bettess, P. (1983). 'A novel boundary infinite element.' *Intl. J. Num. Meth. Eng.*, **19**, 393–404.

# 7 Constitutive Models of the Behaviour of Jointed Rock Masses

## 7.1 INTRODUCTION

Numerical methods are very powerful and effective in the analysis of jointed rock structures. The solutions can be improved by a finer discretization of the domain of analysis. However, the key common factor in all numerical methods is the characterization of the behaviour of the jointed rock masses, i.e. the constitutive law.

In Chapter 6 we saw how discrete zones of weaknesses, like faults and shear zones, can be modelled. If the number of joints in the rock mass is very large so that modelling each and every joint is impossible, one has to have a constitutive law for the rock mass which would take into account the pattern of fabric of joints. In general, the fabric of joints will govern the deformational as well as strength properties of rock masses.

This chapter is devoted to the discussion of a number of constitutive laws of varying complexities which can be used for the analysis of jointed rock masses. After discussing 'anisotropic elastic' and 'no-tension' models, a general framework for developing constitutive models of jointed rock masses is presented. Stress–strain relations of the intact rock and that of rock joints are incorporated in this framework to obtain a specific model. The choice of the model depends on the problem to be solved.

## 7.2 JOINTED ROCK MASS AS A LINEAR ELASTIC ANISOTROPIC MATERIAL

This is one of the simplest of models for jointed rock mass. It obviously has limited applications.

Generally speaking, anistropy in rocks is of two types: (a) anisotropy of elastic moduli and (b) anisotropy of strength. If we characterize a jointed rock mass as a linear elastic material, the strength parameters do not influence the analysis. The strength parameters are useful in simply identifying the zones where strength would be exceeded.

For a general anisotropic material, 21 independent elastic constants are required to describe the complete stress–strain relationship in three dimensions. This number of constants is so large that it has been impossible to obtain them for any rock.

In many cases it is possible to assume that there are three mutually perpendicular directions or symmetry. This may be, for example, due to three sets of mutually perpendicular joint sets (Figure 7.1). The rock mass in such a case would behave orthotropically. If we assume a set of local axes $x'$, $y'$, $z'$ to coincide with the orientation of the normals to the joint planes, the inverse of elasticity matrix ($D_e'^{-1}$) can be written as

$$[D_e']^{-1} = \begin{bmatrix} \dfrac{1}{E_{x'}} & -\dfrac{\nu_{y'x'}}{E_{y'}} & -\dfrac{\nu_{z'x'}}{E_{z'}} & 0 & 0 & 0 \\ & \dfrac{1}{E_{y'}} & -\dfrac{\nu_{z'y'}}{E_{z'}} & 0 & 0 & 0 \\ & & \dfrac{1}{E_{z'}} & 0 & 0 & 0 \\ & & & \dfrac{1}{G_{x'y'}} & 0 & 0 \\ & \text{Symmetric} & & & \dfrac{1}{G_{y'z'}} & 0 \\ & & & & & \dfrac{1}{G_{z'x'}} \end{bmatrix} \qquad (7.1)$$

so that

$$\boldsymbol{\epsilon}' = [D_e']^{-1}\boldsymbol{\sigma}'$$

where

$$\boldsymbol{\sigma}' = [\sigma_{x'}, \sigma_{y'}, \sigma_{z'}, \tau_{x'y'}, \tau_{y'z'}, \tau_{z'x'}]^T$$

and

$$\boldsymbol{\epsilon}' = [\epsilon_{x'}, \epsilon_{y'}, \epsilon_{z'}, \gamma_{x'y'}, \gamma_{y'z'}, \gamma_{z'x'}]^T$$

are the stress and strain components respectively in the local $x'$, $y'$, $z'$ system. The prime of $D_e$ is used to remind that the matrix relates the

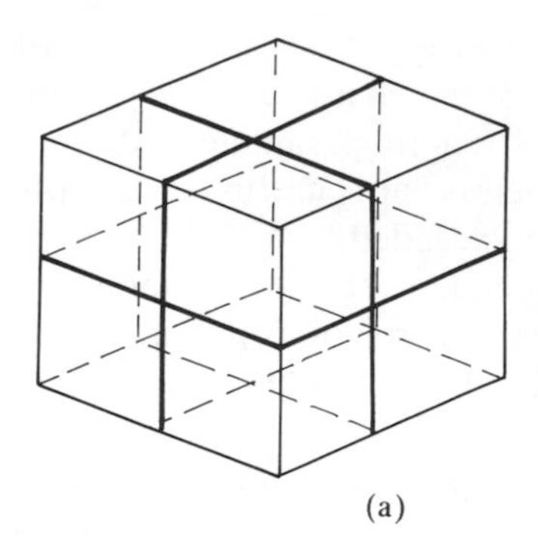

(a)

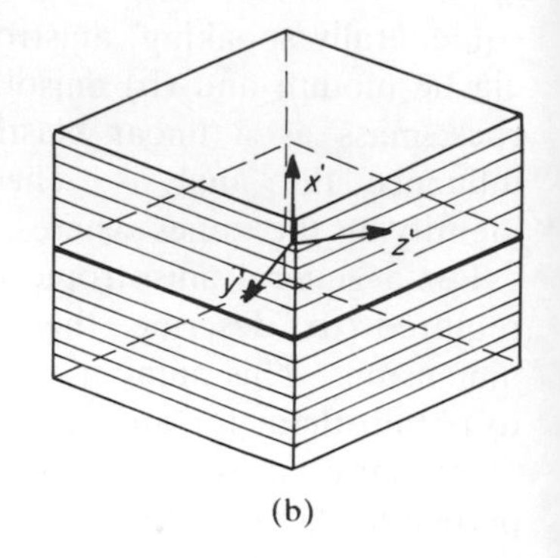


(b)

**Figure 7.1**
Anisotropy of jointed rock masses. (a) Three sets mutually perpendicular joint sets; (b) one set of joint

stresses and strains in the local coordinate system which is fixed on the basis of the orientation of the axes of anisotropy.

In equation (7.1) $E_{x'}$, $E_{y'}$ and $E_{z'}$ represent Young's moduli along $x'$, $y'$ and $z'$ axes respectively. The Poisson's ratios $\nu_{ij}$ are the strains produced in the $i$ direction when a unit strain is applied in the $j$ direction.

A further simplification in the $D_e'^{-1}$ matrix arises if the rock mass has only one dominant joint set. In this case there is a symmetry of elastic properties about an axis (parallel to the normal to the joint set), say $x'$ (Figure 7.1). The $[D_e']$ in this case is given by

$$[D_e']^{-1} = \begin{bmatrix} \frac{1}{E_{x'}} & -\frac{\nu_{y'x'}}{E_{y'}} & -\frac{\nu_{y'x'}}{E_{y'}} & 0 & 0 & 0 \\ & \frac{1}{E_{y'}} & -\frac{\nu_{y'z'}}{E_{y'}} & 0 & 0 & 0 \\ & & \frac{1}{E_{y'}} & 0 & 0 & 0 \\ & & & \frac{1}{G_{x'y'}} & 0 & 0 \\ & \textit{Symmetric} & & & \frac{2(1+\nu_{y'z'})}{E_{y'}} & 0 \\ & & & & & \frac{1}{G_{x'y'}} \end{bmatrix} \tag{7.2}$$

The five independent, elastic constants in equation (7.2) are $E_{x'}$, $\nu_{y'x'}$, $G_{y'x'}$, $E_{y'}$ and $\nu_{y'z'}$. For plane strain situation, equation (7.2) reduces to

$$[D'_e]^{-1} = \begin{bmatrix} \frac{1}{E_{x'}} & -\frac{\nu_{y'x'}}{E_{y'}} & -\frac{\nu_{y'x'}}{E_{y'}} & 0 \\ & \frac{1}{E_{y'}} & -\frac{\nu_{y'z'}}{E_{y'}} & 0 \\ & \textit{Symmetric} & \frac{1}{E_{y'}} & 0 \\ & & & \frac{1}{G_{x'y'}} \end{bmatrix} \quad (7.3)$$

In Chapter 3 a procedure for deriving the complete elasticity matrix for an arbitrary joint fabric was outlined. The final elasticity matrix is a function of the elasticity matrix of intact rock, stiffnesses and orientations of joint sets and spacing of joints in each set. In many practical applications, an elasticity matrix derived on the above basis may represent an adequate model for the designer. It is noted that intact rock itself can be anisotropic. This anisotropy may have a significant influence on the distribution of stresses in certain problems such as tunnels (Schweiger *et al.*, 1986).

In theory there are no problems in adopting anisotropic behaviour of rock masses. Computer application is simple and straight forward. The problems lie in obtaining reliable and meaningful experimental or field parameters to describe the anisotropy.

## 7.3 ROCK MASS AS A 'NO TENSION' MATERIAL

For the solution of many practical problems like design of underground openings, foundations of large structures, dams, etc., the designer is often interested in a stress analysis in which tensile stresses have been either eliminated or an estimate of the tensile stresses which would persist is obtained. This enables the designer to either provide for this tension by reinforcement or introduce a construction joint at the appropriate place. Tensile nonlinearity (incapability to take substantial tensions) is perhaps the most important nonlinearity in soil and rock mechanics problems and any nonlinear analysis must take account of this aspect.

### 7.3.1 Time independent 'no tension' model

Zienkiewicz and Valliappan (1968) introduced the concept of a 'no tension' material. Conceptually it is a material having randomly oriented joints/cracks so that it is incapable of withstanding any tensile principal stresses. An iterative initial load procedure was used to relax tensions present, if any, on initial elastic analysis. Thus, if $\boldsymbol{\sigma}_p$ represents the tensile principal stresses (assuming all principal stresses† are tensile)

† If only one principal stress is tensile, $\boldsymbol{\sigma}_p = \{\sigma_1 \quad 0 \quad 0\}^T$.

$$\boldsymbol{\sigma}_p = \begin{Bmatrix} \sigma_1 \\ \sigma_2 \\ \sigma_3 \end{Bmatrix}, \qquad \sigma_1 > \sigma_2 > \sigma_3 > 0, \tag{7.4}$$

the Cartesian stress components can be obtained by premultiplying $\boldsymbol{\sigma}_p$ by a transformation matrix $[T]$,

$$\boldsymbol{\sigma} = [T]\boldsymbol{\sigma}_p \tag{7.5}$$

The nodal force correction required to nullify tensions is given by (see Chapter 5)

$$F^e = \int_v B^T \boldsymbol{\sigma} \, \mathrm{d}V \tag{7.6}$$

and revised nodal force vector is used to compute the principal stresses again. Iterations are continued till the principal stresses are no longer tensile.

In some problems the convergence is noted to be slow. Chang and Niar (1972) made an empirical modification in which they accounted for Poisson's ratio effect of relaxation of one principal tension on other principal stresses and reported speedier convergence with their modification.

### 7.3.2 TIME DEPENDENT 'NO TENSION' MODEL

The time independent 'no tension' model proposed by Zienkiewicz and Valliappan (1968) two decades ago has, it appears, been used widely in many practical problems. However, this model has a few drawbacks. Firstly, it drops out the time element which is quite crucial for rock soil/structure interaction problems like linings in tunnels, support systems for deep trenches, etc. Secondly, the solution obtained is just one possible solution or in other words the solution is not unique. Thirdly, with the given system of loads and boundary conditions, if no solution is found, the designer in many cases gets no feedback as to what order of magnitude of tensile stresses will persist in the structure so that he could make a decision whether they are permissible or not, or alternatively, provide joints in the structure at suitable places and reanalyse the structure.

These drawbacks appear to get at least minimized when a viscoplastic form of the 'no tension' model is used. Explicitly for this case the yield function can be written as

$$F = \sigma_1 = 0 \tag{7.7}$$

where $\sigma_1$ represents the algebraically largest principal stress.

Tension positive convention is used here. Using the stress invariants $\sigma_m$, $\bar{\sigma}$ and $\theta$ (Appendix I) yield function can be written as

$$F = \sigma_m + \tfrac{2}{3}\,\bar{\sigma}\sin\left(\theta + \frac{2\pi}{3}\right) = 0 \tag{7.8}$$

and the rate of viscoplastic strains is given by

$$\dot{\boldsymbol{\epsilon}}^{vp} = \mu\langle \mathrm{F}\rangle \frac{\partial F}{\partial \boldsymbol{\sigma}} \tag{7.9}$$

where $\mu$ is the fluidity parameter and an associated flow rule has been assumed. Equation (7.9) simply represents that $\boldsymbol{\epsilon}^{vp} \neq 0$ if algebraically largest principal stress is positive (i.e. tensile). Since the expression is written for algebraically largest principal stress and if it is not tensile, then it is assured that other principal stress would also not be tensile. It is noted that an associative flow rule has been applied in equation (7.9) which is essential for tension relaxation. The yield surface in stress space is a trirectangular pyramid (Figure 7.2) and also has corners at $\theta = \pi/6$. However, local smoothing techniques can be easily applied.

This model introduces 'time' element (on an arbitrary time scale) and since it is fashioned after plasticity/viscoplasticity models and has convex, numerically smoothened form of the yield surface and associative flow rule,

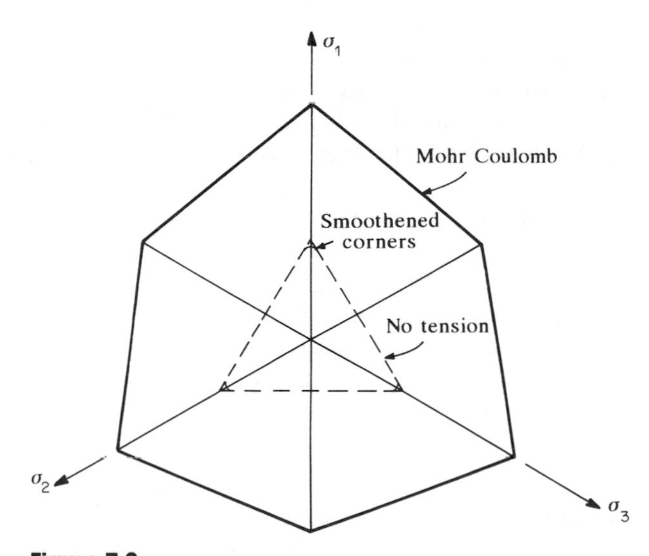


**Figure 7.2**
Trace of Mohr–Coulomb and 'no tension' yield function on the $\pi$ plane

it can be argued, gives unique solutions. The solution, if convergence is not achieved, also provides a steady state of stress field which is quite useful for designers.

The uniqueness however is superficial. For a specified yield function like equation (7.7) the solution is unique. However, equation (7.7) itself does not appear to be the only way of describing 'no tension' behaviour. One could also write

$$F = \langle \sigma_1 \rangle + \langle \sigma_2 \rangle + \langle \sigma_3 \rangle \tag{7.10}$$

where $\langle\ \rangle$ indicates that quantity within brackets to be considered only if positive. Alternatively

$$F = \sigma_1^{2n} + \sigma_2^{2n} + \sigma_3^{2n} \tag{7.11}$$

where $n$ is a positive integer.

Associativity of flow can be evoked with each of such yield functions and corresponding strain-rate equations can be written. Thus, philosphically, the problem of uniqueness has not been solved but circumvented.

## 7.4 MULTILAMINATE FRAMEWORK OF MODELS OF JOINTED ROCK MASSES

The complex behaviour of rock masses is dominated by the presence of fissures, joints and planes of weaknesses. In Chapter 6 a number of formulations of joint elements were discussed.

Joint elements can be used only for modelling a few major discontinuities in the rock mass like a fault or crushed zone. If the number of joints is large, then a constitutive model which accounts for the fabric of joints is more appropriate. Multilaminate framework of models discussed in this section are in this category.

We shall first discuss the framework in simple terms with specific reference to Mohr–Coulomb as a failure function, associated flow rule and no hardening. Next, we look at some more possible forms of yield or failure functions, flow rules and hardening/softening rules which may be more appropriate. The framework is discussed in the context of elasto–viscoplasticity first and closely follows the work of Zienkiewicz and Pande (1977). The alternative of elasto-plastic formulation is given at the end of the section.

### 7.4.1 Assumptions

The following assumptions are implicit in the formulation of multilaminate framework:

(a) All joints in a set are parallel, continuous and unfilled (no gouge material).
(b) The volume occupied by the joint sets is small compared to the total volume of the rock mass.

Certain additional assumptions are required to be made regarding the behaviour of joints for incorporation in the multilaminate framework. The following will be used as an example:

(c) The strength of a joint is represented by Mohr–Coulomb failure criterion, together with a no tension cut-off (Figure 7.3).
(d) The behaviour of the joint is elastic–perfectly plastic, i.e. for stresses within the failure envelope, the joint behaves purely elastically.
(e) The joints have no memory of their opening/closing. This assumption and its implications are explained in Section 7.6.

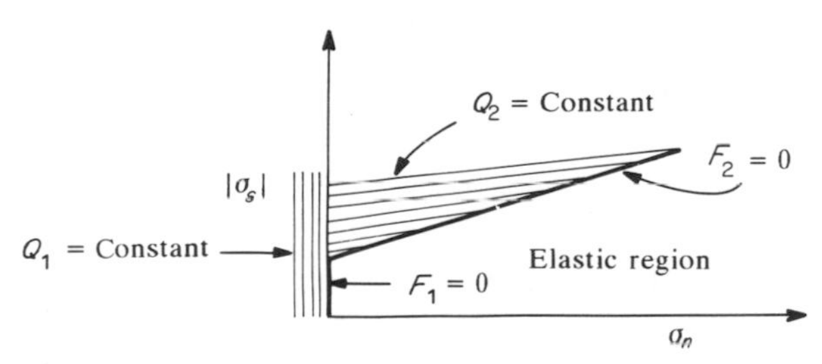


**Figure 7.3**
Mohr–Coulomb yield with no tension cut off for a typical joint set

### 7.4.2 The rheological model

Let us assume that the rock mass is traversed by $n$ sets of discontinuities. In practice $n$ is limited to 1–5. For each of the joint sets we have a failure criterion. The stress acting on the rock mass and the joint sets is the same (resolved on the joint planes). It implies that a rheological model of the jointed rock mass consists of rheological units representing each joint set arranged in series, as shown in Figure 7.4. We have an additional visco–plastic unit to represent intact rock. The elastic behaviour of all joint sets and the intact rock is lumped together in a single spring. The derivation of elasticity matrix of the rock mass (represented by the spring) has been already discussed in Section 4.4.

### 7.4.3 The model

The model will be described in three dimensions. The reduction to two dimensions is straight forward and only appropriate transformation matrices need to be changed.

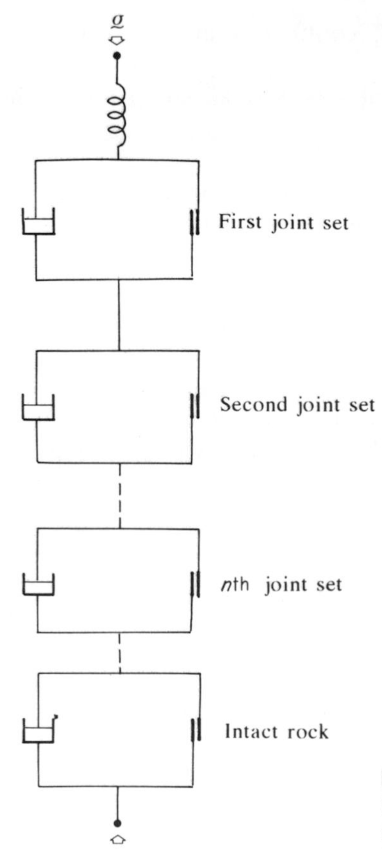


**Figure 7.4**
Rheological analogue of multilaminate ($n$-laminates) model of jointed rock mass

Let $\boldsymbol{\sigma} = [\sigma_x, \sigma_y, \sigma_z, \tau_{xy}, \tau_{yz}, \tau_{zx}]^T$ represent the stress in the rock mass at any point $(x, y, z)$ in a Cartesian coordinate system. Let $\hat{\boldsymbol{\sigma}}_k = [\sigma_n, \sigma_s]_k^T$ represent the normal and resultant shear stress respectively on the $k$th joint set.

$\sigma_{nk}$ and $\sigma_{sk}$ can be found by transforming the stress $\boldsymbol{\sigma}$ into a local system of coordinates $n - s - t$ where $n$ is in the direction of the normal to the joint plane and $s$ and $t$ are on the joint plane.

To model the joint behaviour given in Section 7.4.1 we have to set up two failure criteria. They are:

$$F_1 = [\sigma_n]_{k=1,n} = 0 \tag{7.12}$$

and

$$F_2 = [|\sigma_s| + \sigma_n \tan \phi_k - c_k]_{k=1,n} = 0 \tag{7.13}$$

Readers are reminded that tension positive convention has been used throughout this chapter. $\phi_k$ and $c_k$ are friction angle and cohesion, respectively, of the $k$th joint set. Assuming a general nonassociated flow rule, the plastic potential function for a typical joint plane can be written as

$$Q_2 = [|\sigma_s| + \sigma_n \tan \psi_k - \text{constant}]_{k=1,n} = 0 \tag{7.14}$$

where $\psi_k$ is the dilation angle for the $k$th joint set. Note that the model has the versatility to have different strength and dilation parameters for each of the joint sets. The plastic potential function for 'no tension' on the joint planes has to be associative from a physical view point. Thus,

$$Q_1 = [\sigma_n - \text{constant}]_{k=1,n} = 0 \tag{7.15}$$

The plastic potential and failure functions for a typical joint set have been plotted in Figure 7.3.

The flow equation for a typical joint set can now be written as

$$\dot{\boldsymbol{\epsilon}}_k^{vp} = \gamma_k \langle \mathrm{F}_k \rangle \frac{\partial \mathrm{Q}_k}{\partial \hat{\boldsymbol{\sigma}}_k} \frac{\partial \hat{\sigma}_k}{\partial \boldsymbol{\sigma}} \tag{7.16}$$

$\dot{\boldsymbol{\epsilon}}_k^{vp}$ represents the contribution of the $k$th joint set to the Cartesian components of the viscoplastic strain rate. If any particular joint set is not yielding ($F \leqslant 0$) its contribution would be zero. Here $\gamma_k$ represents the fluidity parameter of the $k$th joint set. Thus, though conceptually it is possible to have different characteristics of flow for each of the joint sets, it will hardly be possible in practice to have material parameters and $\gamma$'s for all joint sets and they will have to be assumed the same. Moreover, in quasi-static applications, where viscoplasticity is used purely as an alternative algorithm for computational convenience, all $\gamma$'s can be assumed as simply one unit.

The contribution of the intact rock, if it yields,, to the viscoplastic strain rates of the rock mass ($\dot{\boldsymbol{\epsilon}}_i{}^{vp}$) will be decided by its yield/failure function and flow rule. The failure of intact rock is likely to take place only in very highly constrained situations, particularly if there are more than two sets of joints. However, for completeness, let $F_i$ and $Q_i$ represent the failure and plastic potential functions, respectively, of the intact rock. The flow equation for intact rock will have the form

$$\dot{\boldsymbol{\epsilon}}_i^{vp} = \gamma_i \langle F_i \rangle \frac{Q_i}{\partial \boldsymbol{\sigma}} \tag{7.17}$$

where $\gamma_i$ represents the fluidity parameter of the intact rock. The failure criterion of intact rock could, for example, be Hoek–Brown criterion or any other suitable criterion and it will largely depend on the type of rock.

The global rate of viscoplastic strains of the rock mass ($\dot{\boldsymbol{\epsilon}}^{vp}$) will be the sum of the contributions of the joint sets and the intact rock as suggested by the rheological model of Figure 7.4. Thus

$$\dot{\boldsymbol{\epsilon}}^{vp} = \sum_{k=1}^{n} \dot{\boldsymbol{\epsilon}}_k^{vp} + \dot{\boldsymbol{\epsilon}}_i^{vp} \tag{7.18}$$

which on substitution of equations (7.16) and (7.17) becomes

$$\dot{\boldsymbol{\epsilon}}^{vp} = \sum_{k=1}^{n} \gamma_k \langle \mathrm{F}_k \rangle \frac{\partial \mathrm{Q}_k}{\partial \hat{\boldsymbol{\sigma}}_k} \frac{\partial \hat{\boldsymbol{\sigma}}_k}{\partial \boldsymbol{\sigma}} + \gamma_i \langle F_i \rangle \frac{\partial Q_i}{\partial \boldsymbol{\sigma}} \tag{7.19}$$

With the viscoplastic strain rate for the rock mass defined by equation (7.19), the model for jointed rock mass is fully defined and solution procedure of the viscoplastic algorithm given in Chapter 2 can be adopted.

Multilaminate framework is extremely modular. The complete description of the model involves

(a) Intact rock: Elastic properties (isotropic or anisotropic), strength parameters and the failure criterion, flow rule and fluidity parameter.
(b) For each joint set: orientation, elastic stiffnesses ($k_n$, $k_s$), spacing of joints, strength parameters, failure criterion, flow rule and fluidity parameters.

The fluidity parameters are required only if true time dependence is to be modelled. An arbitrary value can be assumed for pseudo viscoplastic computations.

### 7.4.4 Limitations of the multilaminate framework

The multilaminate framework is based on the assumption of the distribution of joints throughout the rock mass. It is therefore implied that this assumption is valid only when the spacing of joints in any set is comparatively much smaller than the critical dimension of the structure to be analysed. For example, in analysing the foundation of an arch dam or the tunnel shown in Figure 7.5, the assumptions of the equivalent material are valid since the spacing of joints is much smaller than the base width of the dam in the first case and much smaller than the diameter of the tunnel in the second case.

## 7.5 IMPROVED MODELS OF BEHAVIOUR OF JOINTED ROCK MASSES

As mentioned in the previous section, the multilaminate framework is extremely modular. Different models of strength and dilatancy of rock joints

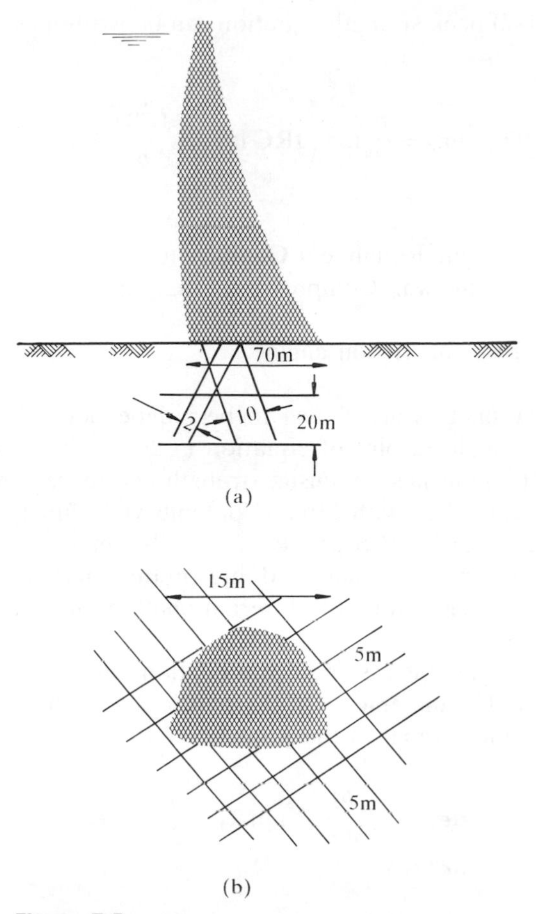


**Figure 7.5**
Situation where assumptions of multilaminate model are valid

can be readily incorporated in the multilaminate framework.

It was pointed out in Chapter 3 that Mohr–Coulomb strength criterion is a poor approximation of the real strength of joints. The main reason is that the strength parameters $c$ and $\phi$ are not truly constants but depend on the normal stress ($\sigma_n$) on the joint. In a boundary value problem, the normal stress on the joints and even on the same joint, can vary several orders of magnitude. The choice of strength parameters ($c$ and $\phi$) for Mohr–Coulomb criterion therefore becomes difficult.

To overcome these problems the empirical relation of Barton and Choubey (see Chapter 3) for peak strength of the joints can be viewed as a failure function and incorporated in the multilaminate framework. Plastic potential functions derived from the experimental results can also be readily incorporated. Pande and Xiong (1982) adopted this approach. Barton and

Choubey's (1977) peak strength equation can be written as a failure function as follows:

$$F = |\sigma_s| + \sigma_n \tan\left(\mathrm{JRC}\log_{10}\left(\frac{\mathrm{JCS}}{-\sigma_n}\right) + \phi_r\right) = 0 \tag{7.20}$$

where

JRC is the Joint Roughness Coefficient
JCS is the Joint wall Compressive Strength

and

$\phi_r$ is the residual friction angle

These three parameters are shown to be independent of normal stress on the joint. A normalized plot of equation (7.20) is shown in Figure 3.9. It is noted that the joint has no tensile strength and there is no need to adopt a 'no tension' cut off as with Mohr–Coulomb yield function.

The parameters JRC, JCS and $\phi_r$ are easily determined by simple field tests. Methods of obtaining them and interpreting their values are contained in the guidelines prepared by the International Society for Rock Mechanics (1978).

If an associated flow is assumed, the dilation angle can be obtained from equation (7.20). The increments of the plastic strains in the directions along the joint ($d\epsilon_s^p$) and normal to the joint ($d\epsilon_n^p$) are given by

$$d\epsilon_n^p = d\lambda \frac{\partial F}{\partial \sigma_n} = \tan\lambda_1 - \frac{\mathrm{JRC}}{2.303}\sec^2\lambda_1 \tag{7.21}$$

$$d\epsilon_s^p = 1$$

where

$$\lambda_1 = \mathrm{JRC}\log_{10}\left(\frac{\mathrm{JCS}}{-\sigma_n}\right) + \phi_r$$

and $d\lambda$ is the usual proportionality constant.

The dilation angle ($\psi$) is then given by

$$\tan\psi = \tan\lambda_1 - \frac{\mathrm{JRC}}{2.303}\sec^2\lambda_1 \tag{7.22}$$

Note that dilation angle is a function of $\sigma_n$ as $\lambda_1$ is a function of $\sigma_n$. Pande and Xiong (1982) noted that the associated flow rule did not fit in with experimental results of dilatancy of joints as given by Barton and Choubey (1977). They have proposed the following relationship (equation (7.23)) for the plastic potential function which models the data on dilatancy of joints, as given by Barton and Choubey.

$$Q = |\sigma_s| + \frac{\sigma_n \tan \lambda_2}{k_1} - \frac{\text{JRC}}{264} \cdot \frac{\sigma_n^2}{\text{JCS}} = \text{constant} \tag{7.23}$$

where

$$\lambda_2 = \lambda_1 - \phi_r$$

$$k_1 = 1 - \tan \lambda_2 \tan \phi_r$$

The rate of dilation ($\tan \psi$) as predicted by equation (7.23) is plotted in Figure 7.6 against ($-\sigma_n$/JCS). The main point is that if experimental data on the dilatancy of joints are available, a plastic potential function can be obtained by curve fitting and incorporated in the multilaminate framework.

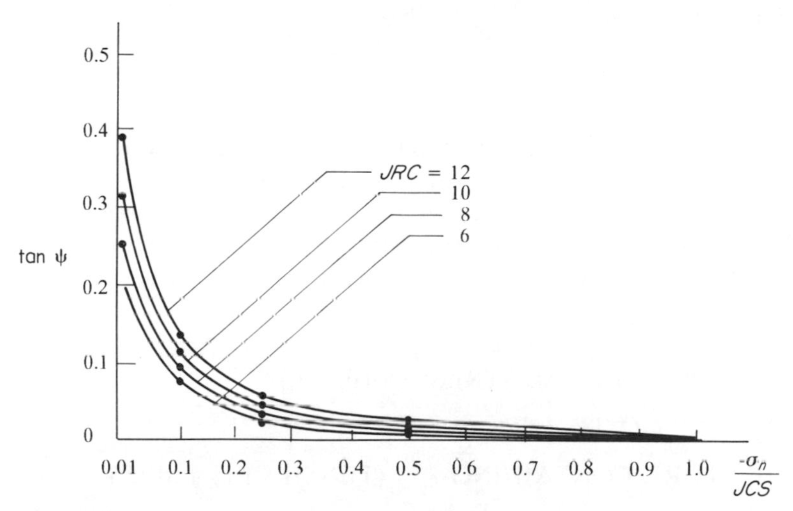


**Figure 7.6**
Variation of dilatancy with changing normal stress

## 7.6 JOINT OPENING/CLOSING MONITOR

In Section 7.4.1, one of the assumptions made for the multilaminate model was that the joints have no memory of their opening/closing.

Due to incorporation of a 'tension cut-off' (equation (7.12)), the presence of tensile stresses normal to any joint set leads to its relaxation and consquent accumulation of plastic strains signifying joint opening. During further loading, should the normal stress across the same joint set become compressive, the compressive stresses are transmitted across the 'open' joint implying a mismatch of joint asperities. In some cases it does not represent the realistic behaviour of the joints, since unless the joints close, compressive stresses may not be transmitted across them. However, the assumptions can be changed to model perfectly matching asperities. Admittedly, the real situation is somewhere in between the two extreme assumptions.

Assuming that joints close perfectly, we characterize the joint behaviour by the following: If the joint has been under tension and $\epsilon_n^{max}$ represents the maximum total tensile strain in the direction normal to the joint in the past loading history then, if the joint is currently under compressive stress, the yield criterion adopted is given by:

$$F_3 = |\sigma_n|_k = 0 \tag{7.24}$$

if

$$(\epsilon_n + \epsilon_n^{max}) > 0$$

and equation (7.12) is used only when

$$(\epsilon_n + \epsilon_n^{max}) \leqslant 0$$

where $\epsilon_n$ represents the current total strain in the direction normal to the joint. Equation (7.24) is thus an additional yield function which ensures that compressive stresses are transmitted across joints only when they are closed.

The computational procedure for obtaining strain rate for each of the joint sets at each Gauss point and at each time-step can be summarized as follows:

1. COMPUTE LOCAL STRESSES $(\sigma_n, \sigma_s)$
2. IF $\sigma_n > 0$ GO TO 7
3. IF STORE CONTAINING $\epsilon_n^{max}$ (TENSILE) IS EMPTY, GO TO 8.
4. GAP$=\epsilon_n + \epsilon_n{}^{max}$ (TENSILE).
5. IF GAP $> 0$, GO TO 9.
6. THE JOINT HAS COMPLETELY CLOSED NOW.
   USE EQN (7.13) TO COMPUTE $\dot{\epsilon}^{vp}$
   GO TO 10.
7. JOINT IS STILL OPENING
   COMPUTE $\dot{\epsilon}^{vp}$ FROM EQN (7.12).
   STORE $\epsilon_n$ AS $\epsilon_n{}^{max}$ IF IT IS GREATER THAN THE LAST TIME-STEP
   GO TO 10.
8. THIS POINT HAS NEVER BEEN IN TENSILE REGION SO FAR,
   USE EQN (7.13) TO COMPUTE $\dot{\epsilon}^{vp}$.
   GO TO 10.

9. THE JOINT IS PARTIALLY CLOSED
   USE EQN (7.24) TO COMPUTE $\dot{\epsilon}^{vp}$.

10. CONTINUE.

## 7.7 MORE SOPHISTICATED MODELS

As mentioned earlier, the multilaminate framework enables assembling the behaviour of the jointed rock mass from that of joints and intact rock. It is therefore possible to introduce quite sophisticated models. For example, the joint behaviour can involve strain hardening as well as strain softening, as discussed in Chapter 3. Here, two surface models can be readily incorporated in the program. Again, description of intact rock behaviour also may include strain hardening and brittle failure as discussed in Chapter 3.

In practice, one has to decide as to how sophisticated the model of jointed rock mass need be. It largely depends on what the designer aims to achieve from thc analysis. In planning open pit mines, a change in the slope by a couple of degrees may mean expenditure of vast sums of money. A thorough investigation, tcsting and a sophisticated numerical modelling is justified in such cases. Sophisticated modcls involve more parameters than simpler models. If one has always a range of the values of these parameters, the number of parametric studies required to be performed soon reaches a large number.

## 7.8 TIME DEPENDENCE IN THE BEHAVIOUR OF JOINTED ROCK MASSES

Multilaminate framework described in Section 7.4 is based on the theory of elasto–viscoplasticity and allows for time dependent behaviour.

Traditionally, the instrumentation and monitoring of geotechnical structures is based on the assumptions of a time dependent behaviour. The forewarning of distress is usually based on movement between 'control' points with time. If the motion is decelerating, the structure is interpreted as stabilising, whilst an acceleration would indicate 'trouble'. However, most of rock engineering practice has been hitherto based on theories which are time independent, viz. theories of elasticity and plasticity. Although multilaminate models based on the theory of elasto–viscoplasticity introduce time element for the behaviour of intact rock as well as rock joints, little experimental or field information is available on time dependent behaviour of jointed rock masses. Applications of the multilaminate model have generally excluded the time dependence by assuming an arbitrary value of 'fluidity parameter'. They, however, retain the sequential history of deformation, albeit on a distorted time scale. This is one of the main advantages of the multilaminate model via viscoplasticity.

## 7.9 PLASTICITY FORMULATION OF MULTILAMINATE FRAMEWORK

Let us consider the $j$th joint set. The yield function associated with it is represented by

$$F(\hat{\boldsymbol{\sigma}}_j, \kappa_j) = 0 \tag{7.25}$$

where $\hat{\boldsymbol{\sigma}}_j = [\sigma_n, \sigma_s]_j^T$ as defined in Section 7.4.3; $\sigma_n$ and $\sigma_s$ being the normal shear stresses respectively and $j$ denoting that the quantities are on the $j$th joint set. $\kappa$ represents a hardening parameter.

The increment of plastic strains contributed by the $j$th plane is given by

$$\mathrm{d}\hat{\epsilon}_j^p = \mathrm{d}\lambda j \frac{\partial Q}{\partial \hat{\boldsymbol{\sigma}}_j} \tag{7.26}$$

where implied summation is not used; $j$'s simply denoting quantities related to the $j$th joint set.

$$\mathrm{d}\hat{\boldsymbol{\epsilon}}_j^p = [\mathrm{d}\epsilon_n^p, \mathrm{d}\epsilon_s^p]_j^T$$

is the vector of normal plastic strain $(\mathrm{d}\epsilon_n{}^p)_j$ and plastic shear strain $(\mathrm{d}\epsilon_s)_j$ increments on the $j$th joint set. $\mathrm{d}\lambda_j$ represents a positive proportionality constant and $\mathrm{Q}(\boldsymbol{\sigma})$ = constant is the plastic potential function.

Assuming for simplicity that $\mathrm{Q} \equiv F$ (associated flow rule) and that the hardening effects are attributed to plastic shear strains $(\epsilon_s{}^p)_j$, the consistency condition for the $j$th joint set is given by

$$\left(\frac{\partial F}{\partial \hat{\boldsymbol{\sigma}}_j}\right)^T \mathrm{d}\hat{\boldsymbol{\sigma}}_j + \left(\frac{\partial F}{\partial \epsilon_s^p}\right)_j (\mathrm{d}\epsilon_s^p)_j = 0 \tag{7.27}$$

Substituting for $(\mathrm{d}\epsilon_s{}^p)_j$ from equation (7.26), we have

$$\mathrm{d}\lambda_j = \frac{1}{H_j}\left(\frac{\partial F}{\partial \hat{\boldsymbol{\sigma}}_j}\right)^T \mathrm{d}\hat{\boldsymbol{\sigma}}_j \tag{7.28}$$

where

$$H_j = -\left(\frac{\partial F}{\partial \epsilon_s^p}\right)_j \frac{\partial F}{\partial \sigma_s} \tag{7.29}$$

The increments of plastic strain in the global system of coordinates contributed by the $j$th joint set are given by

$$d\boldsymbol{\epsilon}_j^p = \begin{Bmatrix} d\epsilon_x^p \\ d\epsilon_y^p \\ d\epsilon_z^p \\ d\gamma_{xy}^p \\ \vdots \end{Bmatrix} = d\lambda_j [T]_j \frac{\partial F}{\partial \hat{\boldsymbol{\sigma}}_j} \tag{7.30}$$

where $[T]_j$ is a transformation matrix given by

$$[T]_j = \left[\frac{\partial \hat{\boldsymbol{\sigma}}}{\partial \boldsymbol{\sigma}}\right]_j = \begin{Bmatrix} \frac{\partial \sigma_n}{\partial \sigma_x} & \frac{\partial \sigma_s}{\partial \sigma_x} \\ \frac{\partial \sigma_n}{\partial \sigma_y} & \frac{\partial \sigma_s}{\partial \sigma_y} \\ \frac{\partial \sigma_n}{\partial \sigma_z} & \frac{\partial \sigma_s}{\partial \sigma_z} \\ \frac{\partial \sigma_n}{\partial \sigma_{xy}} & \frac{\partial \sigma_s}{\partial \sigma_{xy}} \end{Bmatrix} \tag{7.31}$$

To obtain increments of plastic strains for the rock mass having $n$ sets of joints, the contributions by each of the joint sets should be summed up. Thus

$$d\boldsymbol{\epsilon}^p = \sum_{j=1}^{n} d\boldsymbol{\epsilon}_j^p = \sum_{j=1}^{n} d\lambda_j [T]_j \frac{\partial F}{\partial \hat{\boldsymbol{\sigma}}_j} + d\lambda_i \frac{\partial \mathrm{F}_i}{\partial \boldsymbol{\sigma}} \tag{7.32}$$

Here the last term represents the contribution of the intact rock.

It is noted that $d\lambda_j$ will be positive only for those joints which are yielding, and $d\lambda_i$ will be positive if the intact rock is yielding.

The formulation presented above is based on the work of Pietruszczak and Pande (1987) and is quite general and can be adapted to any yield function, flow and hardening rule for the joint sets as well as intact rock.

## 7.10 REINFORCEMENTS IN JOINTED ROCK MASSES

Rock bolts, anchors and dowels, etc. are often used to strengthen jointed rock masses. The analyst then faces the problem of incorporating them in the finite element analysis. There are a wide variety of these devices and they can be 'fully grouted' or partially grouted (grouted at ends only). They may be either passive (without any prestress) or prestressed to induce a precompression in the jointed rock mass. If the number of rock bolts/ anchors is small, it is possible to incorporate them in the analysis by 'bar' (one dimensional) elements. The influence of prestressed anchors can also be accounted for, in an approximate way, by applying the prestress forces at the appropriate nodes.

However, if the number of bolts/anchors is large, modelling them in the analysis by discrete bar elements becomes cumbersome. In this situation, it appears necessary to revise the constitutive law of the jointed rock mass taking into consideration the fabric of rock bolts (anchors, dowel bars, etc.) In other words, the influence of rock bolts is assumed to be distributed (smeared) over the volume of the jointed rock mass and the constitutive relations of the 'equivalent material' are derived taking into account the volume of reinforcement, their mechanical characteristics and orientation. Pande and Gerrard (1983), Gerrard and Pande (1985), Larsson *et al.* (1985), Gerrard *et al.* (1984) and Sharma and Pande (1988) have worked in this area in recent years and the following development of constitutive laws is largely based on their work.

### 7.10.1 Constitutive law of rock masses reinforced by passive fully grouted rock bolts

The multilaminate model of jointed rock mass (Section 7.4) with its assumptions is adopted as the starting point. The following additional assumptions are made regarding the reinforcements:

(a) The reinforcements in their respective sets (there can be more than one set in the rock mass) are continuous, parallel and have spacing much smaller than the critical dimension of the structure.
(b) The passive reinforcement is required to strain in conformity with the jointed rock mass.

The rheological analogue of a reinforced jointed rock mass is shown in Figure 7.7. It is based on an elasto–viscoplastic behaviour of all components, i.e. intact rock, rock joints, reinforcements and interface between the reinforcement and rock mass. The first string of units in series represents jointed rock mass (multilaminate model) as we have seen in Figure 7.4. The units in parallel represent sets of passive reinforcements and their interface. The sum of strains of all units in series is equal and is equal to the strain of the reinforced jointed rock mass. In developing the constitutive relations of the equivalent material, let subscript $i$ denote the $i$th string, subscript $ij$, the $j$th rheological unit in the $i$th string, and the absence of any subscript indicates the equivalent material. Here, $\sigma$ and $\epsilon$ represent stress and strain vectors respectively but the vector sign ($\sim$) has been omitted to avoid confusion. Then

$$\Delta\epsilon = \Delta\epsilon_i = \sum_j \Delta\epsilon_{ij}, \tag{7.33}$$

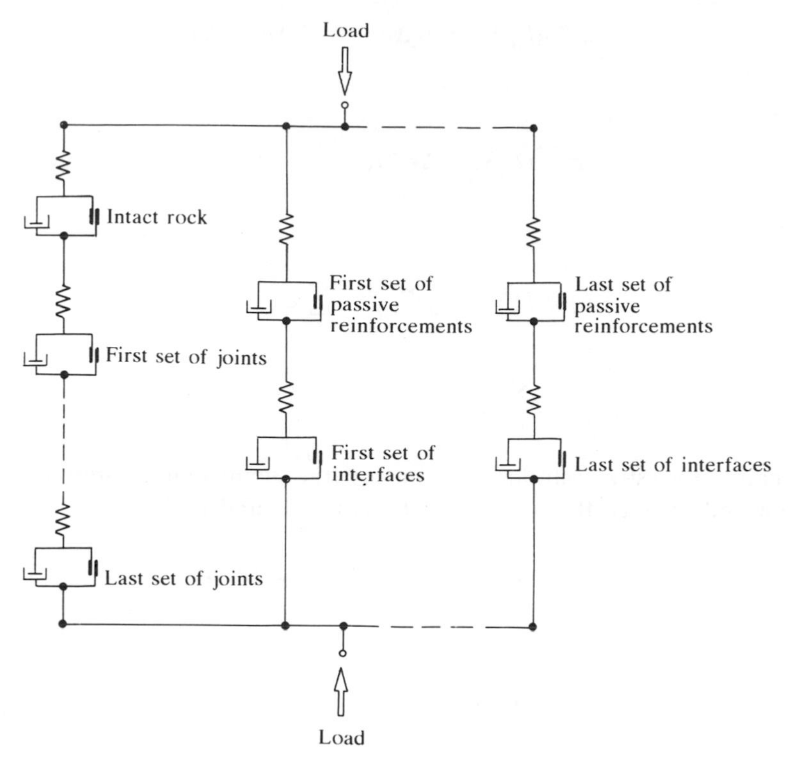


**Figure 7.7**
Rheological model of reinforced jointed rock mass

$$\Delta\sigma = \sum_{i} p_i \, \Delta\sigma_i, \tag{7.34}$$

$$\Delta\sigma_i = \Delta\sigma_{ij}, \tag{7.35}$$

in which $p_i$ is the volumetric proportion of the $i$th string.

The relation between incremental stress and incremental strain vectors for an individual rheological is written as

$$\Delta\epsilon_{ij} = D_{ij}^{-1}\,\Delta\sigma_i + \Delta\epsilon_{ij}^{vp} = C_{ij}\,\Delta\sigma_i + \Delta\epsilon_{ij}^{vp} \tag{7.36}$$

or

$$\Delta\sigma_{ij} = D_{ij}\,(\Delta\epsilon_{ij} - \Delta\epsilon_{ij}^{vp}). \tag{7.37}$$

For the $i$th string, the stress–strain relation is obtained by substituting equation (7.36) into equation (7.33). Thus

$$\Delta\epsilon = D_i^{-1}\,\Delta\sigma_i + \Delta\epsilon_i^{vp} = C_i\,\Delta\sigma_i + \Delta\epsilon_i^{vp}. \tag{7.38}$$

or

$$\Delta\sigma_i = D_i(\Delta\epsilon - \Delta\epsilon_i^{vp}), \tag{7.39}$$

where

$$C_i = D_i^{-1} = \sum_j D_{ij}^{-1} = \sum_j C_{ij} \tag{7.40}$$

$$\Delta\epsilon_i^{vp} = \sum_j \Delta\epsilon_{ij}^{vp}. \tag{7.41}$$

Finally, the stress–strain relation for the equivalent material can be established through the use of equations (7.34) and (7.39).

$$\Delta\epsilon = D^{-1}\,\Delta\sigma + \Delta\epsilon^{vp} = C\,\Delta\sigma + \Delta\epsilon^{vp} \tag{7.42}$$

or

$$\Delta\sigma = D(\Delta\epsilon - \Delta\epsilon^{vp}), \tag{7.43}$$

where

$$D = \sum_i p_i D_i = C^{-1}$$

and

$$\Delta\epsilon^{vp} = D^{-1} \sum_i p_i D_i\,\Delta\epsilon_i^{vp}.$$

The incremental stress vector $\Delta\sigma$ in equation (7.43) represents the total stress loading on the equivalent material due to the loading and/or excavation.

Equations (7.33) to (7.43) are completely general and apply to a three-dimensional situation with six components of stress and strain vectors. Various simplifications can be introduced in this general formulation. For example, for two-dimensional analysis, only four components of stress and strain need be considered. One of the most common simplifications is the assumption that there is no slip between the reinforcement and the surrounding rock mass. Thus, the units in Figure 7.8 representing 'interfaces' are dropped and the strain in reinforcement is assumed the same as in rock mass.

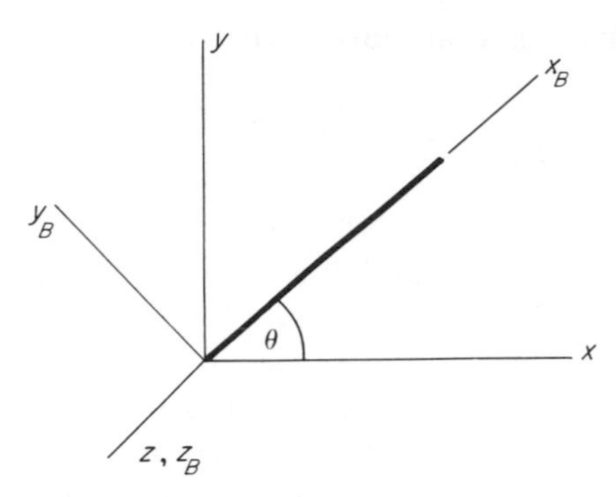


**Figure 7.8**
Local system of coordinate axes for rock bolts

Specific forms of yield/failure functions and plastic potential functions for joint sets have been already discussed in Section 7.4. Here we shall discuss the mechanical characteristics of reinforcements.

*Properties of reinforcement sets*

The mechanical properties of a typical set of reinforcement are defined with reference to the local axes $x_b$, $y_b$ and $z_b$. The $x_b$-axis coincides with the direction of reinforcement and $z_b$ with the global axis $z$ and $y_b$ is perpendicular to $x_b$ as shown in Figure 7.8.

The elastic matrix of the reinforcement set ($D_R$) in the local coordinate system is written as

$$D_R = \begin{bmatrix} E & 0 \\ 0 & G \end{bmatrix}, \tag{7.44}$$

in which $E$ is Young's modulus and $G$ is the shear modulus of the reinforcement material.

As mentioned earlier, the passive reinforcement sets deform to an extent equivalent to the jointed rock mass and hence form part of the reinforced jointed rock mass. Passive reinforcement may be grouted or ungrouted, with only grouted sets capable of resisting shear stress as well as axial stress. Equation (7.44) applies to the grouted passive reinforcement sets; whereas, for the ungrouted sets, $G$ is taken as zero in equation (7.44).

The elastic matrix of the reinforcement set with respect to the global axes ($D_R^G$) is given by

$$D_R^G = T_R^T D_R T_R, \tag{7.45}$$

where $T_R$ is the transformation matrix written as

$$T_R = \begin{bmatrix} c^2 & s^2 & cs & 0 \\ -2cs & 2cs & c^2 - s^2 & 0 \end{bmatrix}, \tag{7.46}$$

in which $c = \cos\theta$, $s = \sin\theta$ and $\theta$ is the angle between $X_b$-axis and $X$-axis.

Since the reinforcement material is usually steel, its yielding is assumed to be governed by Von Mises yield criterion with associated plastic flow. After the initial yield stress $\sigma_{y0}$ is reached, strain hardening can occur with the yield stress being assumed to increase in a specified manner with the viscoplastic octahedral shear strain $\gamma^{vp}$. At a defined level of this shear strain $\gamma_u^{vp}$, the ultimate yield stress $\sigma_{yu}$ is reached and the reinforcement fails. For the reinforcement set which experiences only axial ($\sigma_a$) and shear ($\sigma_s$) stresses, Von Mises criterion can be written as

$$F = \sqrt{\sigma_a^2 + 3\sigma_s^2} - \sigma_y(\gamma^{vp}), \tag{7.47}$$

where $\sigma_y$ is the uniaxial yield stress which is a function of the viscoplastic octahedral shear strain $\gamma^{vp}$. For example, a linear hardening rule with ultimate yield stress as $\sigma_{yu}$ is given by

$$\sigma_y = \sigma_{y0} + \gamma^{vp}\,(\sigma_{yu} - \sigma_{y0})/\gamma_u^{vp}\ . \tag{7.48}$$

and

$$\sigma_y = \sigma_{yu} \quad \text{for } \gamma^{vp} > \gamma_u^{vp}$$

The local viscoplastic strain increment $\Delta\epsilon_b^{vp}$ is calculated by using viscoplastic flow equations (7.47) and (7.48). It is transformed into the global system by the expression

$$\Delta\epsilon_G^{vp} = T_R^T\,\Delta\epsilon_R^{vp}\ . \tag{7.49}$$

### 7.10.2 Constitutive law of rock masses reinforced by active rock bolts

Active reinforcement sets are defined as those that are prestressed during a particular loading stage. As such they form a part of the total loading on the reinforced jointed rock mass when considered together with external loading.

Figure 7.9 shows a rock mass which is reinforced by one set of prestressed rock bolts. If the prestress applied to the bolts is $\alpha$ times the initial yield

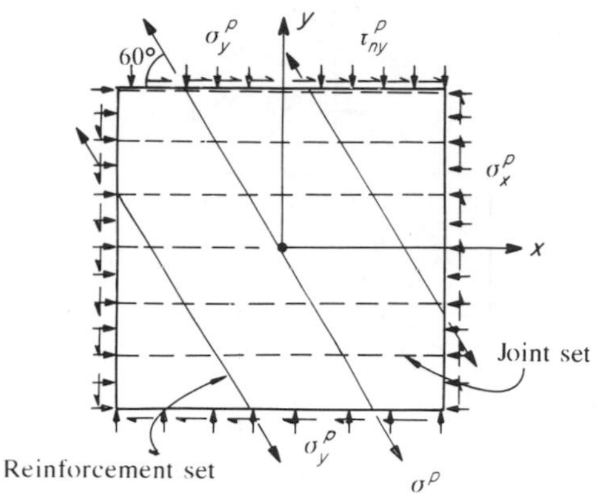


**Figure 7.9**
System of initial stress induced in the jointed rock mass due to a set of prestressed rock bolts

stress of the bolt material, where $\alpha$ is constant less than 1, the compressive stress ($\sigma_p$) induced in the rock mass in the direction of rock bolts can be computed. It is given by

$$\sigma_p = -p\alpha\, \sigma_{y0} \tag{7.50}$$

$\sigma_p$ can be transformed to give components of prestressing stress ($\sigma_x^p$, $\sigma_y^p$,$\sigma_z^p$, $\tau_{xy}^p$, $\tau_{yz}^p$, $\tau_{zx}^p$) in the global coordinate system. In the multilaminate model of rocks, these stresses are simply treated as initial stresses i.e. these stresses are always added to the stresses due to external loading. Obviously, yielding of joints and intact rock is checked taking the prestressing stress ($\boldsymbol{\sigma}_p$) into consideration.

## 7.11 STRENGTH OF REINFORCED JOINTED ROCK MASS UNDER SIMPLE LOADING CONDITIONS

Some numerical examples to illustrate the influence of the orientation of passive rock bolts on the strength of jointed rock mass. We shall consider the jointed rock mass under simple loading situations like uniaxial compression, uniaxial tension and simple shear. The case of uniaxial compression, is quite important as it has practical applications in mine pillars which are subjected primarily to uniaxial compressive forces. Here, the conflicting requirements of safety and maximum extraction of the ore require the pillars with different fabrics of joint sets to be reinforced and designed rationally.

The constitutive equations of reinforced jointed rock mass ((7.43)–(7.45)) can be numerically integrated to compute strains due to an applied stress

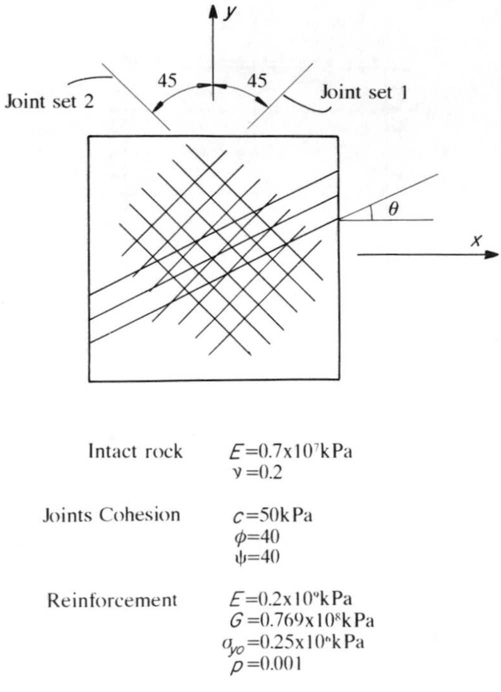


**Figure 7.10**
Assumed fabric of joints and other parameters for reinforced jointed rock mass

path or to compute stresses due to an applied strain path. In the following paragraphs, the results of this numerical integration are presented for some simple loading conditions.

For all cases, the assumed parameters for the jointed rock mass are shown in Figure 7.10. It is noted that the isotropic elastic parameters have been assigned though one could derive anisotropic elasticity matrix based on the theory given in Chapter 4.

### 7.11.1 Uniaxial compression and tension

Figure 7.11 shows the normalized uniaxial stress–strain relationship ($\sigma/c$ − $\epsilon$) in compression for various orientations of the reinforcement. These relationships are trilinear with points $A_1$, $A_2$, $A_3$ representing the initiation of sliding of both joint sets. Points $B_1$, $B_2$, $B_3$ correspond to the failure of reinforcement and consequently the failure of reinforced jointed rock mass.

Figure 7.12 shows a rose diagram of normalized uniaxial failure strength of the reinforced jointed rock mass having the assumed fabric of joints. From this it is obvious that the best orientation for passive fully grouted rock bolts for this case is given by $\theta = 0°$. In Figure 7.12 rose diagram is also shown for the case when linear hardening of reinforcement takes place.

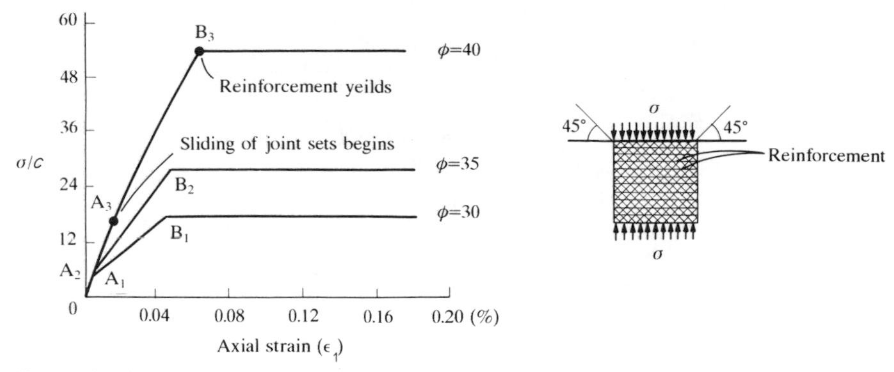


**Figure 7.11**
Uniaxial stress–strain curve for reinforced jointed rock mass-reinforcement perpendicular to the direction of uniaxial stress

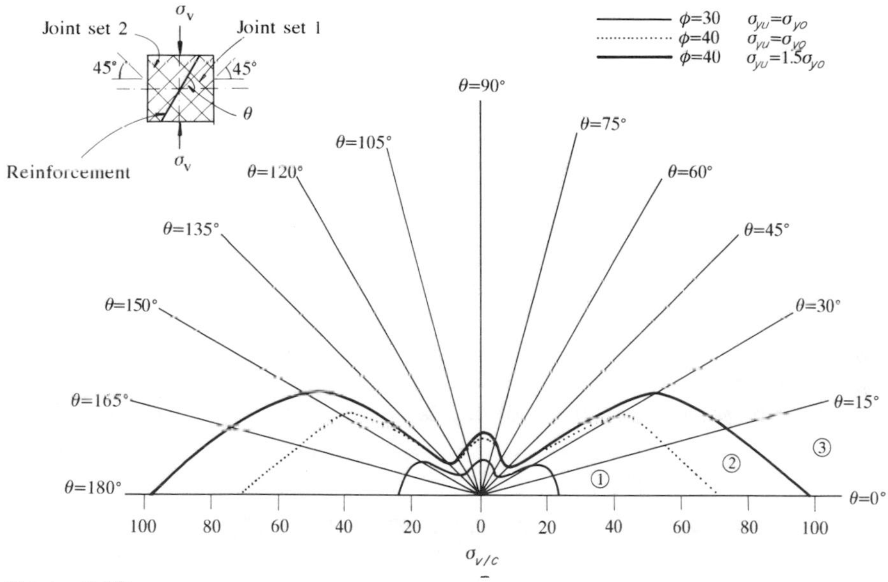


**Figure 7.12**
Rose diagram for reinforced jointed rock under uniaxial compression

For the case of uniaxial tension, the best orientation is represented by $\theta = 90°$ as seen from the rose diagram in Figure 7.13. For the case when the jointed rock mass is subjected to pure shear, the best orientation for reinforcement is along the diagonal ($\theta = 45°$) which is increasing in length due to shear.

Numerically integrated constitutive equations for reinforced jointed rock mass can be used in conjunction with an elastic analysis to obtain the best orientation of reinforcement for practical problems. An example to illustrate this is presented in Chapter 8.

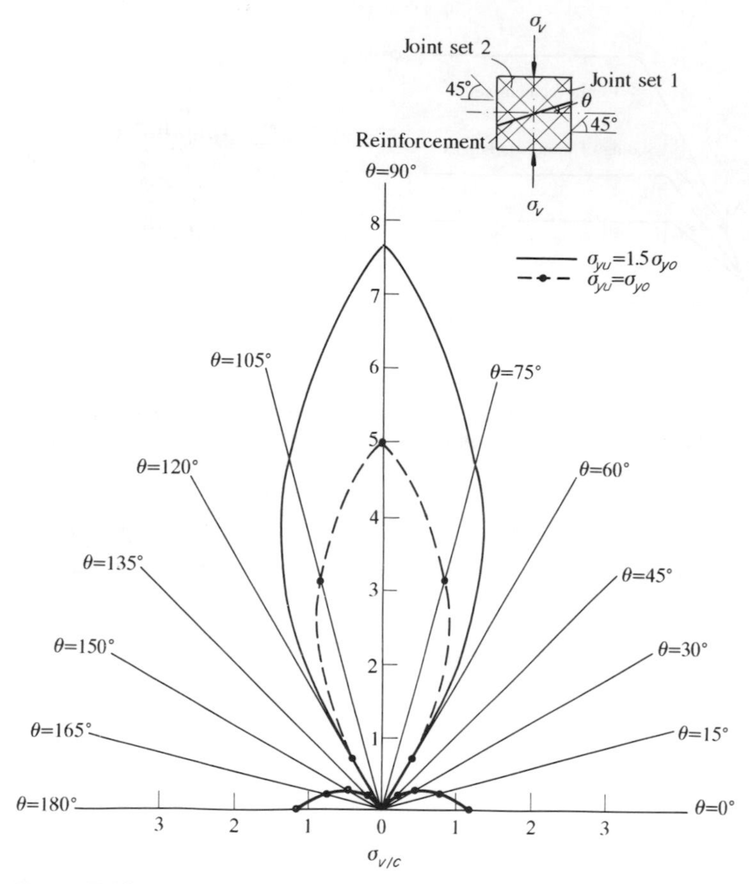


**Figure 7.13**
Rose diagram for reinforced jointed rock mass under uniaxial tension

## 7.12 WATER PRESSURE IN JOINTED ROCK MASSES

The presence of joints, discontinuities, fissures, etc. in a jointed rock mass leads to flow of water through them. Percolation of water through jointed rock masses plays an important role in their stability. Generally the main flow takes place through the network of joints as the surrounding intact rock is comparatively impermeable. The average overall permeability of a jointed rock mass largely depends on the fabric, openness/tightness and orientation of joints and is highy anisotropic. The problem of steady-state flow in rock masses is governed by

$$\frac{\partial}{\partial x'}\left(K_{x'}\frac{\partial U}{\partial x'}\right)+\frac{\partial}{\partial y'}\left(K_{y'}\frac{\partial U}{\partial y'}\right)+\frac{\partial}{\partial z'}\left(K_{z'}\frac{\partial U}{\partial z'}\right)=0 \tag{7.51}$$

where $U$ is the hydraulic potential, $x'$, $y'$ and $z'$ are principal directions of anisotropy of permeabilities with $K_{x'}$, $K_{y'}$, $K_{z'}$ being the principal components of permeability tensor.

A number of numerical as well as analytical techniques exist for solving equation (7.51) with appropriate boundary conditions and we shall not discuss them here.

Our aim in this section is to discuss the influence of water pressure on the deformability and strength of intact rocks and rock joints. It is generally accepted that the strength of intact rocks and rock joints is governed by 'effective stress' and not by total stress. All equations of yield/failure criterion written in this Chapter are strictly in terms of effective stress. Using compression positive convention, the total stress ($\boldsymbol{\sigma}$) and effective stress ($\boldsymbol{\sigma}'$) are related by the equation

$$\boldsymbol{\sigma} = \boldsymbol{\sigma}' + \mathbf{m}p \tag{7.52}$$

where $m = [1\ 1\ 1\ \ 0\ 0\ 0]^T$ and $p$ is the interstitial water pressure called the pore pressure. The concept of effective stresses is well established in Soil Mechanics and readers not familiar with this are referred to standard texts. Equation (7.52) can also be written in the incremental form which is useful for nonlinear behaviour:

$$\Delta\boldsymbol{\sigma} = \Delta\boldsymbol{\sigma}' + \mathbf{m}\,\Delta p \tag{7.53}$$

It is obvious that if there is no water in the rock mass, then distinction between total stress and effective stress disappears as equation (7.53) reduces to

$$\Delta\boldsymbol{\sigma} = \Delta\boldsymbol{\sigma}' \tag{7.54}$$

In the presence of water, when a load is applied the pore pressure rises and two extreme situations can be visualised: (a) undrained—no water escapes/enters the pores. This happens immediately after the application of load (excavation) as water takes time to escape depending on the permeability (b) drained—after a long time, the excess pore pressure generated by the load is either completely dissipated or a state of steady seepage is attained. During the time pore pressures dissipate from undrained to drained state, the rock mass consolidates under the applied load.

For the drained state, there are no excess pore pressures to be considered in the analysis and for the steady seepage case, the pore pressures are known from the seepage analysis (solution of equation (7.51)). They should be accounted for in computing nodal forces as discussed in Section 5.2.8.

The procedure for undrained case is more complex. Undrained case may be critical in some cases such as stability of gravity dams. Here we have to compute total as well as effective stresses and stiffness matrix of the rock structure in 'undrained' state is different from that of the 'drained' state.

### 7.12.1 Effective stress analysis of undrained rock masses

Consider an element of unit volume of rock skeleton and pore fluid. Let this element be subjected to a change of effective stress ($\Delta\boldsymbol{\sigma}'$), and pore pressure change of $\Delta p$. Assume that the element is undrained, i.e. no pore fluid can escape (or enter) from the unit volume.

Now we work out the various components of volume change due to stress.

(a) Change in volume of pore water due to change in pore pressure ($\Delta p$)

$$= \frac{\eta \Delta p}{K_w}$$

where $\eta$ is porosity and $K_w$ is the bulk modulus of water. Rock is assumed to be fully saturated.

(b) Change in volume of rock grains due to change in pore pressure

$$= \frac{(1 - \eta)\Delta p}{K_s}$$

where $K_s$ is the bulk modulus of rock grains.

(c) Change in the volume due to change of effective stress ($\Delta\sigma'$)

$$= \frac{m^T \Delta\boldsymbol{\sigma}'}{3K_s}$$

The total change in the volume is thus equal to

$$\frac{n\Delta p}{K_w} + \frac{(1 - n)\,\Delta p}{K_s} + \frac{m^T\,\Delta\boldsymbol{\sigma}'}{3K_s}$$

which must be equal to the change in the volumetric strain of the rock skeleton ($\Delta\epsilon_v$). Thus

$$\Delta\epsilon_v = \mathbf{m}^T\boldsymbol{\Delta\epsilon} = \frac{n\,\Delta p}{K_w} + \frac{(1 - n)\,\Delta p}{K_s} + \frac{\mathbf{m}^T\,\Delta\boldsymbol{\sigma}'}{3K_s} \tag{7.55}$$

Introducing

$$\frac{1}{K_f} = \frac{n}{K_w} + \frac{1 - n}{K_s} \tag{7.56}$$

equation (7.55) can be written as

$$\mathbf{m}^T \Delta\boldsymbol{\epsilon} = \frac{\Delta p}{K_f} + \frac{\mathbf{m}^T \Delta\boldsymbol{\sigma}'}{3K_s} \tag{7.57}$$

Equation (7.57) can be rearranged to yield

$$\Delta p = K_f \mathbf{m}^T \left( \Delta\boldsymbol{\epsilon} - \frac{\Delta\sigma'}{3K_s} \right) \tag{7.58}$$

Now the total stress change is related to the strain changes by a modulus matrix $D^*$, i.e.

$$\Delta\boldsymbol{\sigma} = D^* \,\Delta\boldsymbol{\epsilon} \tag{7.59}$$

while the effective stress changes are related to the strain changes by a drained modulus matrix $D$, i.e.

$$\Delta\boldsymbol{\sigma}' = D\,\Delta\boldsymbol{\epsilon} \tag{7.60}$$

Substituting equations (7.58), (7.59) and (7.60) in equation (7.53) leads to

$$D^* = D + K_f\,\mathbf{mm}^T - \frac{K_f}{3K_s}\,\mathbf{mm}^T D \tag{7.61}$$

It is noted that no restriction has been placed on $D^*$, $D$ and $K_f$. They represent tangent modulus matrices and equation (7.61) is valid for any nonlinear law.

The stiffness of an element ($K^e$) in the undrained state is given by (see Chapter 5)

$$K^e = \int_V B^T D^* B \,\mathrm{d}V \tag{7.62}$$

Thus, for the 'undrained' analysis of rock mass, the global stiffness equations are set up in terms of total stresses based on equation (7.62), displacements and strains are then computed. Effective stresses and pore pressures are computed using equations (7.60) and (7.58) respectively.

## REFERENCES

Barton, J., and Choubey, V. (1977). 'The shear strength of rock joints in theory and practice.' *Rock Mechanics,* **10**, 1–54.

Chang, C. Y., and Niar, K. 'A theoretical method of evaluating stability of openings in rock.' Final report, April 72, Woodward–Lundgren & Associates, Oakland, Ca. 94607.

Gerrard, C. M., Macindoe, L., and Pande, G. N. (1984). 'Theoretical analyses of practical cases of rock reinforcement in two and three dimensions.' International Conference on *In situ* Soil and Rock Reinforcement, Paris.

Gerrard, C. M., and Pande, G. N. (1985). 'Numerical model of reinforced jointed rock masses.' *I. Theory. Comput. Geotech.* **I**, 293–318.

Larsson, H., Olofsson, T., and Stephansson, O. (1985). 'Reinforcement of jointed rock mass-*a* non-linear continuum approach.' *Proc. Int. Symp. on Fundamental of Rock Joints*, 567–75, Bjorkliden, Sweden.

Pande, G. N., and Xiong, W. (1982). 'An improved multilaminate model of jointed rock masses.' *Numerical Models in Geomech.* (eds, R. Dungar, G. N. Pande, and J. A. Studer, A. A. Balkema), Rotterdam, 218–26.

Pande, G. N., and Gerrard, C. M. (1983). 'The behaviour of reinforced jointed rock masses under various simple loading states.' *Proceedings of the Fifth Congress of the International Society for Rock Mechanics*, Melbourne, F217–F223.

Pietruszczak, S., and Pande, G. N. (1987). 'Multilaminate framework of soil models—plasticity formulation.' *Int. J. Num. Anal. Meth. Geomech.* **11**, 651–58.

Schweiger, H. F., Aldrian, W. and Haas, W. (1986). 'The influence of joint orientation and elastic anisotropy in the analysis of tunnels in jointed rock masses.' *2nd Int. Symp. on Numerical Models in Geomechanics*, 375–80, Jackson, Redruth, Cornwall.

Sharma, K. G., and Pande, G. N. (1988). 'Stability of rock masses reinforced by passive, fully-grouted rock bolts.' *Int. J. Rock Mech. Min. Sci & Geomech. Abstr.*, **25**(5), 273–85.

Zienkiewicz, O. C. and Valliappan, S., and King, I. P. (1968). 'Stress analysis of rock as a no-tension material.' *Geotechnique,* **18**, 56–66.

Zienkiewicz, O. C., and Pande, G. N. (1977). 'Time dependent multilaminate model of rocks—a numerical study of deformation and failure of rock masses.' *Int. J. Numer. Analyt. Meth. Geomech.,* **1**. 219–47.

# 8 Applications of the Finite Element Method

## 8.1 INTRODUCTION

In this chapter we shall discuss a number of applications of the finite element method and material models discussed in Chapter 7. The finite element method has been used extensively in the design and analysis of rock structures in many countries particularly West European countries. We have chosen five illustrative examples of application. The first example is that of a two-dimensional slope in jointed rock mass. The second, third and fourth examples relate to the stability of large caverns and have been contributed by practising engineers in industry. The fifth example is a repetition of the first example illustrating the modelling of passive rock bolts as discussed in Sections 7.10 and 7.11.

The main objective of these examples is to illustrate as to how the finite element method applied to rock engineering problems can lead to a considerable insight into the problem mainly by conducting parametric studies. Once a finite element model is set up, it is quite convenient to obtain results using a range of parameters for strength, deformability etc. It is also possible to successively reduce the strength parameters to induce collapse of the rock structure. In such studies it is possible to identify the mechanism of failure.

## 8.2 A JOINTED ROCK SLOPE

A jointed rock slope 70 m in height has two sets of planes of weakness. Figure 8.1 shows the relevant details of the problem. This two-dimensional problem is not very realistic but has, nevertheless, been chosen to illustrate the possibilities of parametric studies. The following state of initial stress

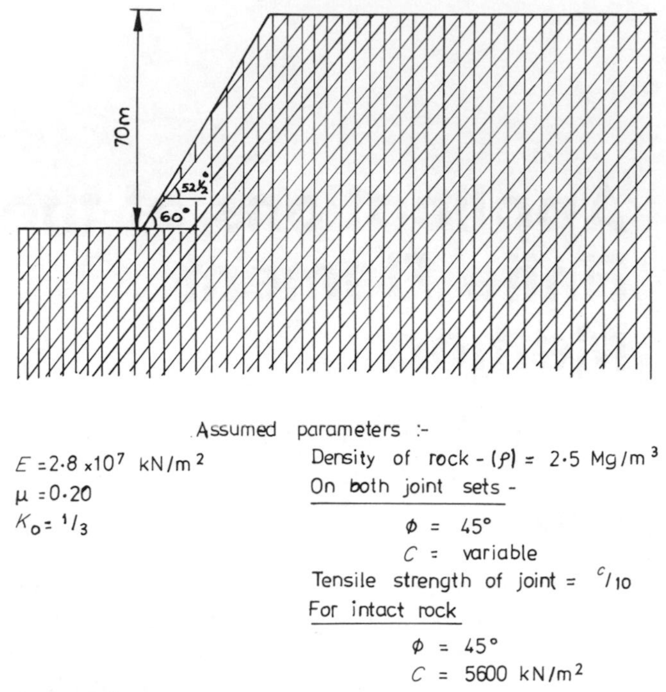


**Figure 8.1**
Details of rock slope problem

was assumed before the excavation (in one step—admittedly, a gross simplification) took place.

$$\sigma_y = \rho g h \qquad (8.1)$$
$$\sigma_x = \sigma_z = K_0 \sigma_y$$

where

- $\rho$ is the mass density of rock
- $g$ is the acceleration due to gravity
- $h$ is the depth of the point under consideration below the virgin rock level (assumed horizontal)
- $K_0$ is the coefficient of initial stress

It may be noted that consideration of initial stress does not arise in conventional limit equilibrium analysis. For our problem, the effect of excavation was obtained by applying equivalent nodal forces on the excavated boundary. First, purely linear analysis was done with a number of finite element meshes to make a choice of the mesh to be used in detailed studies. This is an important step in analysis. The analyst should appreciate that the numerical methods are approximate methods and the accuracy of results

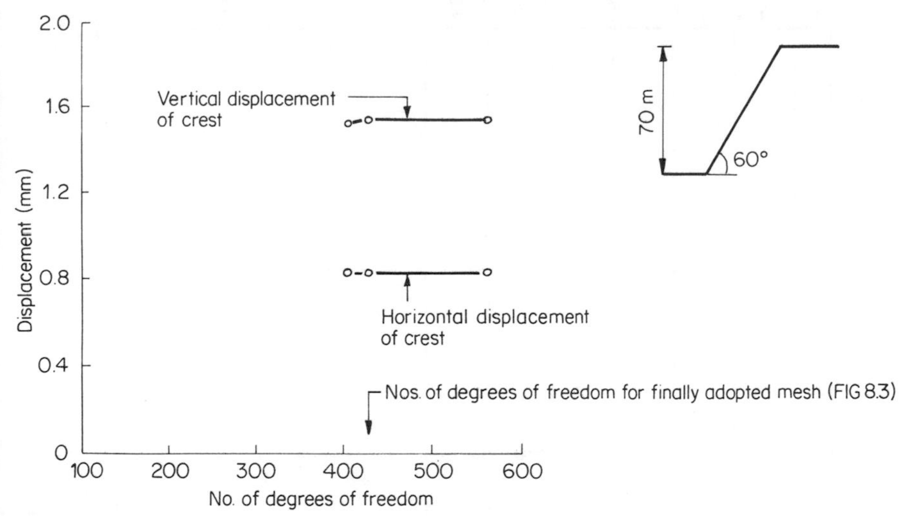


**Figure 8.2**
Variation of displacement with number of degree of freedom (elastic analysis)

obtained depends on the discretization (number of degrees of freedom used to approximate an infinite degree of freedom problem). Figure 8.2 shows the variation of the horizontal and vertical displacements of the crest of the slope with the number of degrees of freedom. A mesh with 450 degrees of freedom was finally adopted for the nonlinear analysis. This mesh is shown in Figure 8.3.

The coefficient of friction on both the joint planes (Figure 8.1) was assumed as unity, $\phi = 45°$) and was taken as fixed, while the value of '$c$' (cohesion) was successfully dropped to achieve collapse of the slope. Figure 8.4 shows the variation of the horizontal displacement of some nodes (refer to Figure 8.3 for node numbering) as the cohesion value was dropped in successive analyses. Figure 8.5 compares the horizontal displacements of a few nodes with those of corresponding nodes on a fine mesh (560 degrees of freedom and 87 isoparametric elements). Displacements are noted to be the same for up to values of $c_1 = c_2 = 50$ kN/m$^2$. However, as values of $c_1$, $c_2$ are further reduced small discrepancies appear. No convergence for $c_1 = c_2 = 27.50$ kN/m$^2$ was achieved within reasonable computer time. Since we are interested in a comparative study only, all further computations were done with the mesh shown in Figure 8.3. Both associative and nonassociated flow rules were used. The difference in the final value of '$c$' for collapse is small—21 kN/m$^2$ for associative law and 23 kN/m$^2$ for nonassociative law. Figure 8.6 shows the effective shear strain contours for $c_1 = c_2 = 23$ kN/m$^2$ when solution converged and for $c_1 = c_2 = 21$ kN/m$^2$ when no convergence took place within 450 time-steps signifying collapse. In Figure 8.7 similar results are shown when a nonassociative flow rule was used. Here the solution converged for $c_1 = c_2 = 25$ kN/m$^2$ and failed to converge

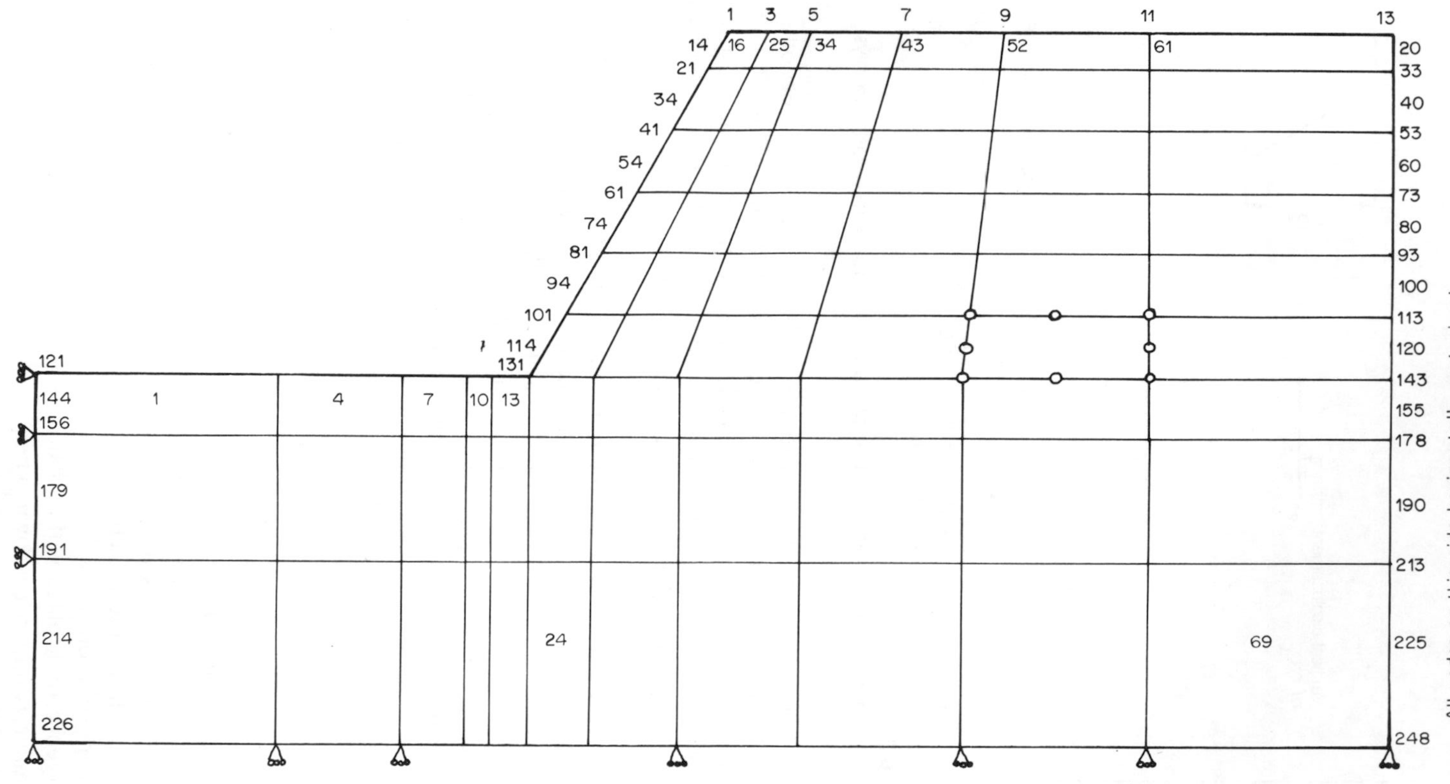


**Figure 8.3**
Finite element idealization—69 isoparametic parabolic elements, 248 nodes, 447 degrees of fredom

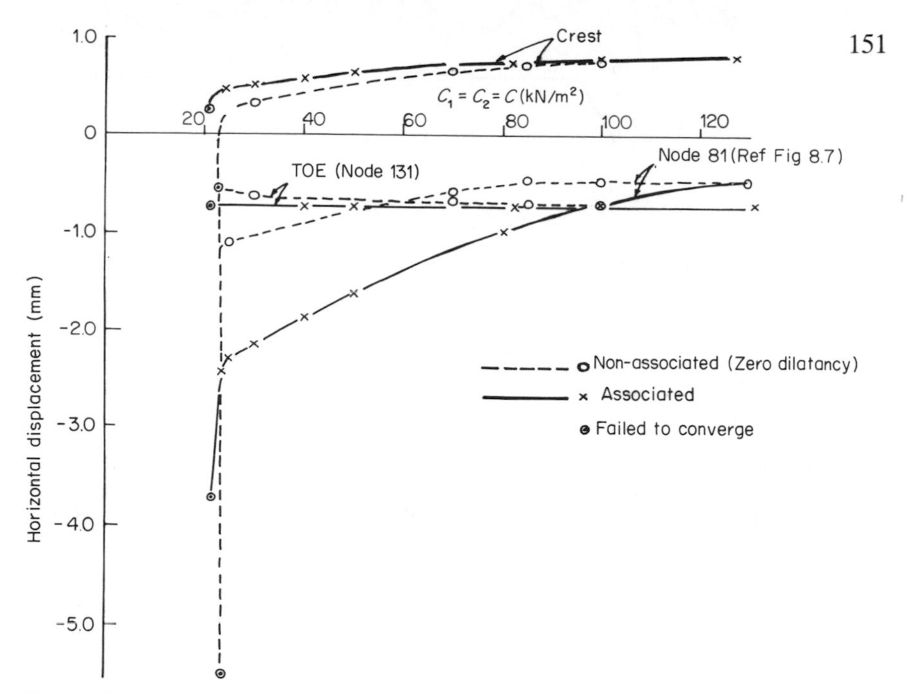


**Figure 8.4**
Variation of horizontal displacement with decreasing value of cohesion on joints

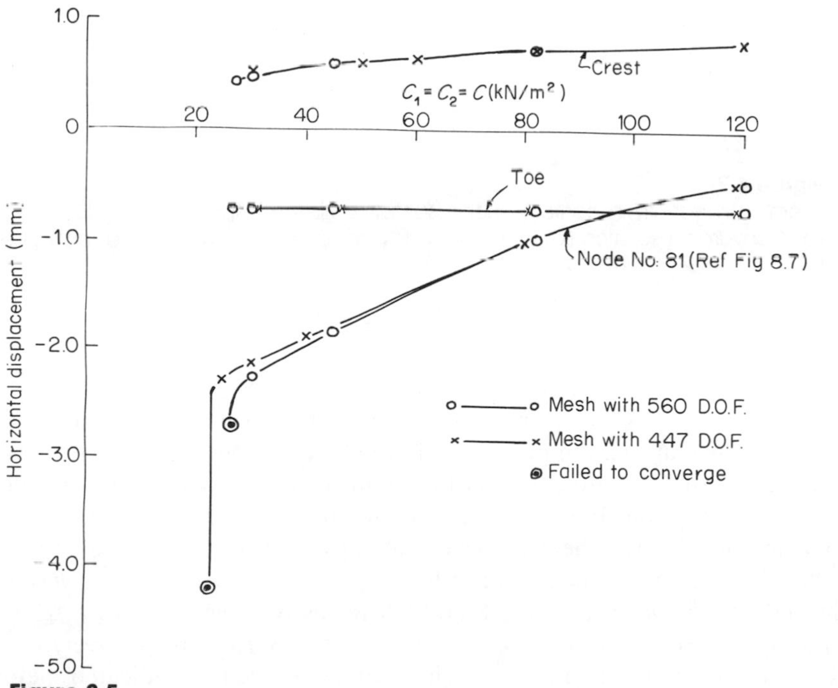


**Figure 8.5**
Comparison with a finer mesh

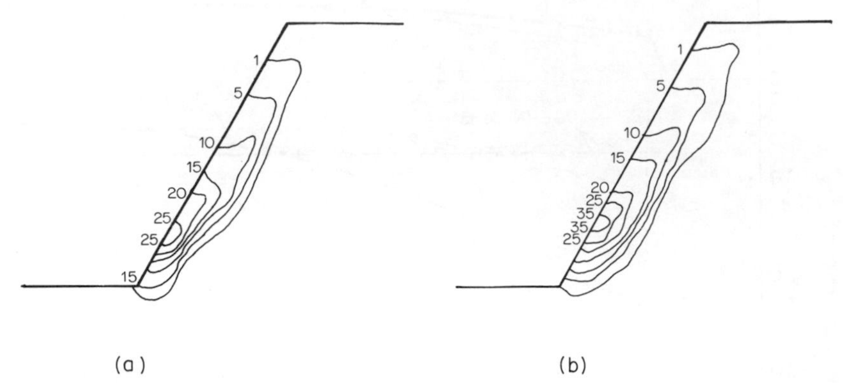


**Figure 8.6**
Effective shear strain contours—associative flow rule. (a) Converged solution ($C_1 = C_2 = 23$ kN/m$^2$, $\phi_1 = \phi_2 = 45°$); (b) at collapse ($C_1 = C_2 = 21$ kN/m$^2$, $\phi_1 = \phi_2 = 45°$)

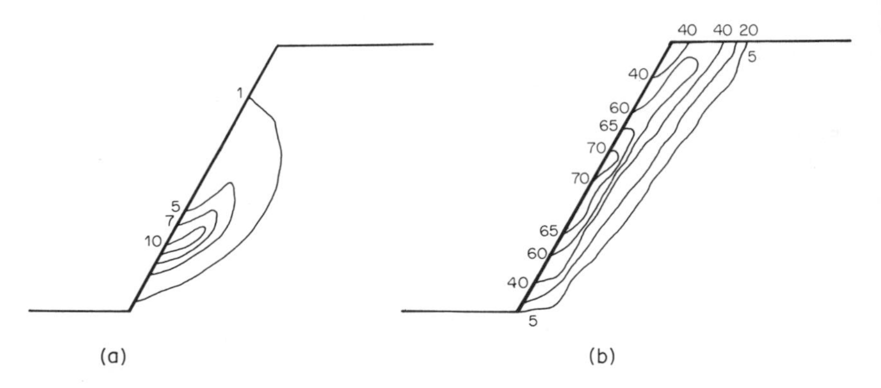


**Figure 8.7**
Effective shear strain contour—nonassociative flow rule ($\psi = 0$, no dilatation). (a) Converged solution ($C_1 = C_2 = 25$ kN/m$^2$, $\phi_1 = \phi_2 = 45°$); (b) at collapse ($C_1 = C_2 = 23$ kN/m$^2$, $\phi_1 = \phi_2 = 45°$)

for $c_1 = c_2 = 23$ kN/m$^2$. In contrast to a circular failure surface for a homogeneous material the plane of failure is seen to be stepped. Moreover, the contours of high effective plastic strain do not pass through or below the toe of excavation. In both associative and nonassociative cases the region of maximum effective shear is located about 0.1 H above the toe.

The Gauss points which were still creeping when convergence was not achieved are shown in Figure 8.8 and 8.9 for associative ($c_1 = c_2 = 21$ kN/m$^2$) and nonassociative ($c_1 = c_2 = 23$ kN/m$^2$) flow rules respectively.

Presence of tension cracks on the surface is clearly indicated before collapse takes place.

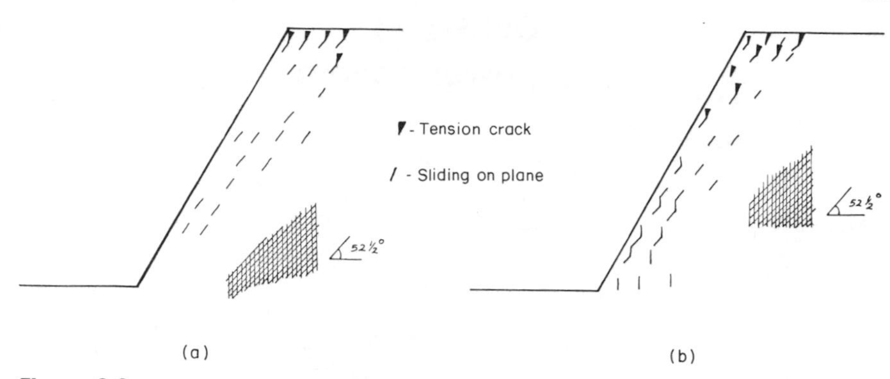


**Figure 8.8**
Gauss points still creeping—nonassociative flow rule. (a) After 450 time steps; (b) after 1000 time steps

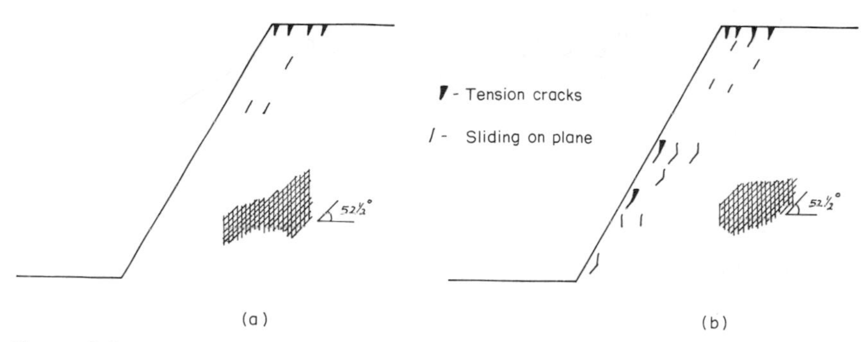


**Figure 8.9**
Gauss points still creeping—nonassociative flow rule. (a) After 450 time steps; (b) after 1000 time steps

## 8.3 ANALYSIS OF POWERHOUSE CAVERNS

Many consulting engineers in West Germany, Austria, Switzerland, etc. have over ten years of experience of applying the multilaminate model to practical problems of tunnels and underground cavities. Honish (1988) has presented case histories of analysis of four powerhouse caverns. Further, there are a number of applications reported in the proceedings of ICONMIG (International Conferences on Numerical Methods in Geomechanics) and NUMOG (International Symposia on Numerical Models in Geomechanics). Here, we have chosen two examples from Honish (1988).

### 8.3.1 Bakun underground powerhouse, Malaysia

For the feasibility study of an underground powerhouse at Bakun, Sarawak/ Malaysia, a cavern with the dimensions 32 × 50 × 300 m was designed to be situated in regularly jointed greywacke and intercalated shales. The

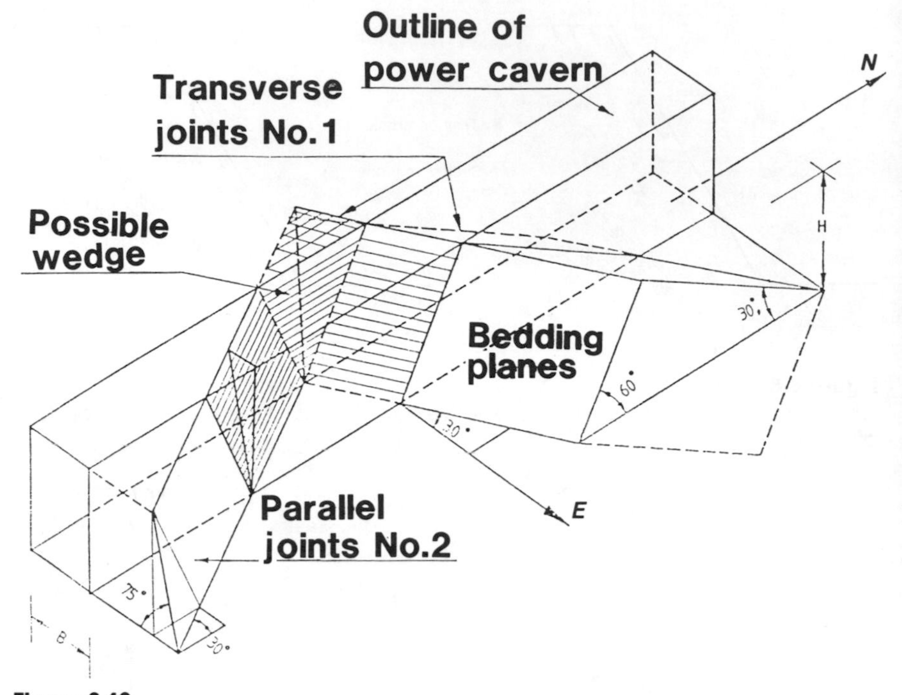


**Figure 8.10**
Schematic diagram of Bakun underground powerhouse, Malaysia showing the pattern of joints

overburden is 140 m. Bedding and longitudinal joints are dipping at 60° and 30° respectively and transverse joints at 75°. (See Figure 8.10.)

The deformability and strength parameters were adopted as given below.

*Rock mass*

Young's Modulus of rock mass, $E_m$ = 6000 MPa
Poisson's ratio of rock mass $V_m$ = 0.25
Alternatively, (transversely anisotropic rock mass)

$E_1 = 8000$ MPa
$E_2 = 4000$ MPa
$\upsilon_1 = 0.25$
$\upsilon_2 = 0.05$
$G = 2250$ MPa

*Rock joints*

| | | |
|---|---|---|
| Bedding planes | N60° | E/60°E |
| Joint set 1 | N240° | E/30°W |
| Joint set 2 | N330° | E/75°W |

Shear strength parameters: For both joint sets, $c = 1$ MPa, $\phi = 40°$
For bedding planes, $c = 0$, $\phi = 32°$

*Initial state of stress*

| | |
|---|---|
| Density of rock | 0.027 MN/m$^3$ |
| Maximum overburden | 140 m |
| Ko | 0.33 |

Figure 8.11 shows the lower hemisphere projection of joint sets and bedding planes. Three-dimensional analyses were performed with the multilaminate model taking various angles between cavern axis and bedding strike into account. The considered values were $\alpha = 60°$, 30° and 0° (cases

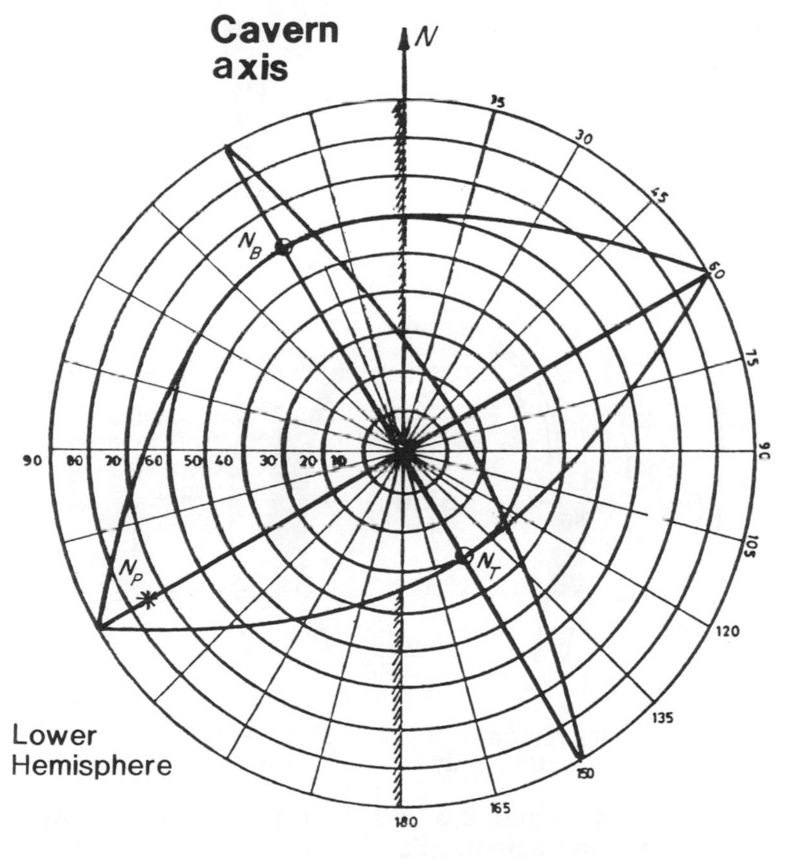


$N_B$ , $N_P$ , $N_T$ : Poles of bedding, parallel and transverse joints

**Figure 8.11**
Lower hemisphere projection of joint sets and bedding planes

1, 2 and 3). The moderate anisotropy of $n = E_1/E_2 = 2$ observed for either greywacke and shales was included for the most critical case of parallel strike only.

The different orientations yielded considerable differences in wall deformations, extent of overstressed zones and maximum stresses between case 1 and the similar cases 2 and 3. These effects are further enlarged when the anisotropy is taken into account (case 4) and are illustrated in Figures 8.12 and 8.13.

The resulting stresses and deformations clearly indicate the influence of bedding orientation on the required cavern support and the capability of model to assure the necessary design.

Representative results for the four cases discussed above are as follows:

| Case No. | Total wall displacement | Thickness of overstressed zone | Maximum tensile stress | Maximum compressive stress |
|---|---|---|---|---|
| 1 | 0.55 mm | 4 m | 87 kPa | 7.16 MPa |
| 2 | 0.82 mm | 9 m | 251 kPa | 6.83 MPa |
| 3 | 0.92 mm | 9 m | 236 kPa | 7.64 MPa |
| 4 | 1.27 mm | 19 m | 288 kPa | 8.08 MPa |

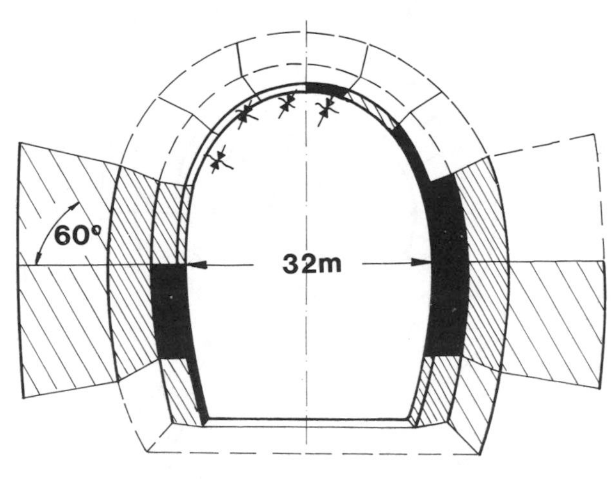


Shear failure,for case 4 (anisotropic,α=0° ) only

Tensile failure, for cases2,3and4

Shear failure, cases2,3 and4 (isotropic and anisotropic)

Note:Failures are at start of iterations only

Shear failure,cases 1to4 No tensile failure for case1

**Figure 8.12**
Failure zones around the cavern

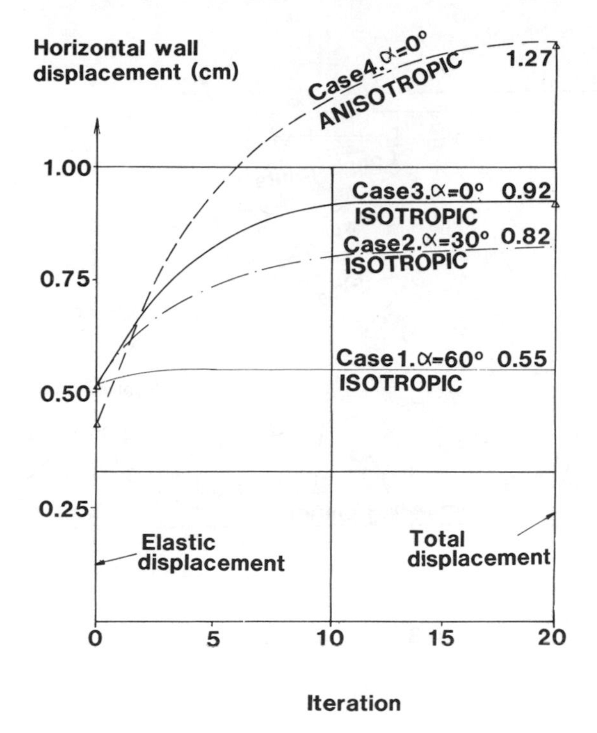


**Figure 8.13**
Horizontal wall displacements in elastoplastic analysis

It is noted that when anisotropy is considered, the total wall displacement increases by 40 percent and the thickness of initially overstressed zone is doubled.

### 8.3.2 Powerhouse cavern for Siah Bishe project in Iran

For the tender design of the powerhouse cavern of the above project pseudo-viscoplastic analysis using the multilaminate material model were performed for different rock types and shear strength parameters. The powerhouse cavern has the dimensions 25 × 40 × 110 m and parallel to the long axis of this cavern, a transformer cavern is proposed which has the dimensions 16.5 × 17.5 × 116.5 m. Figure 8.14 shows schematically the two caverns to be analysed together with the system of joints and bedding planes.

The most prominent discontinuities are bedding and joint set 1, both strike at 50° to the cavern axis and dipping at 60° in opposite directions. A second joint set is vertical and strikes nearly parallel to the powerhouse axis. Figure 8.15 shows the lower hemisphere projection of joint sets and bedding planes. The different rock types observed are quartzitic sandstones, shaly siltstones, limestones and volcanics such as melaphyre.

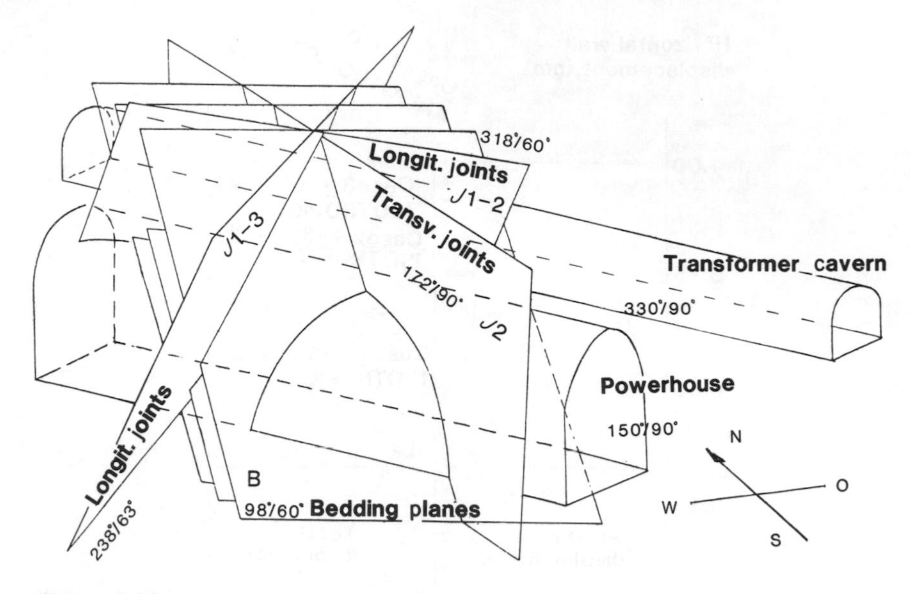


**Figure 8.14**
Schematic diagram of Siah Bishe powerhouse caverns showing the pattern of joints

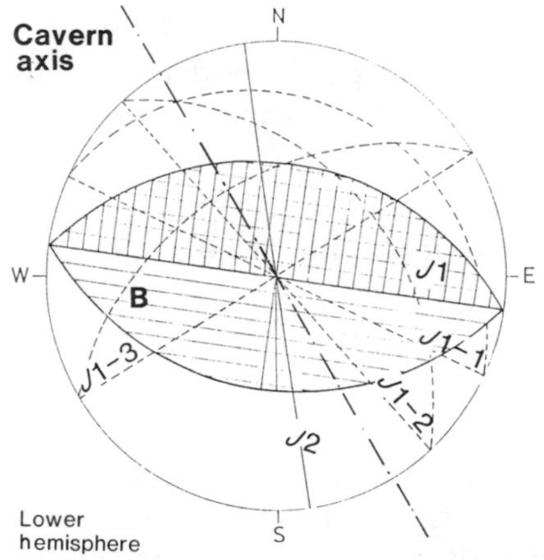


**Figure 8.15**
Lower hemisphere projection of joint sets and bedding planes

Significant analysis features are: considerably high lateral *in-situ* stresses; different strengths of joints and bedding planes; anisotropy of intact rock; and temporary stages of excavation. The presence of the transformer cavern parallel to the powerhouse cavern is also considered to have significant influence on the stability of both the structures.

The deformability and strength parameters adopted are as follows:

### Rock mass

Young's modulus of rock mass, $E_m$ = 7500 MPa
Poisson's ratio of rock mass, $\mu_m$ = 0.3

### Rock joints

| | | |
|---|---|---|
| Bedding planes | N 100° E/60° W | |
| Joint set 1 | N 280° E/60° E | |
| Joint set 2 | N 170° E/90° | |
| Shear strength parameters: | For both joint sets, | $c$ = 0.05 MPa |
| | | $\phi$ = 27.5° |
| | Bedding planes | $c$ = 0 |
| | | $\phi$ = 25° |

### Initial state of stress

| | |
|---|---|
| Density of rock | 0.027 MN/m$^3$ |
| Maximum overburden | 280 m |
| Ko | 1.00 |

The joints were assumed to have no tensile strength and their dilatancy angle was assumed as a half of the friction angle $\phi$.

From Figures 8.16 and 8.17 it can be seen that shear failure on beds is most probable on the upper eastern (right) wall and the lower western (left) wall. The shear failure for joint set 1 takes place on the upper western (left) wall and the lower eastern (right) wall. Shear failure along joint set 2 can be expected symmetrically at walls. Due to the limited shear strength available on bedding planes the respective failure zones are more extended and cover the full rock pillar width between the two caverns under the present strength and support assumptions (rock bolts only).

Although the fabric of joint sets and bedding planes require the closest possible simulation of the discontinuity behaviour, supplementary linear elastic calculations were performed using the boundary element method (see Chapters 9 and 10). The stress failure check was done with artificial shear strength parameters for a 'homogeneous' rock mass. The necessary parameters $m$ = 0.50 and $s$ = 0.2 (see Hoek-Brown criterion, Chapter 3) were derived via rock mass classification results of $RMR$ = 55 and $Q$ = 3

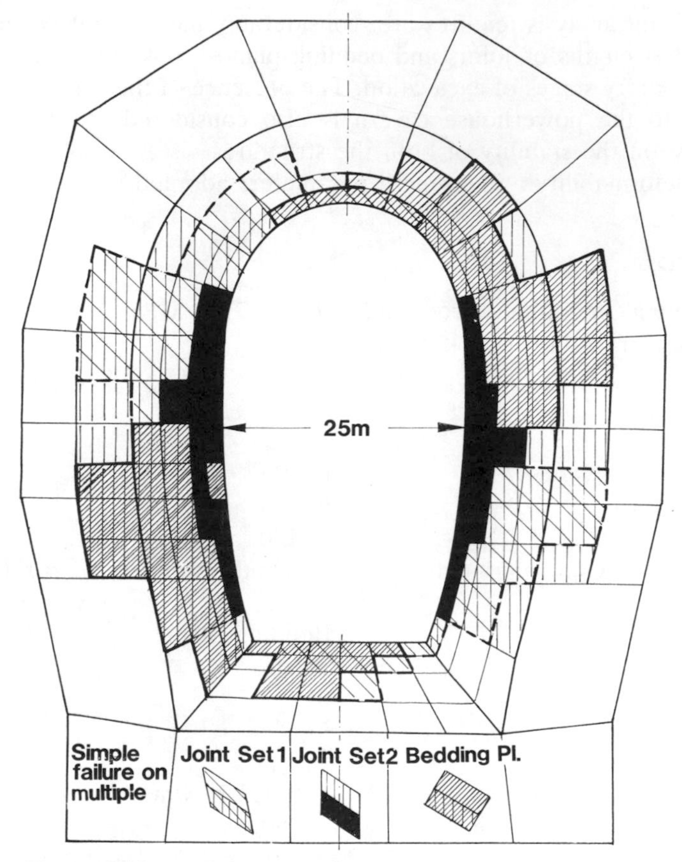


**Figure 8.16**
Failure zones around the powerhouse cavern—without the transformer cavern

for the shaly siltstone. A reasonable agreement for the extent of overstressed zones with visco–plastic results was observed.

### 8.3.3 Hope Brook Gold Mine, Canada

The Hope Brook Gold Mine is situated in southwestern Newfoundland, Canada, and is managed by British Petroleum Resources Canada. Commercial operations commenced in August 1987 with ore produced from an open pit. Production from the underground workings started in March 1989. Mineralisation is lenticular in form and contained within a silicified zone.

The purpose of the finite element analysis performed was to study the stability of open stopes at any given time, during the mining sequences planned by the Canadian mine engineers. The work was carried out using

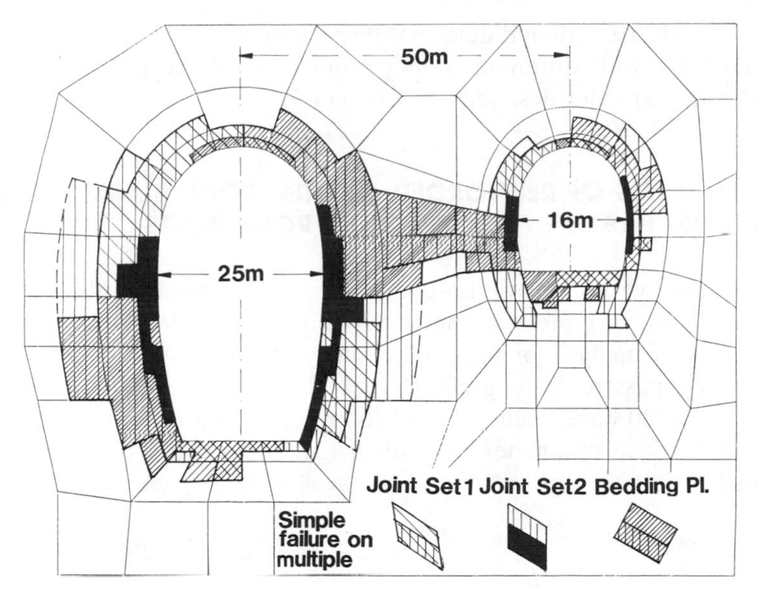


**Figure 8.17**
Failure zones around the powerhouse and transformer caverns

the program 'MINER'. 3-D conditions were assessed because of the complexity of the mining sequences and the mine geometry.

Figure 6.18(a) shows half the finite element mesh used for the analysis purposes, comprising of 3420 20-noded brick elements, which give rise to 16,231 nodes. The stopes to be excavated and their relationship to the open pit position are shown in Figure 6.18(b).

The material properties used in the analysis were:
(a) For the silicified zone:

$$E = 20 \text{ GPa}, \ \nu = 0.14, \ \varphi = 41°, \ \psi = 30°, \ c = \text{MPa and } \rho = 28 \text{ kN/m}^3$$

(b) For the hanging wall rock:

$$E = 18 \text{ GPa}, \ \nu = 0.14, \ \varphi = 39°, \ \psi = 30°, \ c = 2 \text{ MPa and } \rho = 28 \text{ kN/m}^3$$

(c) For the backfill material:

$$E = 1\text{GPa}, \ \nu = 0.26, \ \varphi = 33°, \ \psi = 15°, \ c = 1.1\text{MPa and } \rho = 20\text{kN/m}^3$$

Initially, the analysis has been performed assuming all materials involved in the analysis as elastoplastic obeying Mohr-Coulomb yield criterion with a non-associated flow rule. Overstress zones have been computed for various mining sequences. The pattern of jointing has recently been identified together with strength and deformability parameters of joints. It is expected that analyses with different mining sequences will be repeated using the multilaminate model described in Chapter 7.

## 8.4 ANALYSIS OF REINFORCED JOINTED ROCK MASSES AND DESIGN OF PASSIVE FULLY GROUTED ROCK BOLTS

The constitutive law for reinforced jointed rock mass described in Section 7.10 can be incorporated in a finite element program. We shall present here the results of analysis of the rock slope problem discussed in Section 8.2 considering passive fully grouted rock bolts on the face of the slope. Elements 16–30 have been assumed to be reinforced with dowels at various inclinations. The parameters for intact rock, joints and reinforcement are adopted as follows:

| | | |
|---|---|---|
| Intact rock | $E$ | $= 2.8 \times 10^7$ kPa |
| | $K_0$ | $= 1/3$ |
| | $\nu$ | $= 0.2$ |
| | $\rho$ | $= 25$ kN/m$^3$ |
| Joints | $C_n$ | $= 1 \times 10^{-7}$ m/kPa |
| | $C_s$ | $2 \times 10^{-7}$ m/kPa |
| | Cohesion $c$ | $= 50$ kPa |
| | $\phi$ | $= 45$ ° |
| | Spacing | $= 1$ m |
| Reinforcement | $E$ | $= 0.2 \times 10^9$ kPa |
| | $G$ | $= 0.769 \times 10^8$ kPa |
| | $\sigma_y$ | $= 0.25 \times 10^6$ kPa |
| | $p$ | $= 0.003$ |

The parameters adopted for intact rock and the strength of rock joints are the same as in Section 8.2 but for calculation of elasticity matrix ($D^e$) of rock mass, normal and shear compliances ($C_n$, $C_s$) of joints and their spacing is specified. The procedure given in Chapter 4 has been adopted to calculate $D^e$. No strain-hardening has been assumed for rock bolts and their percentage is 0.3 percent which corresponds to 45 mm $\phi$ bolts at 750 mm centre to centre. In Section 8.2 the cohesion $c$ was progressively reduced to induce collapse in the slope. The friction angle $\phi$ was however kept unchanged. A more rational way is to reduce $c$ as well as tan $\phi$ by a common factor which can be viewed as a factor of safety ($R_f$). The failure criterion of the rock joint sets can then be defined as

$$F = |\tau| + \frac{n \tan \phi}{R_f} - \frac{C}{R_f} = 0 \tag{8.2}$$

Starting with $R_f = 1.0$, its value is increased in successive finite element analyses till collapse takes place. Factors of safety obtained for various inclinations ($\theta$, measured counter clockwise from horizontal) of reinforcement are listed in Table 8.1 below:

**Table 8.1** Factors of safety for different cases using FEM

| | | Factor of safety | |
|---|---|---|---|
| | | Associated flow rule | Nonassociated flow rule |
| Case (i) | no reinforcement | 1.18 | 1.13 |
| Case (ii) | horizontal reinforcement | 2.35 | 2.00 |
| Case (iii) | reinforcement at $\theta = 160°$ | 2.36 | 1.99 |
| Case (iv) | reinforcement at $\theta = 142.5°$ | 2.35 | 2.00 |
| Case (v) | reinforcement at $\theta = 50°$ | 2.15 | 1.97 |

It is apparent that the provision of rock bolts improves the factor of safety dramatically. The use of nonassociated flow rule ($\psi = 0$) results in lower factors of safety as compared to the associated flow rule. There is a marginal difference in the factors of safety when $\theta = 0°$, 142.5° and 160° but it is lower when $\theta = 50°$. Finding the best direction of reinforcement for any given problem requires large amount of computation and implementation of the constitutive law for reinforced jointed rock is also quite involved. There is clearly a need for simplified procedure for practical design of reinforced rock structures. With this aim, a less rigorous 'stress–path method' has been proposed by Sharma and Pande (1988). It is described in the next section using the problem of rock slope as an example.

### 8.4.1 Factor of safety by the 'stress path' method

The stress-path method is well established in soil mechanics but has not been applied in the problems of the reinforced rock structures hitherto. It is usually adopted when nonlinear response of soil has to be simulated without solving nonlinear problem either numerically or analytically. In this method, a linear elastic analysis without reinforcement is carried out by the finite element method and the stresses ($\boldsymbol{\sigma}$) are obtained at the Gauss points. The stresses at the Gauss point due to excavation with respect to initial stresses ($\boldsymbol{\sigma}_{exc}$) are calculated by

$$\boldsymbol{\sigma}_{exc} = \boldsymbol{\sigma} - \boldsymbol{\sigma}_0 \tag{8.3}$$

where $\boldsymbol{\sigma}_0$ are the initial stresses at the Gauss point. The stress-path at any

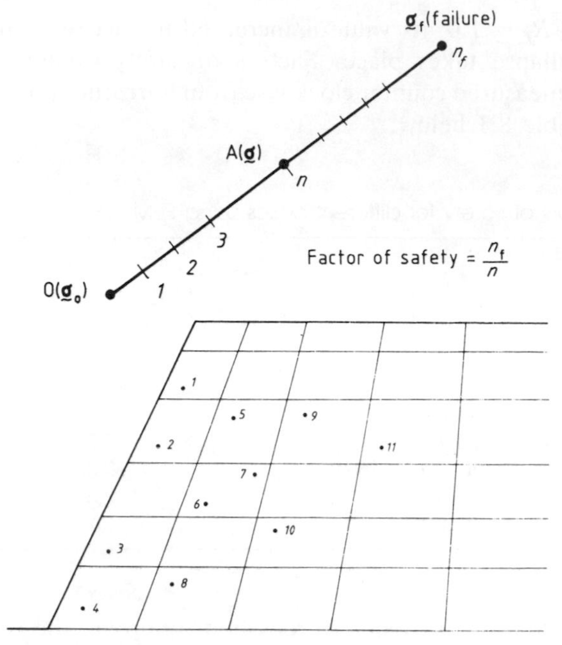


**Figure 8.18**
Gauss points selected for the stress-path method and stress path for a typical Gauss point

chosen Gauss point will be simply the vector $\boldsymbol{\sigma}_{\text{exc}}$ joining the initial stress point $O$ with the total stress point A as shown in Figure 8.18.

In the proposed 'stress-path' method, each Gauss point is considered independently and the steps are as follows:

(a) Stress-path $O$A (Figure 8.18) is calculated using equation (8.3).
(b) Vector length $O$A is divided into a certain number of parts (say '$n$') for incrementation of stress. The incremental stress vector is given by

$$\Delta\bar{\boldsymbol{\sigma}} = \boldsymbol{\sigma}_{\text{exc}}/n \tag{8.4}$$

(c) Starting from the initial stresses ($\boldsymbol{\sigma}_0$) at the Gauss point, the stresses are incremented by $\Delta\bar{\boldsymbol{\sigma}}$ until failure takes place. For each increment, the resulting stresses are given by

$$\hat{\boldsymbol{\sigma}} = \sigma_0 + \Sigma\,\Delta\bar{\boldsymbol{\sigma}} \tag{8.5}$$

The vector $\hat{\boldsymbol{\sigma}}$ is along the stress-path of the Gauss point.
(d) The factor of safety is then calculated by dividing the stress $\boldsymbol{\sigma}_f$ at which failure takes place by the stress $\boldsymbol{\sigma}_{\text{exc}}$ imposed due to excavation. This factor of safety is based on load factor concept, i.e. increasing the load

at a Gauss point and therefore is different from the factor of safety calculated using the constitutive law of reinforced jointed rock mass in the finite element computations. Steps (a)–(d) are repeated for a number of chosen Gauss points and corresponding factors of safety are computed.

For step (c), a simple computer program to integrate the constitutive equations can be developed which computes stress $\boldsymbol{\sigma}_f$ at failure for a single

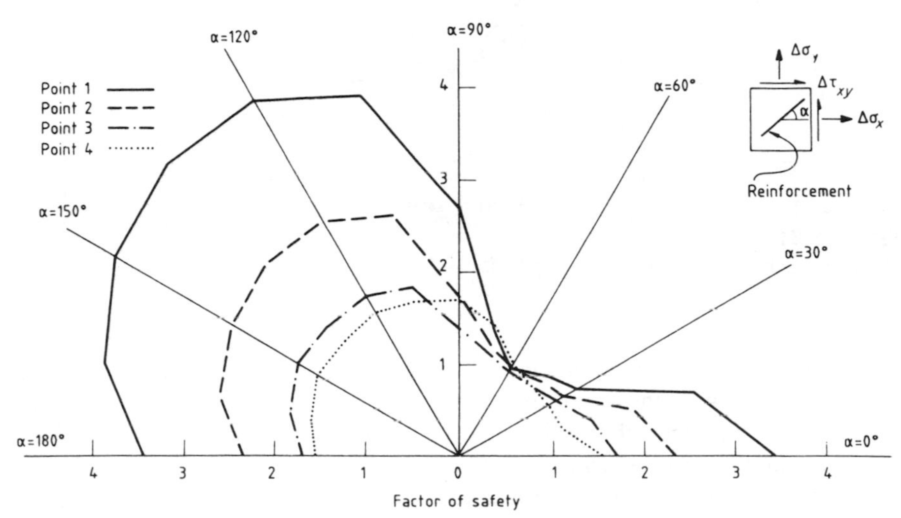


**Figure 8.19**
Rose diagram for factor of safety-points 1–4

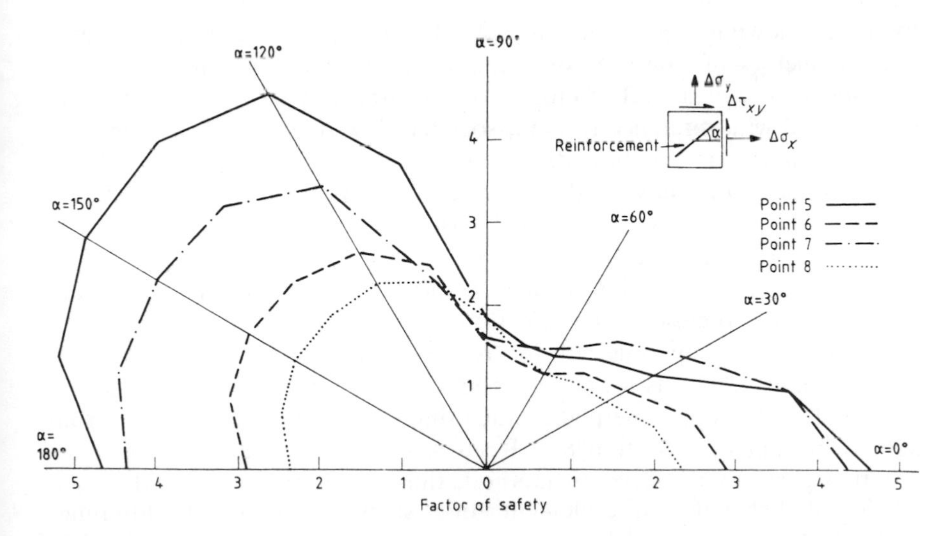


**Figure 8.20**
Rose diagram for factor of safety-points 5–8

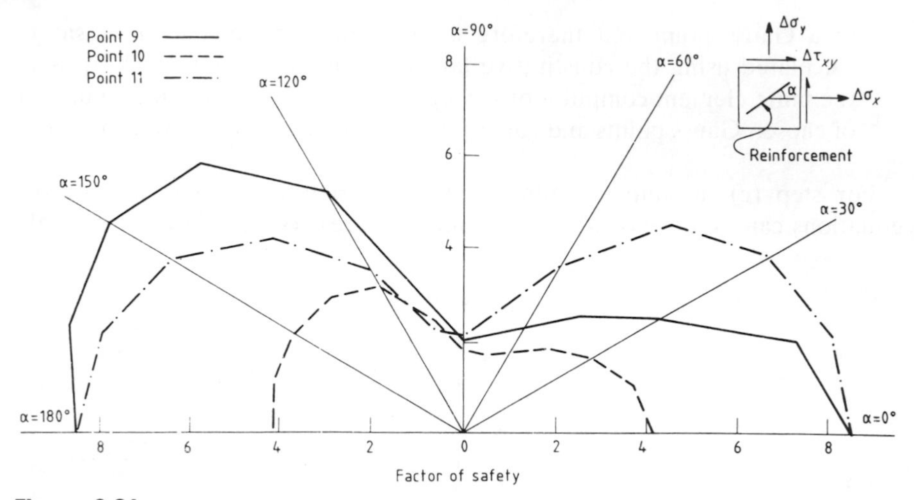


**Figure 8.21**
Rose diagram for factor of safety-points 9–11

stress point (i.e. homogeneous state of stress). The program has the provision for sets of joints and reinforcement and can be used for the plane stress, plane strain or three-dimensional idealization. Both associated and nonassociated flow rules can be used.

To apply this method to the rock slope problem, 11 Gauss points (Figure 8.18) were selected for different orientations of the reinforcement. The proportion of reinforcement was kept fixed at 0.003. Rose diagrams giving the factor of safety for reinforcement along various orientations for the 11 points are shown in Figures 8.19 to 8.21. There is only a marginal difference between factors of safety as obtained by using associated flow rule and nonassociated flow rule. Therefore, rose diagrams of factor of safety have been plotted with reference to the associated flow rule only. It is seen from these diagrams that the best direction of reinforcement, which gives the maximum factor of safety for the selected points, lies between $\alpha = 135°$ and 165°. For the best direction, point 4 has the least factor of safety (=1.8) and point 9 has the largest (=9).

The best direction of reinforcement coincides with the normal to the face of slope. It is also close to the direction of the normal to the inclined joint set; Figure 8.3. Comparing with the results of Section 8.4 above, the least factor of safety of 1.8 is lower than the values obtained using the rigorous method. The stress-path method is much simpler as compared to the rigorous method presented in Section 8.4. It gives a factor of safety which is less than that given by the rigorous method, thus erring on the safe side.

It is noted that the finite element study showed considerable difference in the factors of safety when the 'nonassociated flow rule' was used in comparison to 'associated flow rule'. In the stress-path method adopted in this section, since only one 'Gauss point' is considered, complex interaction of stress does not take place as in the case of the finite element method. This explains why dilatancy of joints influences the factor of safety only marginally in this case.

## References

Hönish, K. (1988). 'Rock mass modelling for large underground powerhouses.' *Numerical Methods in Geomechanics*. (eds, G. Swoboda, A. A. Balkema), **3**, 1517–22.

# 9 Boundary Element Methods

## 9.1 INTRODUCTION

One of the distinct characteristics of a rock mass is that it is very large and, for practical purposes, can be assumed to be of infinite extent. Because of its volume discretization the Finite Element Method is not very well suited for problems with a low ratio of boundary surface to volume because a large number of elements are needed to model the response of the domain.

For analyses in rock mechanics a method where only the surface of an excavation has to be discretized becomes immediately attractive. The amount of input data required to describe a problem is greatly reduced and the influence of the infinite rock mass is automatically considered in the analysis.

One such method is the Boundary Element Method (BEM).

In this chapter we shall discuss the theoretical background of the various BEMs which can be used for problems in rock mechanics.

The explanation of the Boundary Element Method will be kept simple. It will be shown that behind the fancy mathematics sometimes used to explain the method is a very simple and fundamental engineering principle: the method of superposition.

The Boundary Element Methods differ from the Finite Element Method by the fact that *approximations only occur on the boundary of the problem domain*. The solution *inside the domain* will *always satisfy the equations of equilibrium and compatibility exactly*.

For the method to work, however, a *fundamental solution* (i.e. one which satisfies the differential equation of the problem) is required. The particular solution is then obtained by *superposition* of the fundamental solutions in such a way that the given boundary conditions are satisfied either pointwise or in an average way.

A number of different approaches exist. The *modified Trefftz* methods are based on a method Trefftz used in 1926 to solve torsion problems. These methods are the simplest Boundary Element Methods and serve well to demonstrate the basic principles of Boundary Elements but have severe shortcomings with respect to user-friendliness and accuracy. The *indirect*

BEM overcomes most of these disadvantages but requires the solution for 'fictitious forces' before the unknowns on the boundary can be determined. Finally, the *direct Boundary Element* Method eliminates the need to solve for 'fictitious forces' as the unknowns are determined directly. The *advanced* BEM, where Finite Element shape functions are used to describe the geometry and the variation of the knowns and unknowns is most accurate and will be discussed in more detail here.

## 9.2 MODIFIED TREFFTZ METHODS

The simplest Boundary Element Methods are methods based on the basic ideas by Trefftz (1926) who used a method superposition for solving torsion problems as an alternative to the method by Ritz on which the FEM is based.

To explain his method consider, for example, the experiment shown in Figure 9.1. The experimental set up consists of a membrane which is attached to a rigid frame. A facility exists to apply concentrated forces at various locations on the membrane. A closed contour is marked on the membrane and strain gauges are attached at various locations on the contour and these are used to measure the stress perpendicular to the contour.

Assume that for a unit weight applied at a point $P$ the stress measured at point $Q$ is $T(Q,P)$. Consequently the stress at $Q$, $t(Q)$ due to a weight of magnitude $W(P)$ at $P$ can be written as:

$$t(Q) = T(Q,P)W(P) \tag{9.1}$$

Most engineers will recognise $T(Q,P)$ as *influence coefficient*, i.e. a coefficient which gives the effect of a *unit force* on the stress at $Q$.

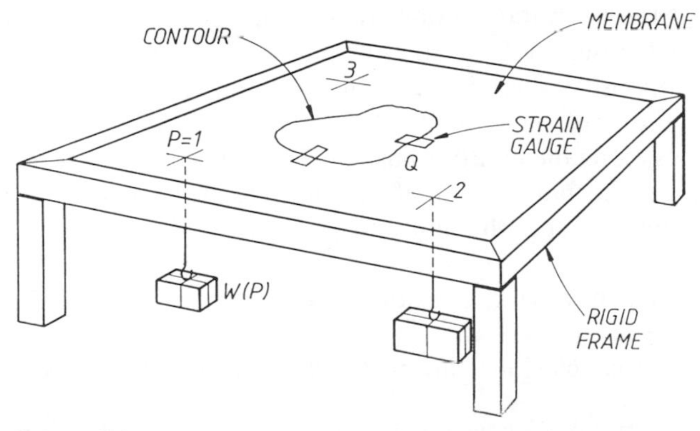


**Figure 9.1**
Experimental set up to explain the principles of the Boundary Element Methods

Assuming that the deformations are elastic and small the method of superposition may be used and hence for a series of loads applied at various points $P = 1, 2, 3 \ldots, N$:

$$t(Q) = \sum_{P=1}^{N} T(Q,P)W(P) \tag{9.2}$$

The consequence of the above is that given a sufficient number of load points $P$ and load values $W(P)$ we can produce any desired value of stress at any point $Q$.

Let us carry this exposé further with a thought experiment.

Assume now that the membrane is of infinite extent and that the point forces are applied in the plane of the membrane instead of perpendicular to it. Assume further that a particular distribution of point loads produces a certain distribution of tractions **t**, on the contour as shown in Figure 9.2.

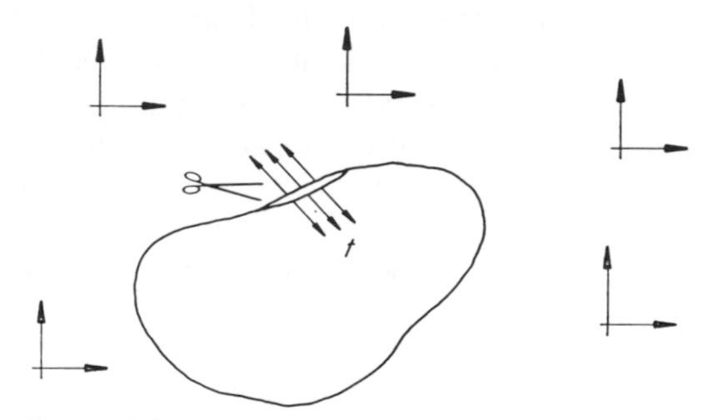


**Figure 9.2**
A thought experiment. The membrane is now of infinite extent and the forces are applied in the plane of the membrane

We now cut the membrane along the contour but as the cut proceeds we apply tractions **t**, resulting from the internal stresses on each surface so that they act in opposite directions. After having completed the cut the displacement and stress field should have remained unchanged because the internal stresses along the contour have been replaced by equivalent external forces acting on the boundaries of the infinite domain (i.e. the material outside the contour) and the finite domain (i.e. the material inside the contour).

We now may, however, consider the exterior and interior domain separately.

Two important conclusions can be made from this thought experiment:

provided the influence coefficients are known any desired stress can be reproduced at any point on the contour by superposition of point load effects;

either the domain inside or outside the contour may be considered separately.

Consequently the method is applicable to determining the displacements and stresses for finite and infinite domains subjected to a given loading on the boundary.

The most important step in the Boundary Element Method is to determine the influence coefficients. As mentioned previously these are essentially solutions of the problem due to concentrated unit forces in an infinite domain.

Such solutions are also called *fundamental solutions*. They can be visualized as special global shape functions—which, in contrast to those used for Finite Elements—exactly satisfy the differential equation governing the problem.

### 9.2.1 Fundamental solutions

For elasticity problems the well known Kelvin (1879) solution for point loads in an infinite elastic, homogeneous and isotropic domain can be used. The solution for the displacement field at $Q$ can be written in matrix form as:

$$\mathbf{u}(Q) = \mathbf{U}(Q,P)\boldsymbol{\phi}(P) \tag{9.3}$$

where for two-dimensional problems:

$$\mathbf{U}(Q,P) = \begin{bmatrix} U_{xx} & U_{xy} \\ U_{yr} & U_{yy} \end{bmatrix}, \qquad \boldsymbol{\phi}(P) = \begin{Bmatrix} \phi_x \\ \phi_y \end{Bmatrix} \tag{9.4}$$

$U_{xx}(Q,P)$ is an *influence coefficient* which gives the $x$-displacement at point $Q$ (field point) due to a unit load in $x$-direction at Point $P$ (load point); $U_{xy}$ and $U_{yy}$ are the $x$- and $y$-displacements due to a load in y direction. $\boldsymbol{\phi}(P)$ are load intensities at $P$. Following Betti's law $\mathbf{U}$ must be symmetric, i.e. $U_{xy} = U_{yx}$. The coefficients of matrix $\mathbf{U}$ are given in Appendix III.

The stresses at point $Q$ can be obtained from the displacements by:

$$\boldsymbol{\sigma}(Q) = \mathbf{D}\,\hat{\mathbf{B}}\,\mathbf{U}(Q,P)\boldsymbol{\phi}(P) = \mathbf{S}(Q,P)\boldsymbol{\phi}(P) \tag{9.5}$$

where $\mathbf{D}$ is the elasticity matrix and for two-dimensional problems:

$$\hat{\mathbf{B}} = \begin{bmatrix} \dfrac{\partial}{\partial x} & 0 \\ 0 & \dfrac{\partial}{\partial y} \\ \dfrac{\partial}{\partial y} & \dfrac{\partial}{\partial x} \end{bmatrix}, \qquad \boldsymbol{\sigma} = \begin{Bmatrix} \sigma_x \\ \sigma_y \\ \tau_{xy} \end{Bmatrix} \tag{9.6}$$

The matrix **S** is also defined in Appendix III. Solutions are also available for nonhomogeneous and orthotropic materials although expressions become quite complicated. Solutions for semi-infinite domains are useful for analysing halfspace problems because the condition of the traction free surface is automatically satisfied by the solution. Such solutions have been presented by Melan (1932) for 2D problems and Mindlin (1936) for 3D problems. They have been presented in a form suitable for computer programming by Telles (1981). The fundamental solutions are often referred to as *Kernels*.

### 9.2.2 Method of solution

Consider again the example in Figure 9.2 but assume now that the tractions, $\mathbf{t}^0$, on the boundaries of the infinite domain are known and that a solution for the displacement field due to this boundary loading is required. For clarity the region inside the boundary although present has been removed in Figure 9.3. At a point on the boundary, $Q$, with the normal vector $\mathbf{n}(Q)$ the following relationship between boundary tractions $\mathbf{t}$ and stresses $\boldsymbol{\sigma}$ exists.

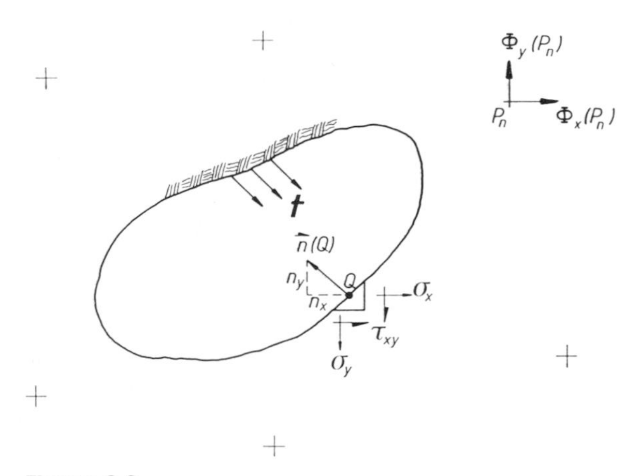


**Figure 9.3**
Modified Trefftz Method: a hole in an infinite domain with applied tractions

$$\mathbf{t}(Q) = \mathbf{n}(Q)\boldsymbol{\sigma}(Q) \tag{9.7}$$

where

$$\mathbf{t} = \begin{Bmatrix} t_x \\ t_y \end{Bmatrix}.$$

The matrix $\mathbf{n}(Q)$ is

$$\mathbf{n}(Q) = \begin{bmatrix} n_x & 0 & n_y \\ 0 & n_y & n_x \end{bmatrix} \tag{9.8}$$

Where $n_x$ and $n_y$ are the $x$ and $y$ components of $\mathbf{n}$.

Note that the tractions acting on the boundary of the interior domain are obtained in exactly the same way. For this case, however, the normal $\mathbf{n}$ acts in the opposite direction.

Following the previous argument a series of concentrated forces of intensity $\boldsymbol{\phi}(P_n)$ are applied at $N$ points $P_n$. The stresses at point $Q$ due to these forces are given by

$$\boldsymbol{\sigma}(Q) = \sum_{n=1}^{N} \mathbf{S}(Q,P_n)\boldsymbol{\phi}(P_n) \tag{9.9}$$

Using the relationship (equation (9.7)) the traction vector at point $Q$ is given by:

$$\mathbf{t}(Q) = \mathbf{n}(Q)\boldsymbol{\sigma}(Q) = \sum_{n=1}^{N} \mathbf{n}(Q)\mathbf{S}(Q,P_n)\boldsymbol{\phi}(P_n) \tag{9.10}$$

Substituting

$$\mathbf{T}(Q,P_n) = \mathbf{n}(Q)\mathbf{S}(Q,P_n) \tag{9.11}$$

we obtain

$$\mathbf{t}(Q) = \sum_{n=1}^{N} \mathbf{T}(Q,P_n)\boldsymbol{\phi}(P_n) \tag{9.12}$$

Satisfaction of the traction boundary condition requires

$$\mathbf{t}(Q) = \mathbf{t}^0(Q) \tag{9.13}$$

at all points $Q$ along the boundary. In general it is not possible to satisfy this condition at all the boundary points. Instead we attempt to satisfy equation (9.13) in an approximate way. Two different approaches will be presented here.

*Approach 1:* Pointwise satisfaction of boundary conditions (Patterson and Sheikh, 1982).

Here we attempt to satisfy the boundary conditions at a discrete number of points as shown in Figure 9.4, i.e. we require that:

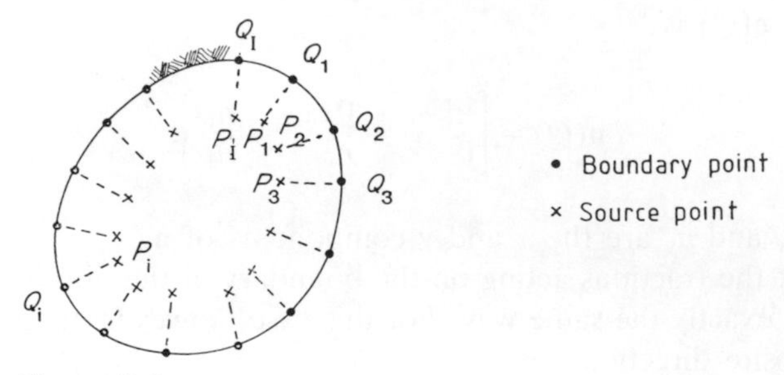


**Figure 9.4**
Modified Trefftz methods: location of boundary and source points

$$\begin{aligned} \mathbf{t}(Q_1) &= \mathbf{t}^0(Q_1) \\ &\cdot \\ &\cdot \\ &\cdot \\ \mathbf{t}(Q_I) &= \mathbf{t}^0(Q_I) \end{aligned} \tag{9.14}$$

For point $Q_i$ we can write

$$\mathbf{t}(Q_i) = \sum_{n=1}^{N} \mathbf{T}(\mathrm{Q}_i,\mathrm{P}_n)\boldsymbol{\phi}(P_n) \tag{9.15}$$

The equations (9.14) can be written in matrix form as

$$\{\mathbf{t}\}^0 = [\mathbf{T}]\,\{\boldsymbol{\phi}\} \tag{9.16}$$

where

$$\{\mathbf{t}\}^0 = \begin{Bmatrix} \mathbf{t}^0(Q_i) \\ \cdot \\ \cdot \\ \cdot \\ \mathbf{t}^0(Q_I) \end{Bmatrix} \qquad \{\boldsymbol{\phi}\} = \begin{Bmatrix} \boldsymbol{\phi}(P_1) \\ \\ \boldsymbol{\phi}(P_N) \end{Bmatrix} \tag{9.17}$$

and

$$[\mathbf{T}] = \begin{bmatrix} \mathbf{T}(Q_1,P_1) & \ldots & \mathbf{T}Q_1,P_N) \\ \cdot & & \\ \cdot & & \\ \cdot & & \\ \mathbf{T}(Q_I,P_1) & \ldots & \mathbf{T}(Q_I,P_N) \end{bmatrix} \tag{9.18}$$

Equation (9.16) represents a system of simultaneous equations which can be solved for unknown fictitious force intensities $\{\boldsymbol{\phi}\}$. To be able to solve the equations we require that [**T**] is a square matrix, that is $N = I$, i.e. the number of locations $P_n$ (source points) is equal to the number of boundary points. It is convenient to place the source points close to the boundary points as suggested in Figure 9.4. The source points must not be too close to the boundary points since as the distance between source and boundary point decreases to zero the coefficients of [**T**] tend to infinity.

A severe shortcoming of the simple method just presented is that the boundary conditions are only satisfied at selected points. As shown in Figure 9.5, significant errors can still occur in between the points if they are not closely spaced.

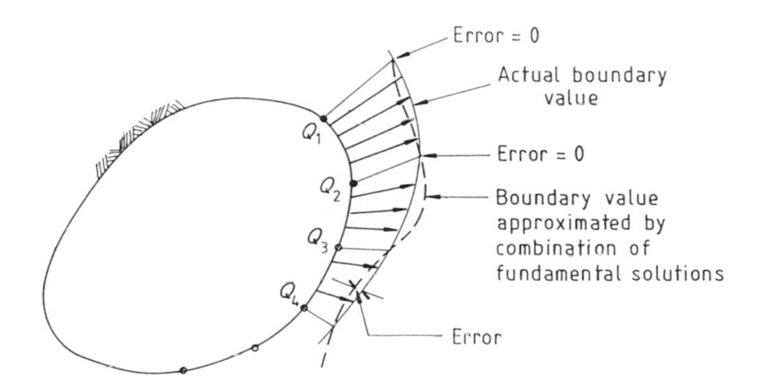


**Figure 9.5**
In Approach 1 (pointwise matching) the deviation between the computed value and the actual value is zero at the nodes but may be large between the nodes

A measure for the error at a particular point can be defined as

$$E = \frac{\sqrt{(t_x - t_x^0)^2 + (t_y - t_y^0)^2}}{\sqrt{(t_x^0)^2 + (t_y^0)^2}} \tag{9.19}$$

where $t_x^0$ and $t_y^0$ are the applied (known) tractions. For the present method $E$ is zero at the points $Q_1, Q_2, \ldots Q_I$ but may be large in between these points.

*Approach 2:* 'Least Square' matching of boundary conditions (Gioda, 1984).

A better approximation to the actual boundary conditions is obtained by making the energy norm of the errors a minimum. This can be expressed as

$$I = \int_S \left[ (t_x - t_x^0)^2 + (t_y - t_y^0)^2 \right] \mathrm{d}s \Rightarrow \text{minimum} \tag{9.20}$$

The integration is taken over the entire boundary of the problem $S$. Equation (9.20) can be recognized as a 'least square' approximation which does not ensure exact satisfaction of the boundary conditions at any particular point but a better fit to the prescribed values over the entire boundary (Figure 9.6).

The minimum of $I$ is achieved by taking derivatives of $I$ with respect to the fictitious forces $\phi_x(P_n)$ and $\phi_y(P_n)$ and setting them to zero:

$$\frac{\partial I}{\partial \phi_x(P_n)} = 0$$

$$\frac{\partial I}{\partial \phi_y(P_n)} = 0$$

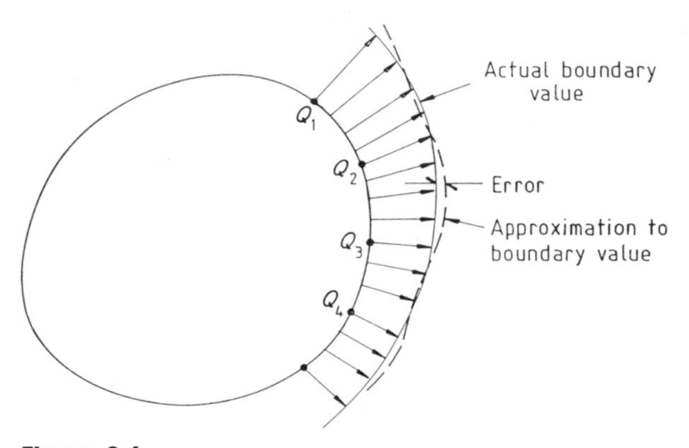


**Figure 9.6**
In Approach 2 (least square matching) the actual value is approximated everywhere on the boundary, not just at the nodes, but the approximation is much better than with point wise matching

This gives $2N$ equations for $2N$ unknown 'fictitious force' components for two-dimensional problems.

The integration is carried out by dividing the boundary into straight line segments and by assuming the given traction value to be constant along each segment.

Whichever method we use a system of equations is obtained which can be solved for the fictitious force intensities $\{\boldsymbol{\phi}\}$ at all the nodes. Once these have been determined the displacements can be obtained at any point $Q$ on the boundary or inside the continuum by

$$\mathbf{u}(Q) = \sum_{n=1}^{N} \mathbf{U}(Q,P_n)\boldsymbol{\phi}(P_n) \tag{9.21}$$

In a similar manner the stresses at point $Q$ are obtained by

$$\boldsymbol{\sigma}(Q) = \sum_{n=1}^{N} \mathbf{S}(Q,P_n)\boldsymbol{\phi}(P_n) + \boldsymbol{\sigma}^0(Q) \tag{9.22}$$

where $\boldsymbol{\sigma}^0$ are initial stresses.

The modified Trefftz methods which we have discussed here require that the source points (i.e. the points where the 'fictitious forces' are applied) do not coincide with the points on the boundary since the expressions tend to infinity as $r \to 0$. In fact, because of the singular nature of these functions, the selection of the location of the source points becomes critical.

This is illustrated by the example in Figure 9.7. Here the accumulated error $E$ as defined in equation (9.20) for a circular excavation subjected to internal pressure is plotted as a function of the ratio of size of boundary segment (= distance between boundary points) and distance to the source point for three different numbers of boundary points. It can be seen that for $H/D$ ratios smaller than one the solution deteriorates rapidly and that for $H/D < 0.75$ a solution cannot be obtained.

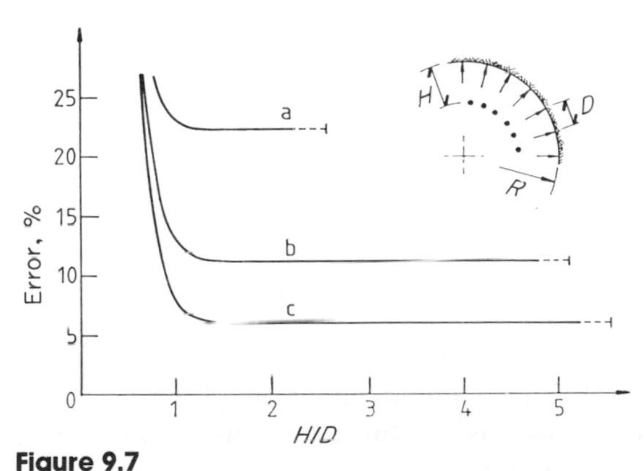


**Figure 9.7**
One of the drawbacks of the modified Trefftz Method is that the points of application of the 'fictitious forces' must not be on the boundary. The figure shows that their location is critical. (After Gioda (1984))

## 9.3 INDIRECT BOUNDARY ELEMENT METHODS

The previous methods had one significant drawback in that the location of the source points had to be different from the boundary points. This occurred because the Kelvin solution is singular near the source point. In the indirect Boundary Element Method presented here we use a different fundamental solution which allows to overcome this limitation and improve the user-friendliness of the method.

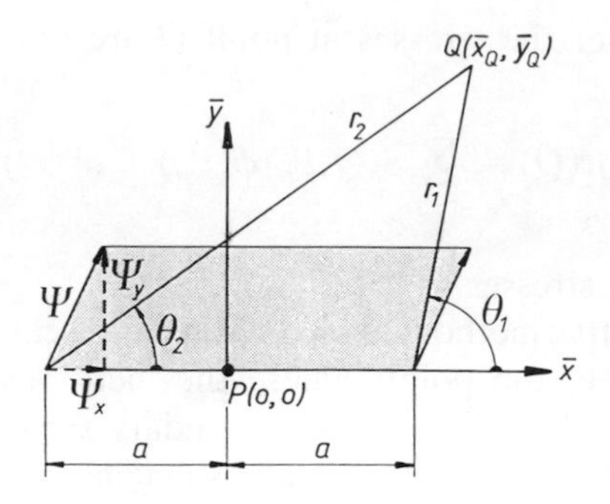


**Figure 9.8**
Constant traction over a line segment

The fundamental solution used is that for constant tractions $\psi_x$ and $\psi_y$ applied to a line segment $|\bar{x}| \leqslant$ a, $\bar{y} = 0$ in an infinite elastic solid (Figure 9.8); $x$ and $\bar{y}$ are local coordinates (relative to the centre of the line segment). The solution is obtained by integrating the Kelvin solution between $-a$ and $+a$.

The solution for the displacements is written as (Crouch, 1983)

$$\bar{\mathbf{u}} = \Delta\bar{\mathbf{U}}\,\boldsymbol{\psi} \tag{9.23}$$

where $\psi$ are distributed fictitious forces and

$$\Delta\bar{\mathbf{U}} = \int_{-a}^{+a} \mathbf{U}\,(Q, P(\bar{x}))\,\mathrm{d}\bar{x} \tag{9.24}$$

The integrated Kelvin solutions $\Delta\bar{\mathbf{U}}$ are given in Appendix III. The solution for the stresses is obtained by integrating the stress solutions presented previously and is given by:

$$\bar{\boldsymbol{\sigma}} = \Delta\bar{\mathbf{S}}\,\boldsymbol{\psi} \tag{9.25}$$

where components of $\Delta\bar{\mathbf{S}}$ are also given in Appendix III.

As can be seen, the integrated Kelvin solution is much better behaved than the original Kelvin solution. For example the solution does not tend to infinity as the centre of the line segment ($\bar{x} = 0$, $\bar{y} = 0$) is approached.

Consider the case where $|\bar{x}| < a$ and $\bar{y} = 0$ is approached from the positive side. Then from Figure 9.9a $\theta_1 \Rightarrow \pi$, $\theta_2 \Rightarrow 0$ and therefore the stress in $y$ direction due to distributed sources in $y$-direction is given by (see Appendix III).

$$\sigma_{\bar{y}} = -\tfrac{1}{2}\,\text{(compression)}.$$

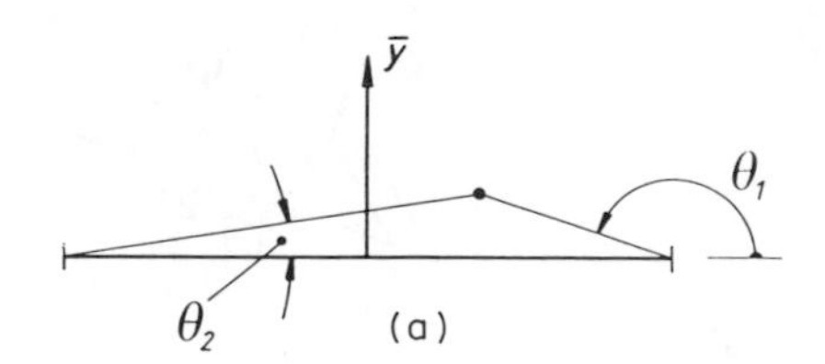


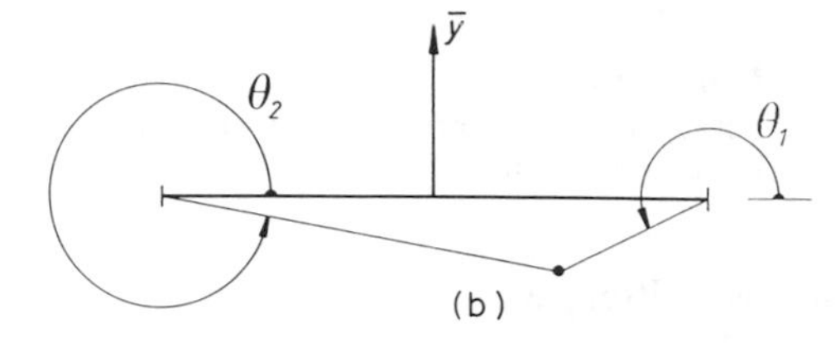


**Figure 9.9**
For the integrated Kelvin solution, different results are obtained depending on from which side the point is approached

If $\bar{y} = 0$ is approached from the negative side (Figure 9.9b) then

$$\theta_1 \Rightarrow \pi, \qquad \theta_2 \Rightarrow 2\pi \quad \text{and} \quad \sigma_{\bar{y}} = \tfrac{1}{2}\,(\text{tension})$$

From the solution matrices given in the Appendix it can be seen that the solutions are well behaved on the line segment $|\bar{x}| < a$ except at the points $|\bar{x}| = a$ where a singularity occurs for some stress components. The consequence of this is that we can now locate the sources of 'fictitious loads' on the boundary.

### 9.3.1 Method of solution

Before presenting the method of solution for the indirect BEM the integrated Kelvin solutions have to be expressed in global coordinates. Following Figure 9.10 the local coordinates $\bar{x}$, $\bar{y}$ are expressed in global $x$, $y$ coordinates by

$$\begin{aligned} \bar{x} &= x_0 + x\cos\alpha - y\sin\alpha \\ \bar{y} &= y_0 + x\sin\alpha + y\sin\alpha \end{aligned} \tag{9.26}$$

Similarly the displacement solution can be written in global coordinate directions as:

$$\Delta\mathbf{U} = \mathbf{T}^{T}\Delta\bar{\mathbf{U}} \tag{9.27}$$

where $\mathbf{T}$ is the transformation matrix as explained in Section 6.2.1.

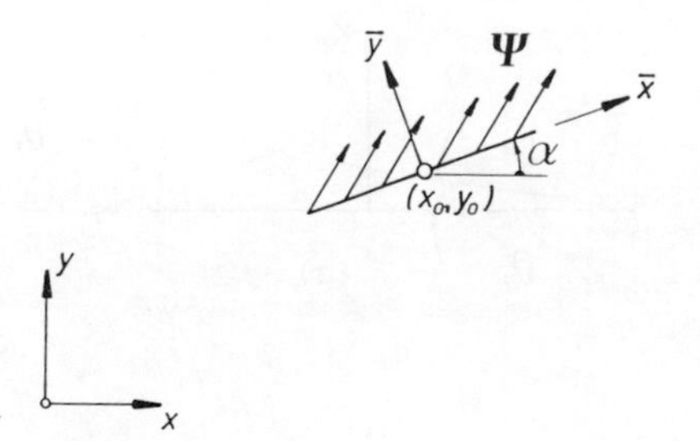


**Figure 9.10**
Loaded line segment in a global coordinate system

The stress solution is written as

$$\Delta \mathbf{S} = \mathbf{T}_{\sigma}\, \Delta \bar{\mathbf{S}} \tag{9.28}$$

where $\mathbf{T}_{\sigma}$ is the stress transformation matrix as shown in Appendix II.

Consider now the example of Figure 9.11 where the boundary is divided into line segments. Assume that segment $j$ is subjected to a distributed 'fictitious force' of $\boldsymbol{\psi}(j)$ as shown. The stresses at the mid-point of segment $i$ due to the loading on segment $j$ can be computed from

$$\boldsymbol{\sigma}(i) = \Delta \mathbf{S}(i,j)\boldsymbol{\psi}(j) \tag{9.29}$$

Using Equation (9.7) the tractions at point $i$ are:

$$\mathbf{t}(i) = \mathbf{n}(i)\, \Delta \mathbf{S}(i,j)\boldsymbol{\psi}(j) \tag{9.30}$$

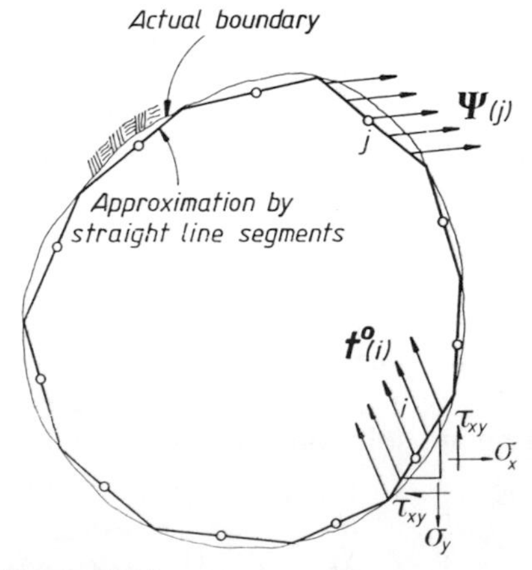


**Figure 9.11**
A hole in an infinite domain: subdivision into constant Boundary Elements

Assume now that the tractions are known on the boundary and that they are approximated by constant values acting on the line segments $\mathbf{t}^0(i)$. The unknown displacements are also assumed to be constant along each line segment. This means, of course, that there is no continuity of displacement where two line segments meet—an absurd situation which is difficult to understand by the practical engineer. The concept of an average displacement on a segment might be more acceptable. A constant Boundary Element is analogous to nonconforming elements in the FEM. and has been shown to perform reasonably well.

The method of solution is now identical to the Trefftz Method presented previously. A distribution of 'fictitious tractions' at all boundary segments, $j$, is determined which gives the given tractions at all the line segments $i$. Using superposition the traction at segment $i$ is

$$\mathbf{t}^0(i) = \sum_{j=1}^{N} \mathbf{n}(\mathrm{i})\, \Delta\mathbf{S}(i,j)\boldsymbol{\psi}(j) \tag{9.31}$$

or

$$\mathbf{t}^0(i) = \sum_{j=1}^{N} \Delta\mathbf{T}(i,j)\boldsymbol{\psi}(j) \tag{9.32}$$

where $N$ is the number of segments and $\Delta\mathbf{T}(i,j) = \mathbf{n}(i)\,\Delta\mathbf{S}(i,j)$. A total of $N$ matrix equations (9.32) can be written resulting in a system of $2N$ simultaneous equations:

$$[\Delta\mathbf{T}]\{\boldsymbol{\psi}\} = \{\mathbf{t}\}^0 \tag{9.33}$$

where

$$[\Delta\mathbf{T}] = \begin{bmatrix} \Delta\mathbf{T}(1,1) & & \Delta\mathbf{T}(1,N) \\ \cdot & & \cdot \\ \cdot & & \cdot \\ \cdot & & \cdot \\ \Delta\mathbf{T}(N,1) & \dots & \Delta\mathbf{T}(N,N) \end{bmatrix} \tag{9.34}$$

and

$$\{\boldsymbol{\psi}\} = \begin{Bmatrix} \boldsymbol{\psi}(1) \\ \cdot \\ \cdot \\ \cdot \\ \boldsymbol{\psi}(N) \end{Bmatrix} \qquad \{\mathbf{t}\}^0 = \begin{Bmatrix} \mathbf{t}^0(1) \\ \cdot \\ \cdot \\ \cdot \\ \mathbf{t}^0(N) \end{Bmatrix} \tag{9.35}$$

Note that it is now possible to obtain the influence coefficient for the case where the points $i$ and $j$ coincide. For example, if the problem in Figure 9.11 is a hole in an infinite domain then

$$\Delta\mathbf{T}(i,i) = \begin{bmatrix} -\frac{1}{2} & 0 \\ 0 & -\frac{1}{2} \end{bmatrix} \tag{9.36}$$

Equation can now be solved for the 'fictitious tractions' $\{\boldsymbol{\psi}\}$. These are 'fictitious' in that they do not really exist but are an artificial device for obtaining the solution due to given tractions.

The unknown displacements at the centre of the line segments are computed indirectly via the 'fictitious forces' by

$$\mathbf{u}(i) = \sum_{j=1}^{N} \Delta\mathbf{U}(i,j)\boldsymbol{\psi}(j) \tag{9.37}$$

Furthermore the displacements and stresses can also be computed at any point $Q$ inside the domain by

$$\mathbf{u}(Q) = \sum_{j=1}^{N} \Delta\mathbf{U}(Q,j)\boldsymbol{\psi}(j) \tag{9.38}$$

$$\boldsymbol{\sigma}(Q) = \sum_{j=1}^{N} \Delta\mathbf{S}(Q,j)\boldsymbol{\psi}(j) \tag{9.39}$$

It has been explained previously that the method is equally applicable to interior problems (i.e. one where the material is inside the contour). All that will be different is that the direction of the normal vector $\mathbf{n}$ will be reversed i.e. it must always point away from the material.

For the case where the displacements are known on some segments and the tractions on others two types of simultaneous equations are written

$$\mathbf{t}^0(i) = \sum_{j=1}^{N} \Delta\mathbf{T}(i,j)\boldsymbol{\psi}(j) \quad \text{For segment } i \text{ with known tractions, } \mathbf{t}^0 \tag{9.40}$$

$$\mathbf{u}^0(k) = \sum_{j=1}^{N} \Delta\mathbf{U}(k,j)\boldsymbol{\psi}(j) \quad \text{For segment } k \text{ with known displacements, } \mathbf{u}^0 \tag{9.41}$$

The Boundary Element Method just presented is called—for obvious reasons—the indirect method. It is much more user-friendly than the Trefftz methods discussed previously because the fundamental solution employed allows the boundary segments to be used as loading locations. This eliminates the need to have two different sets of points: one for the boundary values and one for the 'fictitious forces'.

Further improvements to the method can be made by eliminating the need to compute 'fictitious forces'.

## 9.4 DIRECT BOUNDARY ELEMENT METHOD

In the direct method the reciprocal theorem by Betti (1872) is used to eliminate the 'fictitious forces'. This theorem is valid for any solid in a state of elastic equilibrium. Betti's Theorem states:

'For a linear-elastic solid subjected to quasistatic displacements the work done by the *displacements* of load case I and the *forces* of load case II is equal to the work done by the *forces* of load case I and the *displacements* of load case II.' Or, in short

$$W_{\text{I,II}} = W_{\text{II,I}} \tag{9.42}$$

Consider the example of Figure 9.12 and let load case I be the case where the boundary is loaded by traction $\mathbf{t}$. Select quite arbitrarily load case II to be that of a point force applied at point $P$ in the solid resulting in displacements $\mathbf{U}$ and tractions $\mathbf{T}$.

The work done by the displacements of load case I and the forces of load case II is

$$W_{\text{I,II}}(P_x = 1) = \int_S \Big( u_x(Q) T_{xx}(Q,P) + u_y(Q) T_{yx}(Q,P) \Big) \, \mathrm{d}s + u_x(P) \cdot 1 \tag{9.43}$$

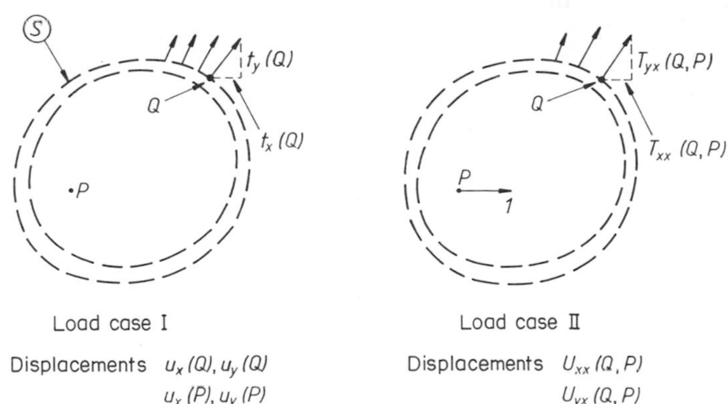


**Figure 9.12**
Betti's Theorem is applied to two load cases: one real, the other fictitious

and that of the forces of load case I and the displacements of load case II:

$$W_{\mathrm{II,I}}(P_x = 1) = \int_S (t_x(Q)\, U_{xx}(Q,P)\; t_y(Q) U_{yx}\,(Q,P)]\, \mathrm{d}s \qquad (9.44)$$

Now we take the load case II to be the same as before except that the unit load is now applied in the $y$-direction. Omitting some values in parentheses for clarity the following is obtained

$$W_{\mathrm{I,II}}(P_y = 1) = \int_S (u_x T_{xy} + u_y T_{yy})\, \mathrm{d}s + u_y(P)\cdot 1 \qquad (9.45)$$

$$W_{\mathrm{II,I}}(P_y = 1) = \int_S (t_x U_{xy} + t_y U_{yy})\, \mathrm{d}s \qquad (9.46)$$

Setting $W_{\mathrm{I,II}} = W_{\mathrm{II,I}}$ for both cases we obtain two simultaneous integral equations which are written in matrix form as

$$\mathbf{u}\,(P) + \int_S \mathbf{T}(Q,P)\mathbf{u}(Q)\, \mathrm{d}s = \int_S \mathbf{U}(Q,P)\mathbf{t}(Q)\, \mathrm{d}s \qquad (9.47)$$

The above equations give the displacement at a point $P$ due to a distribution of tractions and displacements on the boundary $S$. Note that in contrast to the indirect method $Q(s)$ is the integration variable.

If point $P$ is on the boundary a singularity occurs in the integration as the point is approached. For this particular case the integral is evaluated in a special way. The point $P$ is surrounded by a contour as shown in Figure 9.13. The integral is then split into two integrals, one over boundary $S - \epsilon$

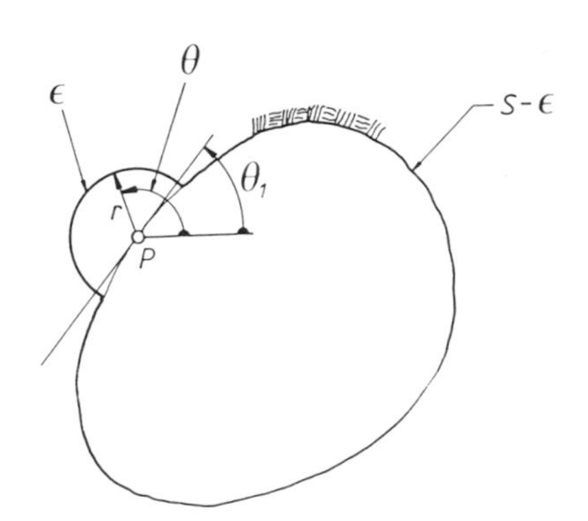


**Figure 9.13**
When $P$ is on the boundary a special integration technique has to be used near $P$

and the other over boundary $\epsilon$. The value of the integral is then obtained by letting $\epsilon \rightarrow 0$. For example for the integral on the left-hand side we obtain:

$$\lim \epsilon \rightarrow 0 \left| \int_{\epsilon} \mathbf{T}\mathbf{u} \, \mathrm{d}s + \int_{S-\epsilon} \mathbf{T}\mathbf{u} \, \mathrm{d}s \right| = \mathbf{c}(P)\mathbf{u}(P) \tag{9.48}$$

$$+ \int_{S} \mathbf{T}'\mathbf{u} \, \mathrm{d}s$$

If the boundary is smooth at point $P$ then we obtain using polar coordinates (Figure 9.13)

$$\mathbf{c}(P) = \lim r \rightarrow 0 \int_{\theta_1}^{\theta_1 + \pi} \mathbf{T} r \, \mathrm{d}\theta = \begin{bmatrix} -\frac{1}{2} & 0 \\ 0 & -\frac{1}{2} \end{bmatrix} \tag{9.49}$$

which is the same result as obtained in equation (9.36). The integral equations on the boundary for the case where $P$ is on a smooth boundary $S$ is written as

$$\tfrac{1}{2} \mathbf{I}\mathbf{u}(P) + \int_{S} \mathbf{T}\mathbf{u} \, \mathrm{d}s = \int_{S} \mathbf{U}\mathbf{t} \, \mathrm{d}s \tag{9.50}$$

where **I** is the unit matrix. It will be shown later that it is not necessary to compute **c** and that the treatment of nonsmooth boundaries is straightforward.

### 9.4.1 Method of solution

The integral equation is solved by discretization of the boundary. In the simplest case we use again linear line segments where the actual tractions and displacements are assumed to be constant along the segment. Then the integrals can be written as a sum of integrals over all the line segments $N$ (Figure 9.14) that is, for example

$$\int_{S} \mathbf{T}(Q,P)\mathbf{u}(Q) \, \mathrm{d}s = \sum_{j=1}^{N} \left[ \int_{S_j} \mathbf{T}\{Q(s),P\} \right] \mathbf{u}(j) \, \mathrm{d}s \tag{9.51}$$

where $\mathbf{u}(j)$ is the displacement at the centre of element $j$.

The discretized form of equation (9.52) is written as

$$\mathbf{c}(P)\mathbf{u}(P) + \sum_{j=1}^{N} \Delta\mathbf{T}(j,P)\mathbf{u}(j) = \sum_{j=1}^{N} \Delta\mathbf{U}(j,P)\mathbf{t}(j) \tag{9.52}$$

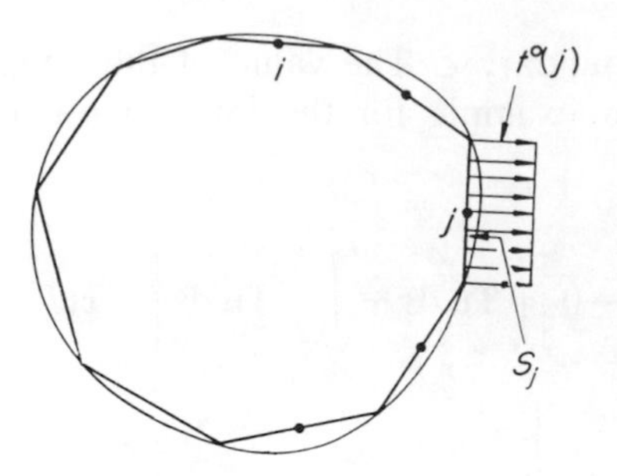


**Figure 9.14**
Direct BE: discretization of an opening using constant Boundary Elements

where

$$\Delta\mathbf{T}(j,P) = \int_{S_j} \mathbf{T}\{Q(s),P\}\,\mathrm{d}s \tag{9.53}$$

and

$$\Delta\mathbf{U}(j,P) = \int_{S_j} \mathbf{U}\{Q(s),P\}\,\mathrm{d}s \tag{9.54}$$

By putting the load point at the centre of each boundary segment in turn i.e. substituting $P = 1, 2, 3, \ldots N$ a system of simultaneous equations of the form

$$\mathbf{Aa} = \tilde{\mathbf{B}}\mathbf{b} \tag{9.55}$$

is obtained. Where

$$\mathbf{a} = \begin{Bmatrix} \mathbf{u}(1) \\ \vdots \\ \mathbf{u}(N) \end{Bmatrix} \qquad \mathbf{b} = \begin{Bmatrix} \mathbf{t}(1) \\ \vdots \\ \mathbf{t}(N) \end{Bmatrix} \tag{9.56}$$

and

$$\mathbf{A} = \begin{bmatrix} \Delta\mathbf{T}(1,1) + \mathbf{c}(1) & \ldots & \Delta\mathbf{T}(1,N) \\ \cdot & & \cdot \\ \cdot & & \cdot \\ \cdot & & \cdot \\ \Delta\mathbf{T}(N,1) & \ldots & \Delta\mathbf{T}(N,N) + \mathbf{c}(N) \end{bmatrix} \tag{9.57}$$

$$\tilde{\mathbf{B}} = \begin{bmatrix} \Delta\mathbf{U}(1,1) & \ldots & \Delta\mathbf{U}(1,N) \\ \cdot & & \cdot \\ \cdot & & \cdot \\ \cdot & & \cdot \\ \Delta\mathbf{U}(N,1) & & \Delta\mathbf{U}(N,N) \end{bmatrix} \tag{9.58}$$

This system of equations can then be solved for either unknown displacements or traction values on the boundary. The displacements at any point $P$ inside the solid can then be obtained by

$$\mathbf{u}(P) = -\sum_{j=1}^{N} \Delta\mathbf{T}(j,P)\mathbf{u}(j) + \sum_{j=1}^{N} \Delta\mathbf{U}(j,P)\mathbf{t}(j) \tag{9.59}$$

The stresses at this point can be found by equation (9.5):

$$\boldsymbol{\sigma}(P) = \mathbf{D}\hat{\mathbf{B}}\mathbf{u}(P) \tag{9.60}$$

Hence

$$\boldsymbol{\sigma}(P) = -\sum_{j=1}^{N} \Delta\mathbf{E}(j,P)\mathbf{u}(j) + \sum_{j=1}^{N} \Delta\mathbf{S}(j,P)\mathbf{t}(j) \tag{9.61}$$

where

$$\Delta\mathbf{E}(j,P) = \int_{S_j} \mathbf{D}\hat{\mathbf{B}}\mathbf{T}(j,P)\,\mathrm{d}S_j \tag{9.62}$$

and

$$\Delta\mathbf{S}(j,P) = \int_{S_j} \mathbf{D}\hat{\mathbf{B}}\mathbf{U}(j,P)\,\mathrm{d}S_j \tag{9.63}$$

## 9.5 ADVANCED BOUNDARY ELEMENT METHOD

While using a constant distribution of tractions and displacements simplifies the solution procedure this is by no means the only possibility.

Watson (1979) seems to have been the first to use Finite Element shape functions to describe the geometry of Boundary Elements and the distribution of both tractions and displacements. This results in a gain in accuracy because complicated boundaries can be modelled more accurately with fewer elements. Figure 9.15 shows various forms of Boundary Elements for two- and three-dimensional problems. These elements can be described as follows:

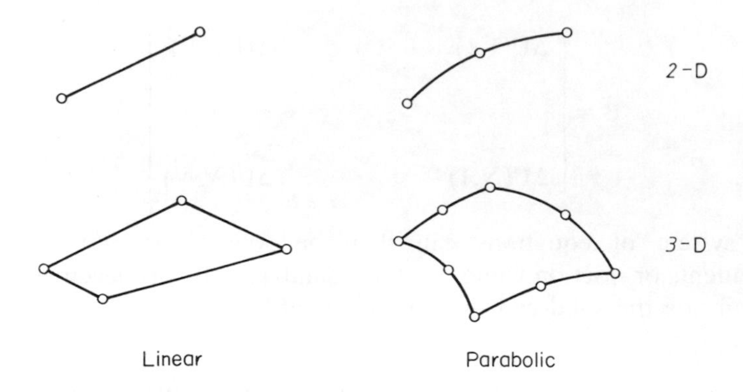


**Figure 9.15**
Isoparametric Boundary Elements: these are essentially the same as Finite Elements except that their dimension is reduced by one

$$\text{GEOMETRY} \qquad \mathbf{x} = \sum_{j=1}^{J} N_j \mathbf{x}_j^e$$

$$\text{TRACTION} \qquad \mathbf{t} = \sum_{j=1}^{J} N_j \mathbf{t}_j^e \tag{9.64}$$

$$\text{DISPLACEMENT} \qquad \mathbf{u} = \sum_{j=1}^{J} N_j \mathbf{u}_j^e$$

where $\mathbf{x}_j^e$, $\mathbf{t}_j^e$ and $\mathbf{u}_j^e$ are coordinates, tractions and displacements at the $j$th node of element $e$. The shape functions $N_j$ are linear or parabolic shape functions as discussed in Chapter 5 and $J$ is the number of nodes per element.

Substitution of relationships equation (9.64) into equation (9.52) gives for $E$ Boundary Elements:

$$\mathbf{c}(P)\mathbf{u}(P) + \sum_{e=1}^{E} \sum_{j=1}^{J} \Delta\mathbf{T}_j^e(P)\mathbf{u}_j^e = \sum_{e=1}^{E} \sum_{j=1}^{J} \Delta\mathbf{U}_j^e(P)\,\mathbf{t}_j^e \tag{9.65}$$

where

$$\Delta\mathbf{T}_j^e(P) = \int_{S_e} \mathbf{T} N_j \,\mathrm{d}s$$

$$\Delta\mathbf{U}_j^e(P) = \int_{S_e} \mathbf{U} N_j \,\mathrm{d}s \tag{9.66}$$

Whereas analytical evaluation of the integrals—although cumbersome—is possible for constant Boundary Elements numerical integration is essential for the integrals in (9.66).

Gauss Quadrature as used for Finite Elements can be employed. However, because of the singular nature of the fundamental solutions special schemes involving a variable number of Gauss points depending on the proximity of the point $P$ must be used. It is the numerical integration of the Kernel-shape function products (equation (9.66)) which requires most of the skill in writing efficient computer programs. The reader is referred to papers by Watson (1979) and Beer (1985) for a more detailed discussion on integration.

### 9.5.1 Collocation

The integral equation (9.65) has to be satisfied for any location of the load point $P$. Obviously it is not possible to write this equation for an infinite number of points and numerical techniques are necessary which involve some degree of approximation. The simplest method is to use a finite number of points which are conveniently taken the nodal points of the Boundary Elements. This method is called point *collocation* and the points used for the location of $P$ *collocation points*.

If the source point $P$ is now placed at each node of the mesh (1 to $NP$) a matrix of influence coefficients is obtained for each element. Since the equation (9.65) has to be satisfied for every location of $P$ the following system of equations can be written which relates the displacements to the tractions of all the nodes of the boundary.

$$\mathbf{Ca} + \sum_{e=1}^{E}\sum_{j=1}^{J}[\Delta\mathbf{T}]_j^e\,\mathbf{u}_j^e = \sum_{e=1}^{E}\sum_{j=1}^{J}[\Delta\mathbf{U}]_j^e\mathbf{t}_j^e \tag{9.67}$$

where

$$[\Delta\mathbf{U}]_j^e = \begin{bmatrix} \Delta\mathbf{U}(P_1)_j \\ \cdot \\ \cdot \\ \cdot \\ \Delta\mathbf{U}(P_{NP})_j \end{bmatrix} \tag{9.68}$$

and

$$[\Delta\mathbf{T}]_j^e = \begin{bmatrix} \Delta\mathbf{T}(P_1)_j \\ \cdot \\ \cdot \\ \cdot \\ \Delta\mathbf{T}(P_{NP})_j \end{bmatrix} \tag{9.69}$$

the vector **a** contains the displacement components at all the nodes of the mesh and matrix **C** is a sparse matrix containing only submatrices of order $2 \times 2$ for the two-dimensional case or $3 \times 3$ for the three-dimensional case on its diagonal.

### 9.5.2 Assembly of influence coefficient matrices

Equation (9.67) can be rewritten in matrix form as:

$$(\mathbf{C} + \mathbf{A}')\mathbf{a} = \tilde{\mathbf{B}}\mathbf{b} \tag{9.70}$$

where $\mathbf{A}$ $(=\mathbf{C} + \mathbf{A}')$ and $\mathbf{B}$ are coefficient matrices and the vector **b** contains the traction components at all the nodes of the mesh.

These coefficient matrices are assembled from element contributions.

To demonstrate the assembly process consider the simple example in Figure 9.16 consisting of three linear Boundary Elements.

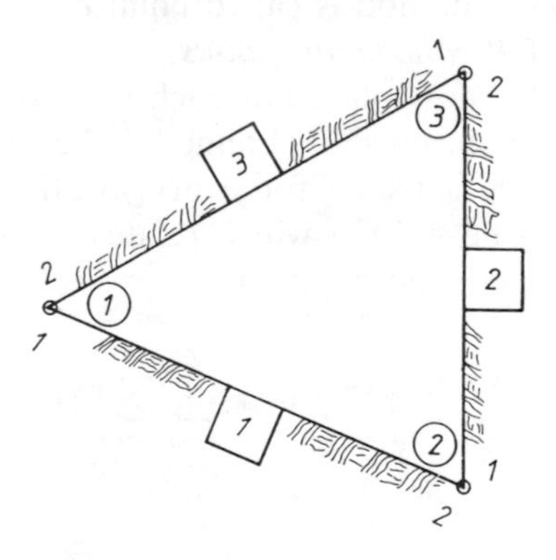


**Figure 9.16**
Three-element problem for demonstrating assembly

The 'connectivity' of each element is:

| Element | First Node | Second Node |
|---|---|---|
| 1 | 1 | 2 |
| 2 | 2 | 3 |
| 3 | 3 | 1 |

The displacements are assumed to be continuous at the nodes that is:

$$\mathbf{a}_1 = \mathbf{u}_1^1 = \mathbf{u}_2^3$$
$$\mathbf{a}_2 = \mathbf{u}_2^1 = \mathbf{u}_1^2 \tag{9.71}$$
$$\mathbf{a}_3 = \mathbf{u}_2^2 = \mathbf{u}_1^3$$

Substituting these relationships into equation (9.67) the coefficient matrix $\mathbf{A}'$ is obtained as:

$$\mathbf{A}' = \left[ [\Delta\mathbf{T}]_1^1 + [\Delta\mathbf{T}]_2^3, [\Delta\mathbf{T}]_2^1 + [\Delta\mathbf{T}]_1^2, [\Delta\mathbf{T}]_2^2 + (\Delta\mathbf{T}]_1^3 \right] \tag{9.72}$$

The matrix $\mathbf{A}'$ is therefore assembled in a similar way as the global stiffness matrix in the Finite Element method. However, now whole columns of coefficients are assembled and that means that the global coefficient matrix will be fully populated and not banded as is the case with FE stiffness matrices.

When integrating over the Boundary Elements, terms $\Delta\mathbf{T}(P)_j$ cannot be obtained if $P$ is the $j$th element node, that is for element contributions to the diagonal submatrices of $\mathbf{A}'$, because the fundamental solution is singular there. The coefficients of the diagonal submatrices $\mathbf{A}_{ii}$ are therefore computed not by assembly of element contributions but using the condition that for an elastic body a rigid body movement must not be associated with any strain energy.

The diagonal submatrices of $\mathbf{A}$ are given by:

$$\mathbf{A}_{ii} = \mathbf{c} + \mathbf{A}'_{ii} \tag{9.73}$$

whereas the off-diagonal submatrices are simply:

$$\mathbf{A}_{ij} = \mathbf{A}'_{ij} \qquad i \neq j \tag{9.74}$$

If a rigid body movement occurs for a finite elastic region then the tractions at all nodes and therefore the right-hand side of equation (9.70) must be zero:

$$\mathbf{A}\mathbf{a}^i = 0 \tag{9.75}$$

Here $\mathbf{a}^i$ is a vector containing unit displacements in the $i$th direction at all the nodes of the mesh. Depending on the dimensionality of the problem, two or three sets of equations (9.75) can be written from which all the missing coefficients of the diagonal submatrices can be determined. The consequence of equation (9.75) is that the coefficients of the diagonal submatrices can be obtained by summing all the off-diagonal coefficients and changing sign.

For an infinite elastic region strictly speaking no rigid body movement is possible. However, the above procedure can still be used to work out the coefficients of the diagonal submatrices by making the infinite domain a finite one by surrounding the finite boundary with a circle (or sphere) of radius $r_0$. Rigid body displacements are then applied as before. An additional term has now to be considered in equation (9.75) which has to do with the circular (spherical) boundary at $r_0$ ($S_0$):

$$\mathbf{A}\mathbf{a}^i + \mathbf{T}_0\mathbf{u}^i = 0 \tag{9.76}$$

The 'azimuthal' integral is defined as:

$$\mathbf{T}_0 = \int_{S_0} \mathbf{T}\, \mathrm{d}S_0 \tag{9.77}$$

and $\mathbf{u}^i$ is a $2 \times 1$ ($3 \times 1$) vector containing unit values of displacement in the $i$th coordinate direction.

Since the region is in fact infinite this means that $r_0$ is also infinite. For both 2D and 3D problems the following result is obtained in the limit as $r_0 \rightarrow \infty$:

$$\lim_{r_0 \rightarrow \infty} \int_{S_0} \mathbf{T}\, \mathrm{d}S_0 = -\mathbf{I} \tag{9.78}$$

The consequence of equation (9.81) is that for infinite regions the off-diagonal coefficients are summed, the 'azimuthal' integral added and the sign is changed to obtain the missing coefficients. Equation (9.76) is identical to stating that the tractions on boundary $S_0$ have to be in equilibrium with the unit force.

Regarding the assembly of coefficient matrix $\tilde{\mathbf{B}}$ the assumption can also be made that the tractions are continuous at the nodes. Analogously to equation (9.72) this yields:

$$\tilde{\mathbf{B}} = \left[ [\Delta\mathbf{U}]_1^1 + [\Delta\mathbf{U}]_2^3, [\Delta\mathbf{U}]_2^1 + [\Delta\mathbf{U}]_1^2, [\Delta\mathbf{U}]_2^2 + [\Delta\mathbf{U}]_1^3 \right] \tag{9.79}$$

This assumption, however, is not necessary. For example, if all the tractions are known at the boundary equation (9.72) can be written:

$$\mathbf{A}\mathbf{a} = \mathbf{F} \tag{9.80}$$

where

$$\mathbf{F} = \sum_{e=1}^{E} \sum_{j=1}^{J} [\Delta\mathbf{U}]_j^e\, \mathbf{t}_j^e \tag{9.81}$$

Since the tractions are defined for each element individually two different traction values can be specified at a particular node and this is particularly useful for analysing problems with corners where traction discontinuities may occur.

### 9.5.3 Solution and postprocessing

The direct and advanced BEM lead to a system of simultaneous equations which relate the displacement components **a**, at each node to the traction components, **b**:

$$\mathbf{A}\mathbf{a} = \tilde{\mathbf{B}}\mathbf{b} \tag{9.84}$$

Equation (9.82) can be solved for unknown nodal displacements **a** or tractions **b** depending if **b** or **a** are known on the boundary.

Excavation problems are solved as follows. (Assume that the initial stress field before excavation is $\boldsymbol{\sigma}^0$ and that the initial displacements are $\mathbf{a}^0$):

Convert the stresses $\boldsymbol{\sigma}^0$ at each nodal point of the Boundary Elements into tractions $\mathbf{t}^0$ using equation (9.7).
Analyse the opening for known tractions $-\mathbf{t}^0$ (i.e. acting in the opposite direction) at the element nodes by solving equation (9.80).
Add the results obtained for the stresses and displacements to the initial conditions.

It can be seen that resulting stress field gives zero tractions at the excavation boundary as required.

Regarding the solution of equation (9.82) it should be mentioned that the matrices **A** and **B** are not symmetric. Also, because of the occurrence of logarithmic and $1/r$, $1/r^2$ functions in the Kernels it is recommended to scale coordinate and load data as suggested by Watson (1979). It is convenient to work not in the normal units of length and force but, for example, to assume the greatest dimension of the structure as the unit of length and adopt a unit of force such that Young's Modulus equals to unity.

Apart from their unsymmetry the matrices in (9.82) differ from the Finite Element matrices in that they are not always positive definite (i.e. the diagonal element may become negative during the elimination). However, if scaling is used, the system of equations is always well conditioned and normal Gauss elimination procedures can be used.

The primary results of a Boundary Element analysis are displacements/tractions at the nodes of the Boundary Element mesh. With the shape functions the distribution of displacements and tractions on the boundary is also defined.

Stresses on the boundary can be computed from this distribution of displacement and traction. For example, for plane strain problems the tangential stress $\sigma_\theta$ on a boundary element is computed as follows:

The displacement tangential to the surface is:

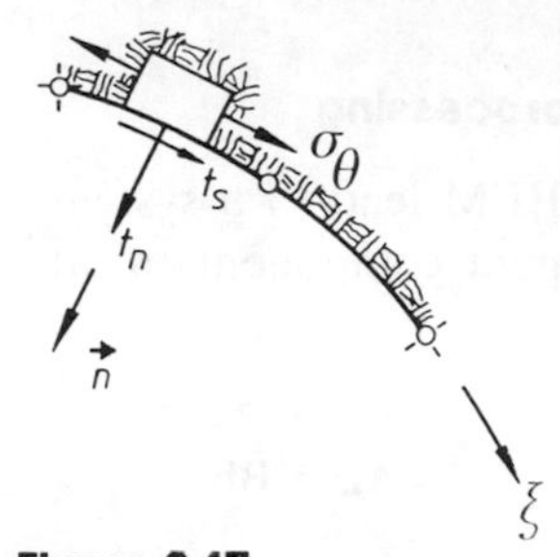


**Figure 9.17**
Boundary stresses for 2D element

$$u_\theta = \boldsymbol{\ell}^T\mathbf{u} = \boldsymbol{\ell}^T\mathbf{N}\mathbf{u}^e \tag{9.83}$$

where $\boldsymbol{\ell}$ is a unit vector in tangential direction and

$$\mathbf{N} = [N_1 \ldots N_j]\,\mathbf{I} \tag{9.84}$$

$$\mathbf{u}^e = \begin{Bmatrix} \mathbf{u}_1^e \\ \cdot \\ \cdot \\ \cdot \\ \mathbf{u}_j^e \end{Bmatrix} \tag{9.85}$$

The strain in tangential direction is:

$$\epsilon_\theta = \frac{\mathrm{d}u_\theta}{\mathrm{d}\xi}\frac{\mathrm{d}\xi}{\mathrm{d}s} = \frac{1}{|J|}\boldsymbol{\ell}^T\frac{\mathrm{d}\mathbf{N}}{\mathrm{d}\xi}\mathbf{u}^e \tag{9.86}$$

or

$$\epsilon_\theta = \mathbf{E}\mathbf{u}^e \tag{9.87}$$

where

$$|J| = \sqrt{(\mathrm{d}x/d\xi)^2 + (dy/d\xi)^2} \tag{9.88}$$

and

$$\mathbf{E} = \frac{1}{|J|}\boldsymbol{\ell}^T\frac{\mathrm{d}\mathbf{N}}{\mathrm{d}\xi} \tag{9.89}$$

Finally, using Hooke's law the tangential stress is obtained for plane strain problems as:

$$\sigma_\theta = \frac{\mathrm{E}}{(1-\nu^2)}\,\epsilon_\theta + \frac{\nu}{(1-\nu)}\,t_n + \sigma_\theta^0 \tag{9.90}$$

where $t_n$ is the traction component normal to the boundary and $\sigma_\theta^0$ is the initial tangential stress.

For three-dimensional problems the boundary stresses are evaluated in a similar way as for the shell elements (Zienkiewicz, 1977). First the strains in the global $x$, $y$, $z$ directions are computed by:

$$\boldsymbol{\epsilon} = \Sigma\, \mathbf{B}_j \mathbf{u}_j^e \tag{9.91}$$

where

$$\mathbf{B}_j = \begin{bmatrix} \dfrac{\partial N_j}{\partial x} & 0 & 0 \\ 0 & \dfrac{\partial N_j}{\partial y} & 0 \\ 0 & 0 & \dfrac{\partial N_j}{\partial z} \\ \dfrac{\partial N_j}{\partial y} & \dfrac{\partial N_j}{\partial x} & 0 \\ 0 & \dfrac{\partial N_j}{\partial z} & \dfrac{\partial N_j}{\partial y} \\ \dfrac{\partial N_j}{\partial z} & 0 & \dfrac{\partial N_j}{\partial x} \end{bmatrix} \tag{9.92}$$

Next the strains are transformed into a local $\mathbf{v}_1$, $\mathbf{v}_2$ coordinate system (as shown in Figure 9.18) which is tangential to the boundary surface. Using the strain transformation matrix, $\mathbf{T}_\epsilon$, (Appendix II) the local strains are given by:

$$\boldsymbol{\epsilon}' = \mathbf{T}_\epsilon \boldsymbol{\epsilon} \tag{9.93}$$

Of the six stress components on the boundary 3 are directly related to the traction components $t_n$, $t_{s_1}$, $t_{s_2}$, i.e.

$$\begin{aligned} \sigma_z &= t_n \\ \tau_{y'z'} &= t_{s_2} \\ \tau_{x'z'} &= t_{s_1} \end{aligned} \tag{9.94}$$

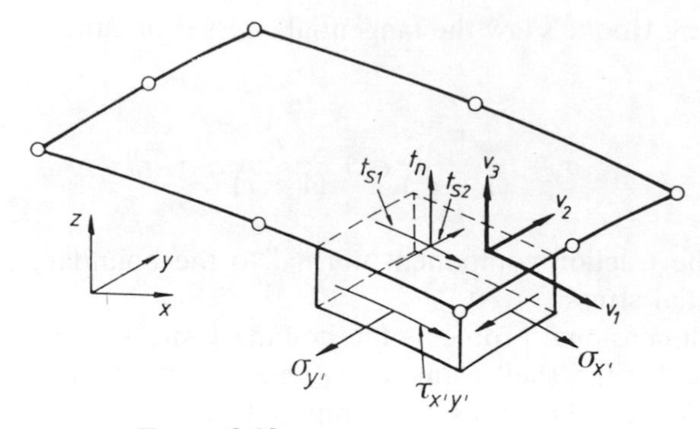


**Figure 9.18**
Boundary stresses for 3D element

The remaining stress components can be computed using Hooke's law:

$$\begin{aligned}\sigma_{x'} &= C_1(\epsilon_{x'} + \nu\epsilon_{y'}) + C_2 t_n \\ \sigma_{y1} &= C_1(\epsilon_{y'} + \nu\epsilon_{x'}) + C_2 t_n \\ \tau_{x'y'} &= C_1\gamma_{x'y'}\end{aligned} \tag{9.95}$$

where

$$C_1 = \frac{E}{(1-\nu^2)'} \qquad C_2 = \frac{\nu}{(1-\nu)} \tag{9.96}$$

The six stress components at the boundary in the local orthogonal system of axes tangential and normal to the surface are

$$\boldsymbol{\sigma}' = \begin{Bmatrix} \sigma_{x'} \\ \sigma_{y'} \\ t_n \\ \tau_{x'y} \\ t_{s_1} \\ t_{s_2} \end{Bmatrix} \tag{9.97}$$

Finally the six stress components in the global $x$, $y$, $z$ axis system can be obtained by multiplying with the stress transformation matrix $\mathbf{T}_\sigma$ and adding the initial stress components.

$$\boldsymbol{\sigma} = \mathbf{T}_\sigma \boldsymbol{\sigma}' + \boldsymbol{\sigma}_0 \tag{9.98}$$

In addition to the values of traction, displacement and stress at the boundary the values of displacement and stress can be computed at any point inside the BE region using the boundary values.

The displacements at a point $P$ (coordinates $x$, $y$, $z$) inside the domain are obtained by

$$\mathbf{u}(P) = \sum_{e=1}^{E} \sum_{j=1}^{J} (\Delta\mathbf{U}(P)_j^e \, \mathbf{t}_j^e - \Delta\mathbf{T}(P)_j^e \, \mathbf{u}_j^e) \tag{9.99}$$

The stresses are obtained by differentiating (equation (9.99)) to obtain the strains and multiplying with the elasticity matrix (see Section 9.3.1). The result is:

$$\boldsymbol{\sigma}(P) = \sum_{e=1}^{E} \sum_{j=1}^{N} (\Delta\mathbf{S}(P)_j^e \, \mathbf{t}_j^e - \Delta\mathbf{E}(P)_j^e \, \mathbf{u}_j^e) + \boldsymbol{\sigma}(P)_0 \tag{9.100}$$

where

$$\Delta\mathbf{S}(P)_j^e = \int_{-1}^{+1} \mathbf{S}(Q(\xi),P)N_j|J| \, \mathrm{d}\xi \tag{9.101}$$

and

$$\Delta\mathbf{E}(P)_j^e = \int_{-1}^{+1} \mathbf{E}(Q(\xi),P)N_j|J| \, \mathrm{d}\xi \tag{9.102}$$

$\boldsymbol{\sigma}(P)_0$ is the initial stress at point $P$.

Indeed one of the distinct features of the BEM is that results of an analysis can be obtained at any point inside the region (rock mass); all one has to do is to specify its $x$, $y$, $z$ coordinates. Any number of points can be specified after a solution has been obtained. If results inside a region are to be plotted then planes ('dummy planes') have to be specified on which results are displayed. This gives the user greater flexibility since one is no longer restricted to Gauss or nodal points of Finite Elements.

## REFERENCES


Beer, G., and Jun, L. (1985). 'Efficient integration of techniques for three-dimensional Boundary Elements.' Invited lecture JSNME meeting Tokyo, Japan Society for Numerical Methods in Engineering.

Betti, E. (1872). 'Teori dell elasicita.' *Il Nuovo Ciemento*, 7–10.

Crouch, S. L., and Starfield, A. M. (1983). *Boundary Element Methods in Solid Mechanics*, London: George Allen & Unwin.

Gioda, G. (1984). 'A simple boundary equation technique for elastic stress analysis of underground cavities.' *Rock Mechanics and Rock Engineering*, **17**, 147–65.

Kelvin and Tait (1879). *Natural Philosophy*, 2nd edn.

Melan, E. (1932). 'Der Spannungszustand der Durch eine Einzelkraft im Inneren Beanspruchten Halb-scheibe.' *Z. Angew. Math. Mech.* **12**, 343–46.

Mindlin, R. D. (1936). 'Force at a point in the interior of a semi-infinite solid.' *Physics*, **7**, 195–202.

Patterson, C., and Sheikh, M. A. (1982). A modified Trefftz Method for Stress Analysis, *Proc. Int. Conf. Finite Element Methods*, Shanghai, China. New York: Gordon and Breach.

Telles, J. C. F., and Brebbia, C. A. (1981). 'Boundary Element solution for half-plane problems.' *Int. J. Solid Structures*, **17**(12), 1149–58.

Trefftz, E. (1926). 'Ein Gegenstück Zum Rizschen Verfahren.' *Proc. 2nd Int. Congress in Applied Mechanics*, Zurich.

Watson, J. O. (1979). 'Advanced implementation of the Boundary Element Method for two- and three-dimensional elastostatics,' Chapter 3 in: *Developments in Boundary Element Methods, 1* (eds, P. K. Banerjee, and R. Butterfield, Applied Science, London, 31–63.

Zienkiewicz, O. C. (1977). *The Finite Element Method* (Chapter 12) (3rd ed.), McGraw-Hill.

# 10 Applications of the Boundary Element Method

## 10.1 INTRODUCTION

The Boundary Element Methods discussed in Chapter 9 permit the analysis of problems in rock mechanics where the rock mass is homogeneous, isotropic and elastic. Such elastic analyses are useful in providing estimates of the stress concentrations induced into the rock mass by excavation of material. Practical examples of such an application are shown here and this serves well to demonstrate the user-friendliness and efficiency of the method which has made possible three-dimensional analyses beyond the capabilities of the Finite Element Method.

The first example is that of five mine excavations whose geometry is typical of the excavation layout in the copper orebodies of the Mount Isa Mine in Australia (Watson and Cowling, 1986). Typical dimensions of these excavations (stopes) are 100 m. Two meshes have been used and the computed stresses at two points are compared with measured values in Figure 10.1. The agreement with measurement is not excellent, possibly due to influences such as jointing and the proximity of other excavations which were present but not considered in the analysis. However, the results of the coarse mesh and the fine mesh agree well indicating that good results can be obtained even with a fairly coarse boundary subdivision. More significantly meshes took only a few hours to prepare whereas an equivalent Finite Element mesh would have taken a substantially greater effort because it would have been necessary to divide the entire rock mass into Finite Elements.

The second example also comes from mining and relates to the excavation strategy for future mining below 21 level at the PASMINCO mine in Broken Hill, Australia using program BEFE (Beer, 1986). Figure 9.2 shows the Boundary Element mesh for one of the 16 excavation stages, which was analysed. The massive mesh in the upper part of Figure 10.2 models the existing mining. Note how the mesh is refined in the area of interest. Also

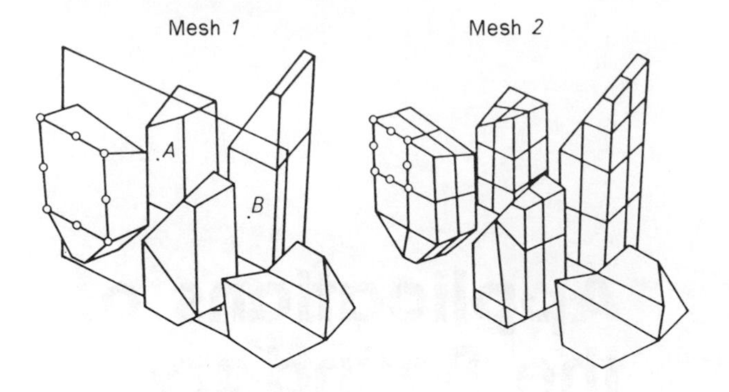


| Point A | Measured MPa | Dip Dir/Dip | Mesh 1 | | Mesh 2 | |
|---|---|---|---|---|---|---|
| $\sigma_1$ | 32·0 | 099/09 | 42·5 | 094/27 | 42·9 | 096/25 |
| $\sigma_2$ | 20·3 | 288/81 | 25·1 | 288/63 | 24·8 | 285/64 |
| $\sigma_3$ | 13·8 | 190/01 | 8·2 | 186/06 | 8·8 | 188/05 |
| Point B | | | | | | |
| $\sigma_1$ | 40·9 | 090/58 | 44·6 | 091/26 | 45·5 | 091/25 |
| $\sigma_2$ | 19·2 | 300/28 | 25·6 | 282/63 | 25·7 | 287/64 |
| $\sigma_3$ | 13·8 | 203/14 | 7·9 | 183/05 | 3·5 | 182/06 |

**Figure 10.1**
Analysis of mining excavations at Mount Isa Mine, Australia

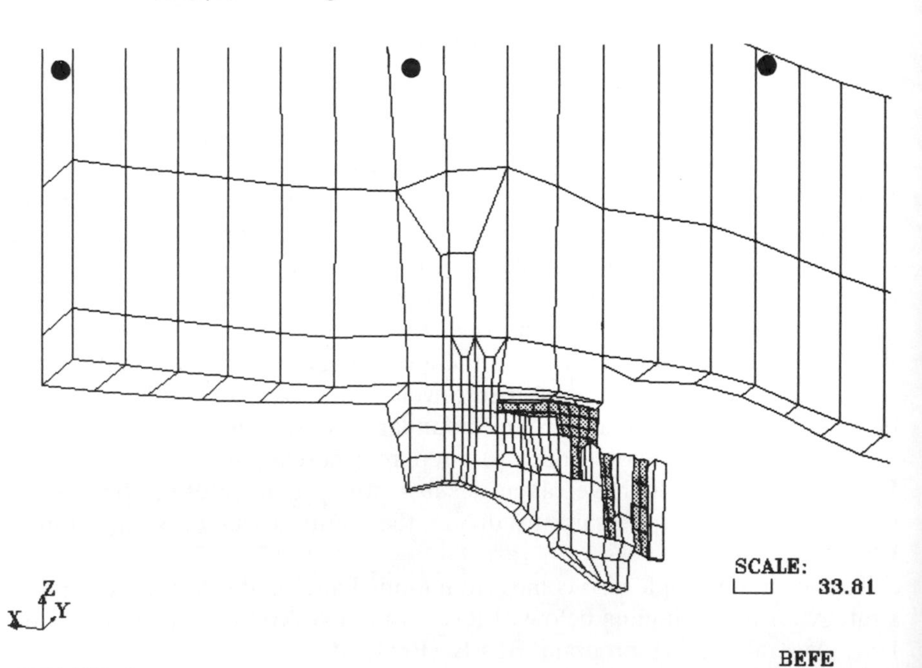


**Figure 10.2**
Boundary Element mesh to analyse mine excavations in Broken Hill, Australia

shown in the perspective view of Figure 10.2 is a mesh of so-called 'dummy Boundary Elements' (shaded) which are used only to plot results and are not part of the BE mesh. For this particular analysis fournoded Boundary Elements with linear shape functions were used. The virgin stress field was assumed to vary linearly and was based on stress measurements. The major principal stress direction is about 48° inclined to the N–S and varies from 32 MPa at 500 m to 38 MPa at 1000 m. The bottom of the mining is about 1000 m underground. The elastic properties of the rock were assumed to be $E = 80\,000$ MPa and $\nu = 0.3$. Figure 10.3(a) (see colour plate) shows a plot of $\sigma_{max} - \sigma_{min}$ (maximum minus minimum principal stress) plotted on 'dummy planes' inside the rock mass. The reason this stress distribution is plotted is because a relationship between rock strength and this stress value was established (Tillmann, 1984) for the rock mass at Broken Hill. Critical stress values $\sigma_{max} - \sigma_{min}$ are in the range of 50–60 MPa, i.e. the red areas in Figure 10.3(a).

The analyses showed that rock pillars are expected to become very highly stressed and highlighted the need for stress relief measures during mining. An example of how the stress values can be plotted on the boundary surfaces is shown in Figure 10.3(b) (see colour plate) for a different excavation stage of the same problem.

As has been stressed throughout this book, real rock in most cases does not behave like an elastic continuum but has distinct features which influence its response to—for example—excavation of material.

Let us recapitulate the distinct features of the rock mass which influence its response, they are:

(1) Major faults.
(2) Closely spaced joints in preferred orientations.
(3) Closely spaced joints in random orientations.

We shall discuss different ways of how we can use the BEM to model joints in more detail.

## 10.2 MODELLING OF JOINTS

There are basically three different approaches of how joints and faults can be modelled with the Boundary Element Method:

(1) By connecting two or more Boundary Element regions.
(2) By using a special fundamental solution for a displacement discontinuity.
(3) By coupling the Finite Element and Boundary Element Method.

### 10.2.1 Boundary Element regions and interfaces

If a major fault is to be modelled then the rock mass can be divided into two regions as shown in Figure 10.4. Initially each region is considered completely independently.

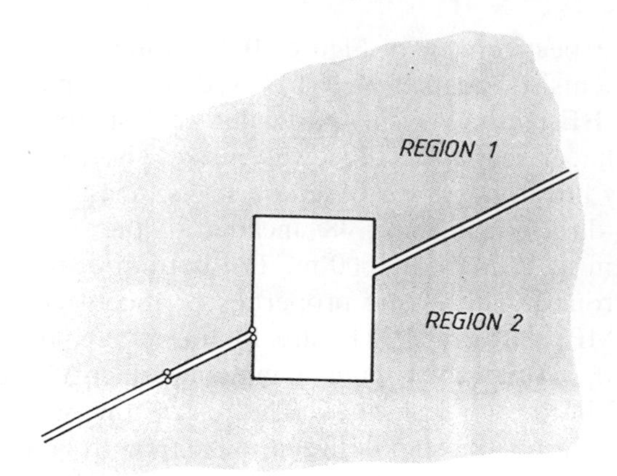


**Figure 10.4**
Excavation in rock mass which is transsected by a fault. The rock mass is modelled by two BE regions which are considered separately

For Regions 1 and 2 we obtain the following system of equations:

$$\mathbf{A}^1\mathbf{a}^1 = \tilde{\mathbf{B}}^1\mathbf{b}^1 \tag{10.1}$$

$$\mathbf{A}^2\mathbf{a}^2 = \tilde{\mathbf{B}}^2\mathbf{b}^2 \tag{10.2}$$

where $\mathbf{a}^1$, $\mathbf{b}^1$ are displacements and traction values at the boundary nodes of Region 1. The corresponding values on the boundary of Region 2 are $\mathbf{a}^2$, $\mathbf{b}^2$.

The two regions are now combined by assuming joint behaviour at the interface between regions.

Firstly we must have equilibrium of forces at the interface, i.e.:

$$\begin{aligned} t_n^1 &= t_n^2 \\ t_s^1 &= t_s^2 \end{aligned} \tag{10.3}$$

where $t_n^1$, $t_n^2$ are tractions normal to the boundaries of Regions 1 and 2 respectively and $t_s$ is the traction tangential to the boundary. Note that the direction of the traction components depend on the direction of the 'outward normal' which is different for Region 1 and 2.

For an elastic joint we can write for a pair of opposing joints (one on Region 1, the other on Region 2):

$$\begin{aligned} t_n &= k_n\,\Delta u_n + t_n^0 \\ t_s &= k_s\,\Delta u_s + t_s^0 \end{aligned} \tag{10.4}$$

where $t_n$, $t_s$ are the normal and shear tractions at a point (either on Region 1 or Region 2), $k_n$ and $k_s$ are normal and shear stiffnesses. $t_n^0$, $t_s^0$ are initial values of tractions. The relative displacements are given by:

$$\Delta u_n = u_n^1 - u_n^2$$
$$\Delta u_s = u_s^1 - u_s^2 \tag{10.5}$$

Where the superscript denote region number.

The joint remains elastic until it slips or separates. For example the onset of slip occurs when (see Section 3.5.2.1 but note that $t$ replaces $\sigma$):

$$|t_s| = |t_s^y| = c + t_n \tan \phi \tag{10.6}$$

The boundary regions are now combined using the constraint equations (10.3) and (10.4). After the first analysis equation (10.6) is checked for each pair of nodes. If

$$|t_s| > |t_s^y| \tag{10.7}$$

then joint slip should have occurred during the time step but because the joint was assumed elastic it has not.

The 'excessive traction'

$$|t_s^p| = |t_s^y| - |t_s| \tag{10.8}$$

has to be redistributed. The iteration continues until equilibrium and yield conditions are satisfied within a given tolerance at all the points on the interface.

Details of the implementation of such a procedure are given by Crotty and Wardle (1985).

An example of application of the above procedure to a practical mining problem using BITEMJ (Crotty, 1982) is given in Figure 10.5 which depicts a pillar between two excavations, with a major fault transecting the entire area. The premining stress field is assumed to be −30 MPa in the horizontal ($x_1$) direction and −20 MPa in the vertical ($x_2$) direction. A Youngs Modulus of 20 GPa and a Poisson's ratio of 0.2 were assumed for the rock mass. For the fault elastic stiffnesses of $k_n$ = 10 GPa/m and $k_s$ = 1 GPa/m are assumed and the angle of friction was 25° and the cohesion zero.

Figure 10.6 shows the deformed mesh for both elastic and elastoplastic material properties of the fault.

In the mesh shown in Figure 10.5 the fault plane was discretized to $x_1$ = 660 and $x_1$ = −600 m. Theoretically the surfaces of the fault can be assumed to extend to infinity in both directions but it has been found that truncation of the mesh at a large distance away from the openings is acceptable.

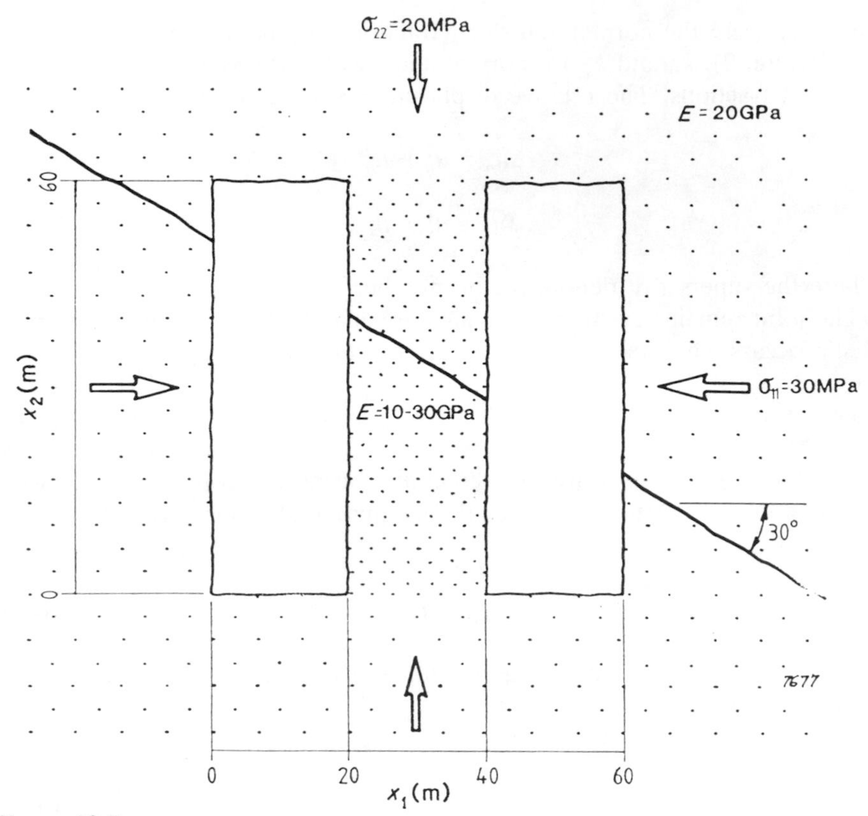


**Figure 10.5**
Underground mining problem of excavations in heterogeneous, faulted rock mass

A more elegant and—especially for three-dimensional analysis—more economic approach is to use infinite Boundary Elements instead of mesh truncation. These infinite elements are similar to the infinite Finite Elements described in Chapter 6 except that line elements are used for plane analysis and surface elements for 3D analyses. Details on infinite Boundary Elements are given by Beer and Watson (1989).

### 10.2.2 Displacement discontinuity modelling

It is possible to derive a fundamental solution, similar to the Kelvin solution, for a displacement discontinuity (DD) in an infinite domain. The displacement discontinuities are defined as relative displacements between top and bottom surface of a crack with zero thickness. For example for plane problems (Figure 10.7) we have:

$$\begin{aligned} &\text{SLIP} && D_x = u_{x,B} - u_{x,T} \\ &\text{OPENING/CLOSING} && D_y = u_{y,B} - u_{y,T} \end{aligned} \tag{10.9}$$

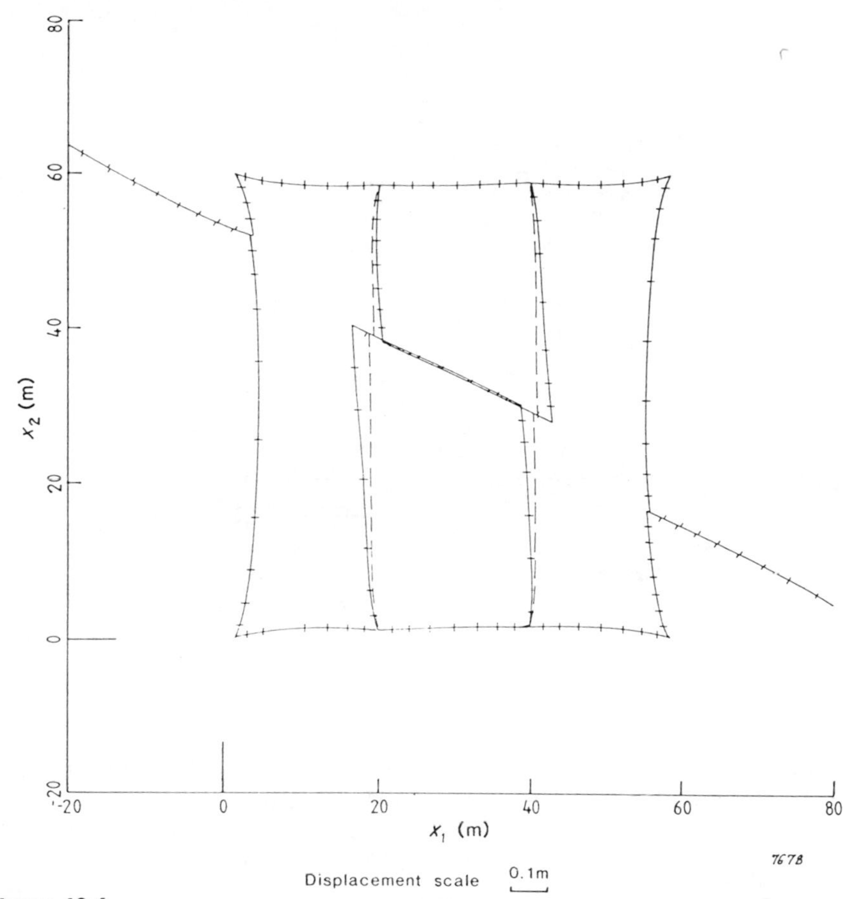


**Figure 10.6**
Deformed mesh for homogeneous rock with fault. ---- elastic fault; ++++ nonlinear fault, $\phi = 25°$

The solution for the displacement at point $Q$ $\mathbf{u}(Q)$ is given by:

$$\mathbf{u}(Q) = \mathbf{U}_{\Delta}(Q,P)\mathbf{D}(P) \tag{10.10}$$

Where $\mathbf{U}$ is an influence coefficient matrix i.e. gives the displacement at $Q$ due to a unit displacement discontinuity $\mathbf{D}$ at $P$ and

$$\mathbf{D} = \begin{Bmatrix} D_x \\ D_y \end{Bmatrix} \tag{10.11}$$

Similarly the stresses at $Q$ can be expressed by:

$$\boldsymbol{\sigma}(Q) = \mathbf{S}_{\Delta}(Q,P)\mathbf{D}(P) \tag{10.12}$$

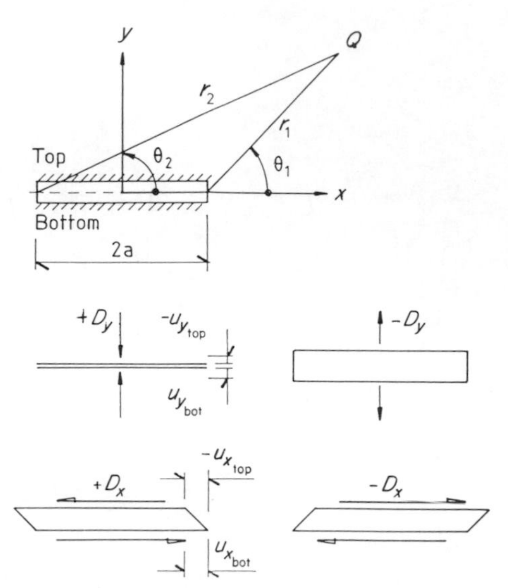


**Figure 10.7**
Displacement discontinuity modes for plane problems

The coefficients of $\mathbf{U}_\Delta$ are given in Appendix III.

Solutions also exist for displacement discontinuities in three-dimensional space. There are now two components of slip (Figure 10.8):

$$\begin{aligned} \text{SLIP 1 } D_x &= u_{x,B} - u_{x,T} \\ \text{SLIP 2 } D_y &= u_{y,B} - u_{y,T} \\ \text{OPENING/CLOSING } D_z &= u_{z,B} - u_{z,T} \end{aligned} \tag{10.13}$$

The fundamental solutions for a three-dimensional displacement discontinuity are fairly complicated expressions which are given by Sinha (1979). The solution proceeds in exactly the same way as for the indirect Boundary Element Method with the exception that instead of one surface, a pair of opposing surfaces is described by one element. Also, instead of 'fictitious forces' we now solve for displacement discontinuities which are no longer fictitious but real normal and shear displacements.

The displacement discontinuity method as presented so far allows for the analysis of excavations where the surfaces are close together (tabular excavations) as they occur in coal mines and some metaliferous mines.

The DD element can also be used to model joints and faults filled with weak material. In this case the DD element surfaces are connected by springs which have shear stiffness $k_s$ and normal stiffness $k_n$. These 'Springs' are similar to the joint stiffness discussed in 6.2.

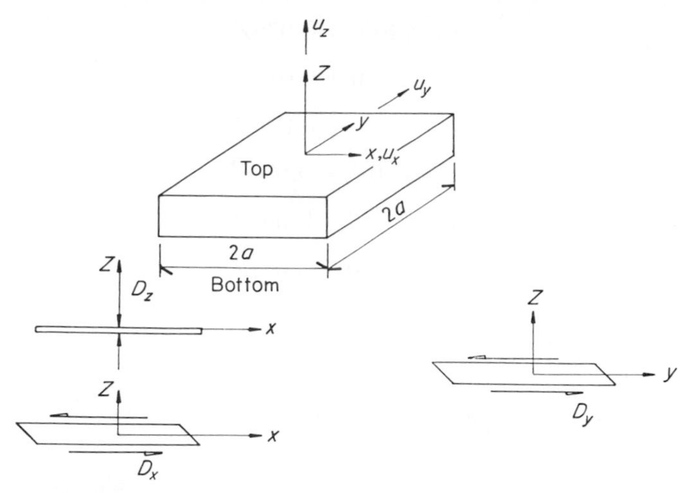


**Figure 10.8**
Displacement discontinuity modes for 3D problem

An example of the application of a three-dimensional Displacement Discontinuity program NFOLD (Sinha, 1979) to tabular excavations at the Mount Isa Mine In Australia is shown.

The modelling of the excavation process with program NFOLD works as follows:

(1) Each tabular orebody is divided into a grid of displacement discontinuity elements whose surfaces are connected by springs. Here the spring constants $k_n$ and $k_s$ represent the normal and shear stiffness of the ore.
(2) Next, excavations are defined by flagging the DD elements inside the excavation. For these elements the springs are removed and a traction equivalent but opposite to the virgin (pre-excavation) stress is applied.
(3) The system is solved. The primary results are the amount of closure in the excavation region. The normal and shear stress inside the orebody can be computed in a similar way as for the direct Boundary Element method.

Figure 10.9(a) shows a cross-section of two lead–silver–zinc orebodies modelled by NFOLD and their approximation by displacement discontinuity elements. The subdivision shown is into macro DD elements. Each of these elements is again subdivided into five DD elements in each direction.

Figures 10.9(b) and (c) show long sections through the two orebodies and the excavated regions (stopes).

The following assumptions were made for the analysis:

The virgin stress field was assumed to be:

$$\sigma_x = 10.0 + 0.026Z$$
$$\sigma_y = 7.0 + 0.012Z$$
$$\sigma_z = 0.0 + 0.028Z$$
$$\tau_{xy} = 0.0 + 0.000Z$$
$$\tau_{yz} = 0.0 + 0.000Z$$
$$\tau_{xz} = 6.0 - 0.001Z$$

where $Z$ is the depth.

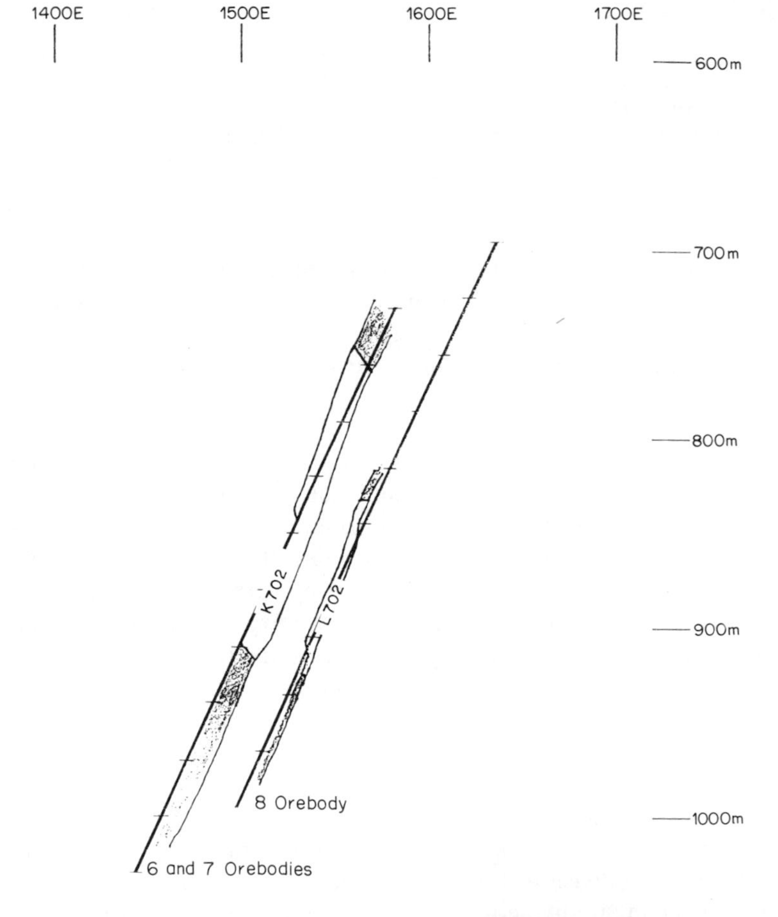


**Figure 10.9**
Idealized cross-section of lead–silver–zinc orebodies at the Mount Isa Mine in Australia (Courtesy Mount Isa Mines Ltd)

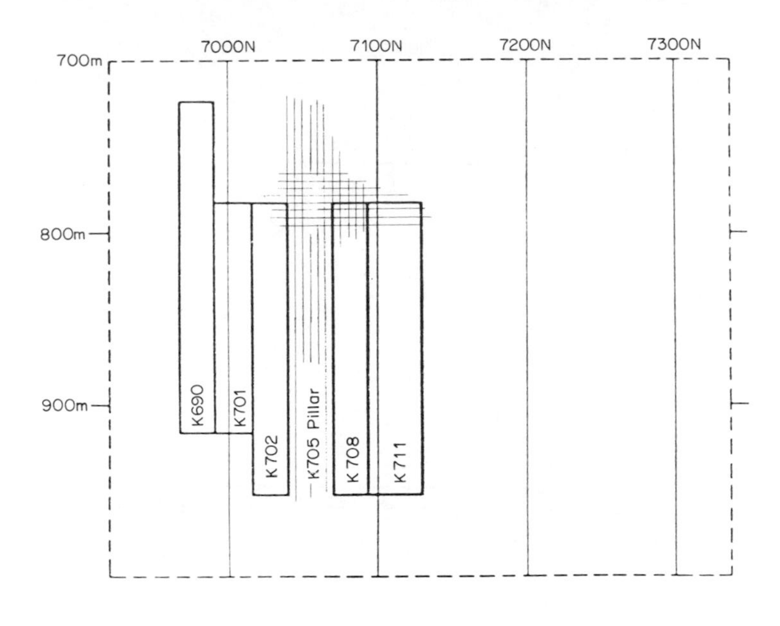


**LONGSECTION**

**Figure 10.9**
Continued

The material properties for both seam and host material were:

$$E = 80\,\text{GPa} \quad \text{and} \quad \nu = 0.25.$$

The results of the analysis are shown in Figure 10.10.
The colour code for the stress levels is:

| | | | |
|---|---|---|---|
| Red | >90 MPa | Green/yellow | 40–50 MPa |
| Red/yellow | 80–90 MPa | Green and yellow | 30–40 MPa |
| Yellow/red | 70–80 MPa | Yellow/green | 20–30 MPa |
| blue | 60–70 MPa | Yellow/white | < 20 MPa |
| White/blue | 50–60 MPa | White | Excavated |

The following conclusions can be made from these plots:

*6 and 7 Orebodies*

(1) The high stresses in K705 pillar;
(2) The high stress levels in the southern abutment of K698;
(3) Low stresses adjacent to this highly stressed abutment; and
(4) High stresses in northern abutment of K711 stope.

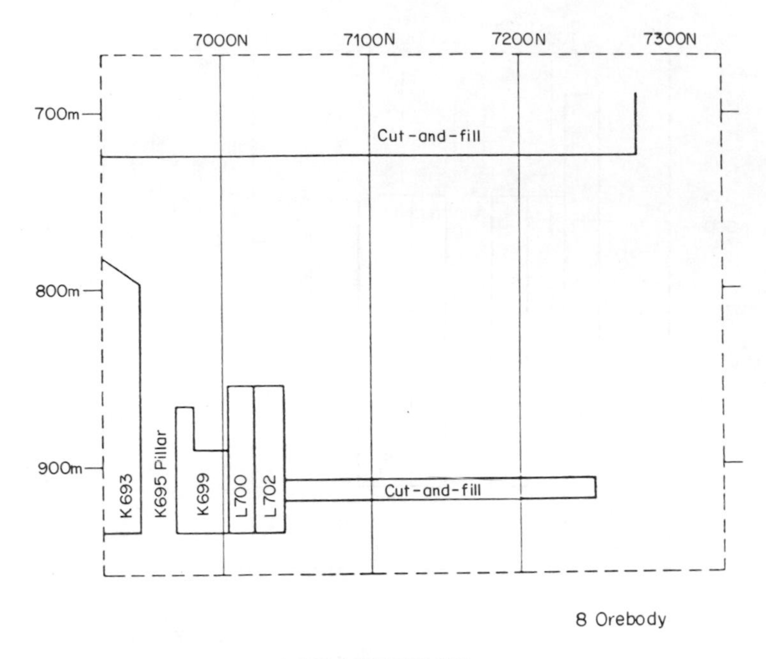


**LONGSECTION**

**Figure 10.9**
Continued

*8 Orebody*

(1) The high stresses in K695 pillar which are being transferred through from the southern abutment of K698 in 6 and 7 orebodies;
(2) The low stresses (shadowing) above the cut and fill area caused by extraction of K708;
(3) The high stresses transferred through from K705, above the Southern portion of the cut and fill; and K711;
(4) Shadowing above K699 and L700 caused by extraction of K698, K701 and K702 stopes in 6 and 7 orebodies.

Mount Isa Mines Ltd. is using the program NFOLD for mine design and has—during the past decade—gained enough experience to be able to correlate the magnitude of normal stress with expected ground condition.

## 10.3 MODELLING OF ELASTO–PLASTIC BEHAVIOUR

A third approach which can be used for the modelling of joints and elastoplastic behaviour is by coupling Boundary Element and Finite Element discretizations.

This allows to use joint Finite Elements, multilaminate, elastoplastic and viscoplastic models, etc. for the part of the rock mass which is jointed.

In addition the coupled procedure allows to consider:

(1) Ground support (rock anchors, shotcrete, etc.)
(2) Sequential excavation and construction.

These features can be best accommodated by the Finite Element Method. Coupling with Boundary Elements will then allow the infinite rock mass to be considered efficiently.

### 10.3.1 Coupled FEM/BEM Analysis-Theory

The FEM leads to a system of simultaneous equations which relate the displacements at all the nodes to *nodal forces*. In the BEM on the other hand a relationship between *nodal displacements* and *nodal tractions* is established.

Consider the two regions in Figure 10.11 which are connected at a common interface. Region $A$ is a Finite Element region and Region $B$ is an infinite Boundary Element region. For the FE Region $A$ the following system of equations is obtained.

$$\mathbf{K}^A \mathbf{a}^A = \mathbf{F}^A \tag{10.14}$$

where $\mathbf{a}^A$ and $\mathbf{F}^A$ are vectors of displacement and nodal forces at the interface of Region $A$ and $\mathbf{K}^A$ is a (condensed) stiffness matrix of the

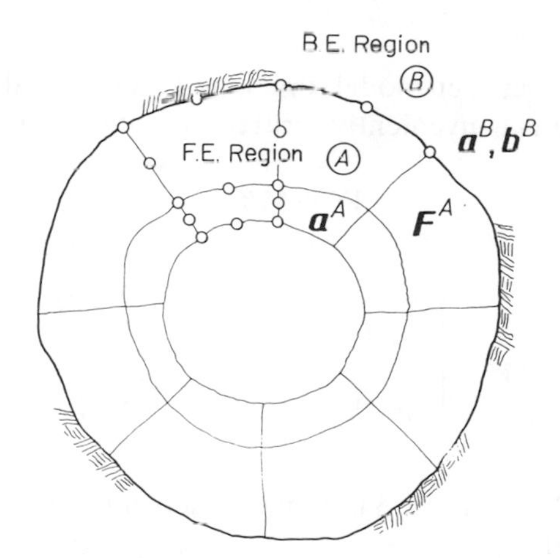


**Figure 10.11**
An excavation is surrounded by Finite Elements which are connected to a Boundary Element region: fully coupled analysis

interface. For the BE Region $B$ the system of equations is written as

$$\mathbf{A}^B\mathbf{a}^B = \tilde{\mathbf{B}}^B\mathbf{b}^B \tag{10.15}$$

where $\mathbf{a}^B$ and $\mathbf{b}^B$ are displacements and tractions at the interface of Region $B$. One approach is to rewrite equation (10.15) into the same form as equation (10.14), i.e. compute a stiffness matrix for the BE region. This requires the conversion of traction into nodal point forces.

A relationship between nodal traction and nodal point forces can be established using the principle of virtual work. The equation for the force $\mathbf{F}^e_l$ at node $l$ of an element $e$ which is subjected to distributed global tractions $\mathbf{t}$ on one side is given by (see Section 5.2.8):

$$\mathbf{F}^e_i = \int N_i \mathbf{t}\, \mathrm{d}S \tag{10.16}$$

where the integration is carried out over the element side. Using the usual interpolation of tractions from nodal values;

$$\mathbf{t} = \Sigma N_j \mathbf{t}^e_j \tag{10.17}$$

substituting into equation (10.16) and considering that $\mathbf{t}^e_j$, being traction values at the nodes of element $e$, can be taken outside the integral we obtain:

$$\mathbf{F}^e_i = \sum_{j=1}^{J} \left( \int N_i N_j \,\mathrm{d}S \right) \mathbf{t}^e_j \tag{10.18}$$

The relationship between nodal tractions $\mathbf{t}^e$ and nodal forces $\mathbf{F}^e$ on all nodes of element $e$ is conveniently written in matrix form

$$\mathbf{F}^e = \mathbf{M}^e\mathbf{t}^e \tag{10.19}$$

where

$$\mathbf{M}^e = \begin{bmatrix} N^e_{11}\mathbf{I} & \dots & N^e_{1J}\mathbf{I} \\ \vdots & & \vdots \\ N^e_{J1}\mathbf{I} & \dots & N^e_{JJ}\mathbf{I} \end{bmatrix} \tag{10.20}$$

In the above $\mathbf{I}$ is a $2 \times 2$ or $3 \times 3$ unit matrix depending on the dimensionality of the problem and

$$N^e_{ij} = \int_{S_e} N_i N_j \,\mathrm{d}S_e \tag{10.21}$$

To obtain the stiffness matrix of the Boundary Element region equation (10.15) is solved for the nodal tractions due to given displacements

$$\mathbf{b} = \mathbf{Q}\mathbf{a} \tag{10.22}$$

Where $\mathbf{Q}$ is a matrix of solutions due to unit displacements, i.e. in matrix notation

$$\mathbf{Q} = \tilde{\mathbf{B}}^{-1}\mathbf{A} \tag{10.23}$$

although in an efficient implementation the inverse of $\tilde{\mathbf{B}}$ need not be computed (Beer, 1983).

Next, the nodal traction values are converted into nodal point forces using equation (10.19).

$$\mathbf{F} = [\mathbf{M}]\mathbf{Q}\mathbf{a} \tag{10.24}$$

where $[\mathbf{M}]$ is a diagonally dominant matrix assembled from element contributions $\mathbf{M}^e$ (Beer, 1986). The stiffness matrix of the boundary element region is then given by

$$\mathbf{K}^B = [\mathbf{M}]\mathbf{Q} \tag{10.25}$$

An alternative way of determining $\mathbf{K}^B$ is to use the principle of minimum potential energy instead of virtual work.

The potential energy at the boundary of Region $B$ subjected to displacements, $\mathbf{a}$, at the nodes can be written as

$$\pi_B = \frac{1}{2}\mathbf{a}^T\mathbf{F} = \frac{1}{2}\mathbf{a}^T[\mathbf{M}]\mathbf{Q}\mathbf{a} \tag{10.26}$$

The minimum of $\pi_B$ is obtained by taking the derivative with respect to $\mathbf{a}$:

$$\frac{\partial \pi_B}{\partial \mathbf{a}} = \frac{1}{2}\left(\mathbf{Q}^T[\mathbf{M}]^T + [\mathbf{M}]\mathbf{Q}\right)\mathbf{a} \tag{10.27}$$

In the absence of external forces on the boundary of $B$:

$$\frac{\partial \pi_B}{\partial \mathbf{a}} = \mathbf{K}^B\mathbf{a} = 0 \tag{10.28}$$

where

$$\mathbf{K}^B = \frac{1}{2}\left(\mathbf{Q}^T\mathbf{M}^T + \mathbf{M}\mathbf{Q}\right) \tag{10.29}$$

is the stiffness matrix of the boundary of Region $B$.

Both derivations must be equivalent because there is an equivalence between virtual work and minimum potential energy principles.

In some cases the stiffness matrix of the Boundary Element region derived using virtual work is not symmetric especially when higher order Boundary Elements and coarse meshes are used.

Since it is convenient and economic to use symmetric equation solvers the second approach using potential energy is preferred by the writer.

At the interface between a Finite Element and a Boundary Element region conditions of compatibility and equilibrium have to be satisfied. Compatibility conditions are satisfied if the same shape functions are used to describe the variation of displacements on the Boundary and Finite Elements. Equilibrium is satisfied if the nodal point forces are equal and opposite. The consequence of this is that we can treat the BE region as a Superelement and assemble its stiffness matrix with the Finite Element stiffness matrices. Although $\mathbf{K}^A$ is a condensed stiffness matrix of the FE regions where all the degrees of freedom of the internal nodes have been eliminated it is not necessary to perform such a condensation explicitly. If a frontal solution program is used and the Boundary Element region is assembled last this condensation is implied in the solution process.

After the coupled system has been solved results can not only be obtained for the Finite Elements but also for points inside the BE region. Since the displacements are known the tractions can be computed on the interface using equation (10.22). With both tractions and displacements known on the interface, stresses can be computed either on the boundary or inside the BE regions using the formulae discussed in the last chapter.

When coupling BE and FE regions it is not necessary to couple all the nodes of the BE region. For example, for the problem of Figure 10.12 some of the BE nodes are on a free boundary. Therefore on a partially coupled

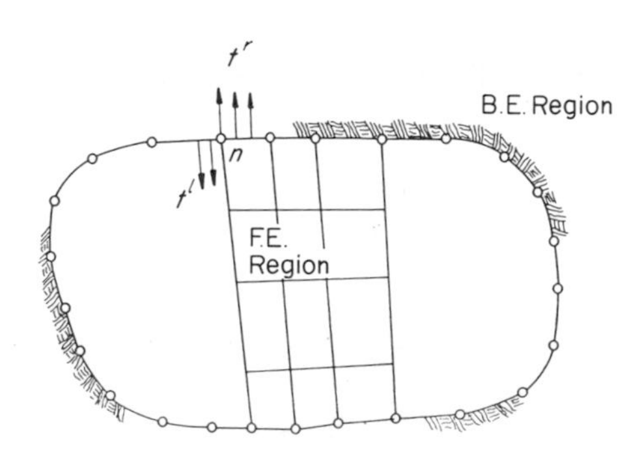


**Figure 10.12**
An excavation is modelled by Boundary Elements which are connected to a Finite Element region on some nodes of the mesh: Partially Coupled analysis

analysis the **Q** matrix contains solutions due to unit displacements at the interface only. Details of the implementation of partially coupled analysis are presented by Beer (1983, 1986). When using this method of coupling care has to be taken at the ends of the FE–BE interface. As shown in Figure 10.12 discontinuous tractions have to be assumed, i.e. the traction at one side of the corner is different to the one on the other side, i.e. for example $t^r$ and $t^l$ at node $n$ in Figure 10.12. This will influence the assembly of the **B** matrix such that only $\Delta\mathbf{U}$ from the element on the right side are assembled since the traction are unknown there. The $\Delta\mathbf{U}$ matrix of the element on the left side of the corner contributes to the right-hand side of the system of equations because the tractions are known there.

If there is a sharp corner inside the interface then there are more unknowns than equations because the traction on both sides of a corner are unknown. Lachat and Watson (1975) have shown that in this case additional equations can be written using the boundary stresses and Hookes' law. However, in 3D the programming is not trivial.

### 10.3.2 Practical applications

Three applications of the coupled FEM/BEM method are shown. One is related to tunnelling, the other two to mining. All problems are three dimensional and could not have been analysed as efficiently with either the BEM or FEM alone.

In tunnel construction the stress state near the tunnel face is of importance. In particular it is of interest to calibrate plane models which have been used to approximate the change in stress and displacement near the face.

The three-dimensional model of a tunnel is shown in Figure 10.13 and consists of four zones:

Zone A. Excavation region
Zone B. Shotcrete lining
Zone C. Nonlinear immediate region
Zone D. Elastic far region.

Zones A to C are described by Finite Elements (Zones A and C consists of 20-node isoparametric quadratic solid elements and Zone B of eight-node shell elements). The elastic region (Zone D) is modelled by eight-node isoparametric Boundary Elements which are connected to the outside of the Finite Element mesh. The virtually infinite extent of the boundary of Zone D is modelled by using infinite Boundary Elements.

Comparison of the three-dimensional results is made with approximate two-dimensional results using the 'stiffness reduction method' (Swoboda *et al.*, 1986) in Figure 10.14. Here the axial forces in the tunnel lining are shown for both 3D and 2D analysis. It can be seen that the approximate plane analysis underestimates the lining forces as computed by the more

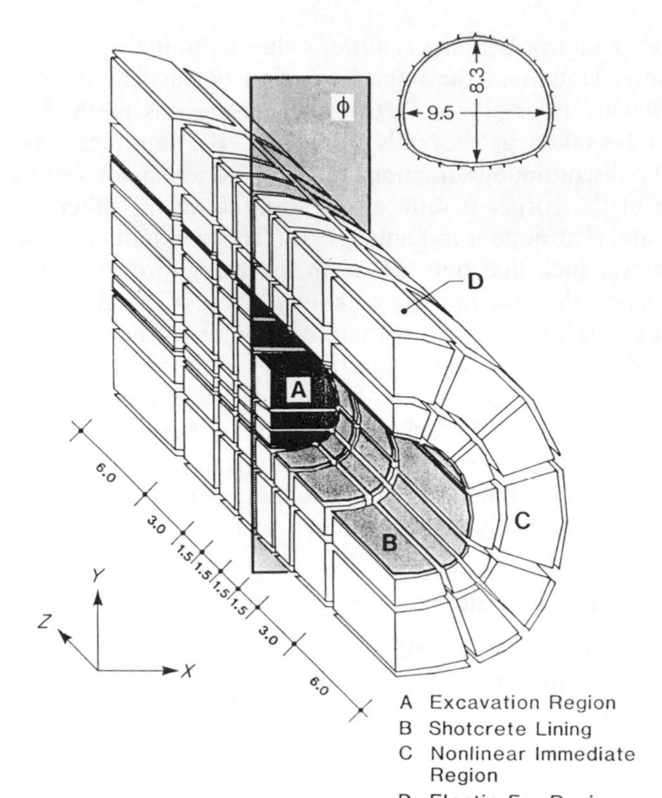


**Figure 10.13**
Three-dimensional coupled mesh for a tunnel excavation problem

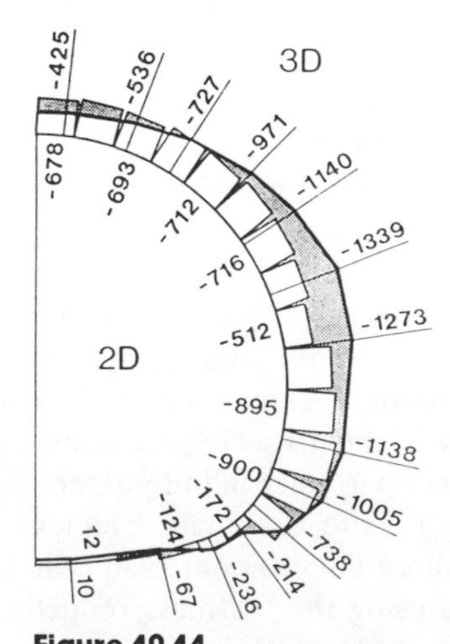


**Figure 10.14**
Comparison of lining forces for simplified plane and three-dimensional analyses

realistic 3D analysis. The coupled model used here results in substantial reduction in effort because the far field can be modelled accurately with the Boundary Elements and there is no need to extend the mesh a large distance away from the opening.

The analysis proceeded as follows:

The stiffness matrix of the Boundary Element region was computed. This had to be only done once. Subsequently the sequential excavation and installation of shotcrete was simulated by activating and deactivating Finite Elements. The rock mass in the vicinity of the tunnel, i.e. the region modelled by Finite Elements, was assumed to behave elasto–plastically and obey the Mohr–Colomb yield condition.

The program FINAL (Swoboda, 1986) to which subroutines from program BEFE for the stiffness computation of the BE region have been added was used for the analysis.

The second analysis relates to a partially coupled analysis of sequential mining at the Elura mine in Australia. Because the sequential excavation and the nonlinear rock mass behaviour had to be modelled the rock mass in new excavation region was modelled by Finite Elements. These Finite Elements are surrounded by Boundary Elements which represent the infinite, elastic rock mass. The Boundary Element mesh also includes the surfaces of existing excavations (Figure 10.15).

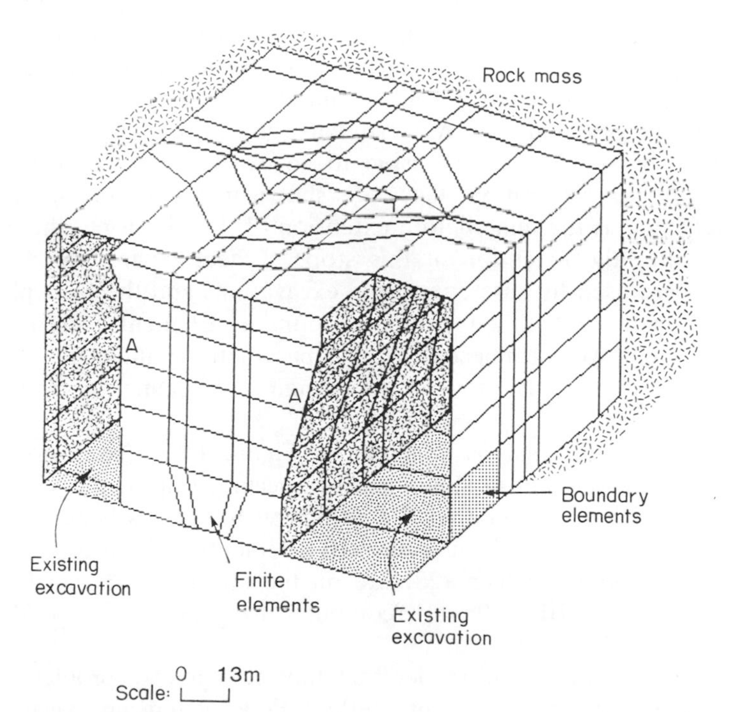


**Figure 10.15**
Coupled mesh of mining excavations

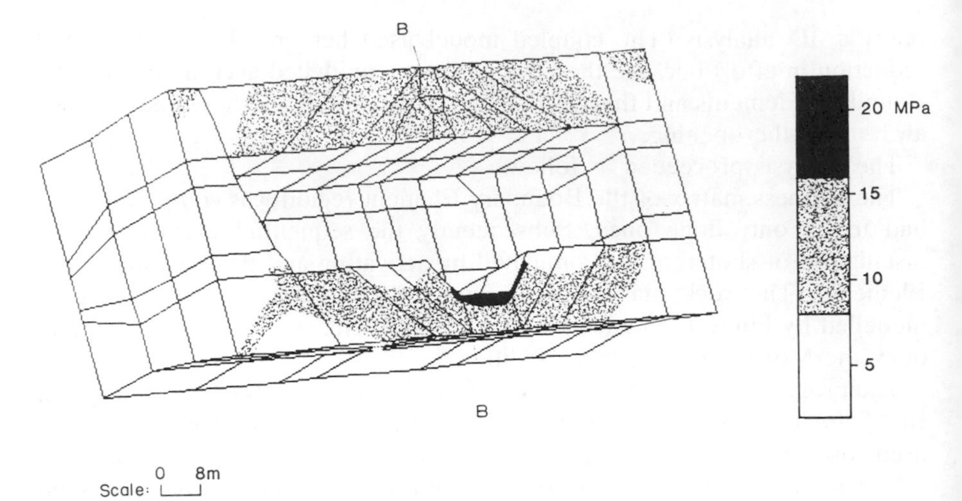


**Figure 10.16**
Deviatoric stress on section A–A for excavation stage 3

For the rock mass a multilaminate material model as explained in Section 7.4 was used but the nonlinear material behaviour is restricted to the Finite Element mesh. Figure 10.16 shows the distribution of $\sigma_{\max} - \sigma_{\min}$ on a plan section for a particular excavation stage. In Figure 10.17 (see colour plate) is shown contours of slip on bedding planes on cross Section B–B.

The last example shows an application of the modelling of faults using two BE regions connected by joint elements. The example is that of the Mt. Charlotte gold mine in Western Australia. The mine experienced a severe rock burst when an extension to an existing excavation was fired. The relevant two excavation stages are shown in Figure 10.18. The mesh on the left models the mine before extension of the G2 stope, the mesh on the right after the extension of this stope. Linear (four-node) boundary elements were used for discretizing the excavations and the fault plane. At the perimeter of the fault plane infinite Boundary Elements were used. The two BE regions were connected with isoparametric joint elements (Beer, 1985) which are similar to the Goodman-type elements discussed in Chapter 6.

One result of the analysis is shown in Figure 10.19 (see colour plate) which depict the amount of slip which has occurred up to the stage where G2 has been extracted and after the extension to G2 has been fired.

The mesh contains 763 linear Boundary Elements and has a total of 3369 degrees of freedom of which 1287 are on the interface. The analysis took 736 seconds on a CYBER 205 supercomputer using a fully vectorised version of program BEFE (Beer, 1986).

It may be obvious to the reader that the analysis of a problem of such complexity by the Finite Element Method alone would have resulted in a much more substantial effort in generating a mesh and much longer

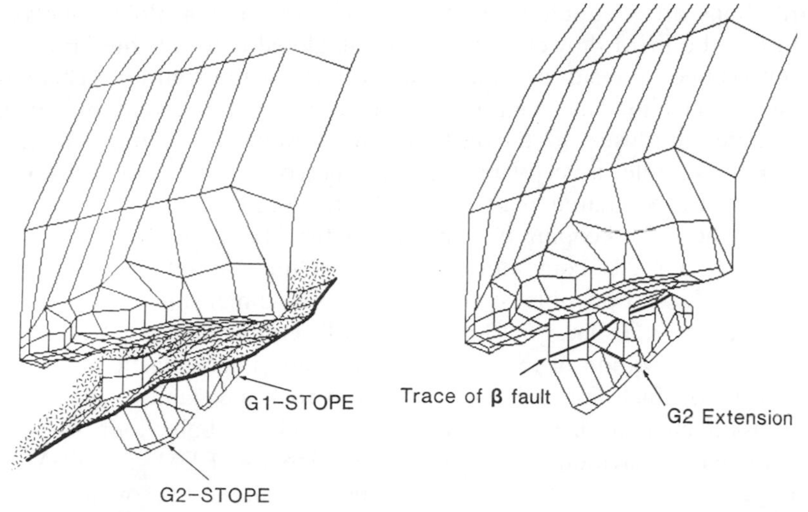


**Figure 10.18**
Boundary Element mesh for modelling excavations in a rock mass containing a fault

computation times. The required effort in generating a volume discretization for the infinite rock mass would make such a Finite Element analysis impracticable. On the other hand, Finite Elements are indispensible when dealing with nonlinear material, sequential excavation and ground support.

## 10.4 CONCLUSIONS AND FUTURE OUTLOOK

It has been attempted in this chapter to convey to the reader the power of the BEM in dealing with problems involving an essentially infinite domain as they occur in rock mechanics. Because the Boundary Element Methods include the effect of an infinitely distant boundary in the solution they are ideally suited for analysing excavations in an infinite rock mass. Furthermore, because only the excavation surfaces need to be discretized the methods are—especially when dealing with three-dimensional problems—much more user friendly.

The BEM can—as has been demonstrated—be explained using fundamental engineering principles: the principle of superposition or Betti's reciprocal theorem. It is not necessary to use either fancy mathematical derivations or tensor notation as has been perpetrated in the literature and which has scared off potential users of the method.

The implementation of the BEM is more complicated than that of the FEM because the integration of the singular fundamental solutions or Kernel functions require special programming techniques. Apart from this, the implementation of BEM follows a similar process to the FEM, i.e. element contributions are computed and then assembled into a global matrix. In contrast to the FEM, however, the resulting global matrix is fully populated and not symmetric resulting in a greater effort for the solution. This has in the past been claimed as a major drawback of the method. With the advance in computer hardware in the last decade, however, this does no longer apply. For example, a vector-processing supercomputer is more efficient in solving a fully populated matrix because the vector length can be made large and the book-keeping is dispensed with. This is particularly true for the CDC Cyber 205 computer which has a long optimal vector length. However, it is also possible to solve large three-dimensional analyses on a microcomputer in a somewhat longer but still reasonable computation time.

With a discretization of the surfaces alone only linear elastic analysis can be performed which limits the application in rock mechanics. Typically the rock mass is characterized by its nonlinear, inelastic behaviour. To model the nonlinear behaviour, it is best to combine the BEM and FEM in a 'marriage a la mode'. This, then, allows us to combine two or more elastic regions—which are discretized into Boundary Elements—within nonlinear (inelastic) regions which are discretized into Finite Elements. Here all the advantages of the FEM as explained in this book are retained, surely a mutually beneficial situation.

The three-dimensional examples presented here indicate the future trend in numerical methods in rock mechanics:

More emphasis on three-dimensional analyses;
Less man-hours spent in inputting data for such analysis;
Greater flexibility in presenting results of complex analyses.

CAD systems and graphical preprocessors are now available which take the pain out of specifying the mesh and drastically reduce the time spent in defining and inputting the mesh. Figure 10.20 (see colour plate) shows a three-dimensional Boundary Element mesh which was generated interactively from mine sections in just one day using a graphical preprocessor developed recently specifically for mining applications (Beer and Mertz, 1989). With user friendly postprocessors and the display of results in colour as shown previously numerical modelling will become a routine tool for civil and mining engineers. We have also found that using these pre- and postprocessors the likelihood of errors is greatly reduced and that analyses are error free the first time they are submitted.

Numerical modelling should be fun!

## REFERENCES

Banerjee, P. K., and Butterfield, R. (1981). *Boundary Element Methods in Engineering Science*, U.K.: McGraw-Hill.

Beer, G. (1983). 'Finite Element, Boundary Element and coupled analysis of

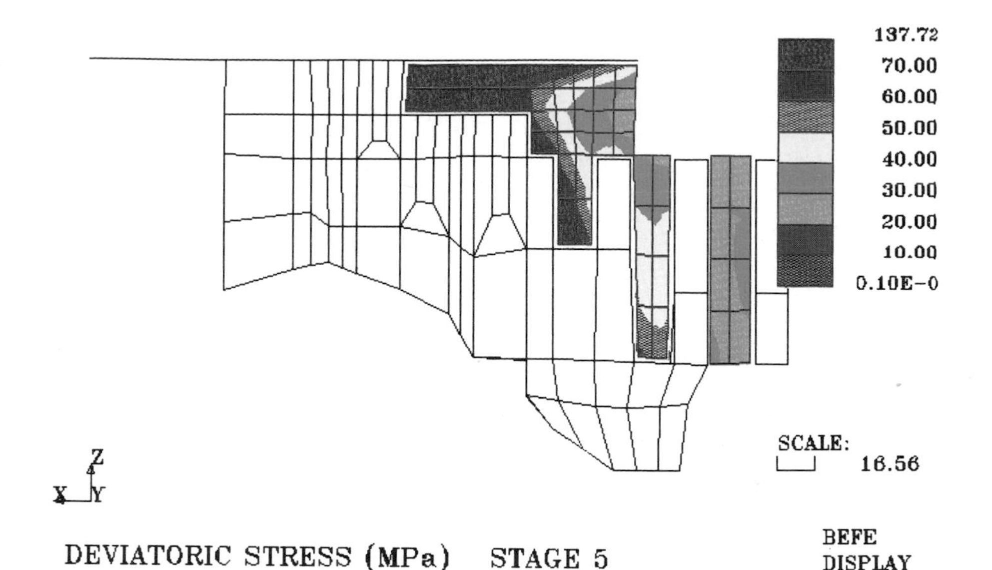


**Figure 10.3(a)**
Contours of $\sigma_{max} - \sigma_{min}$ plotted on 'dummy planes' inside rock pillars

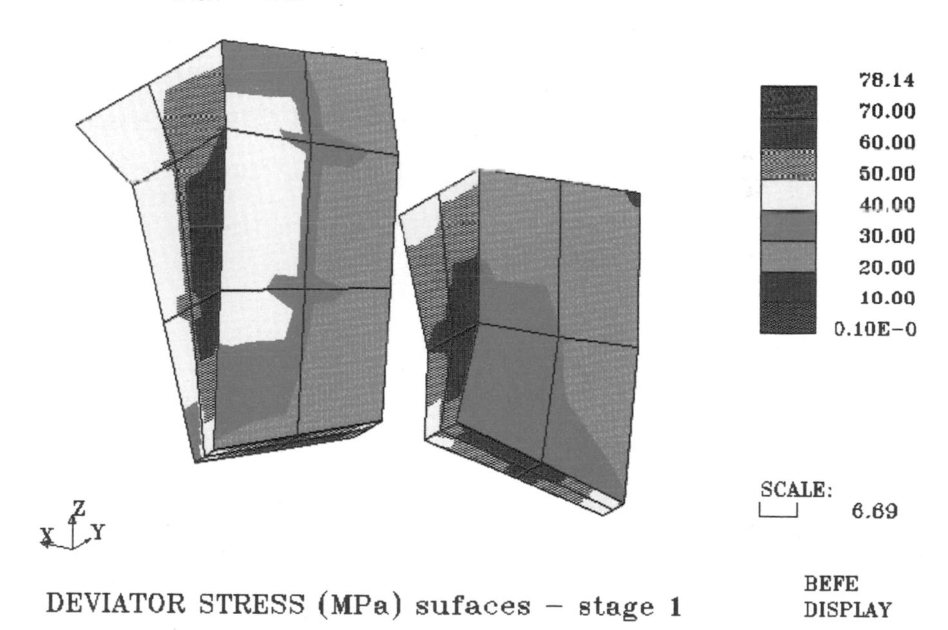


**Figure 10.3(b)**
Contours of $\sigma_{max} - \sigma_{min}$ on surfaces of excavations

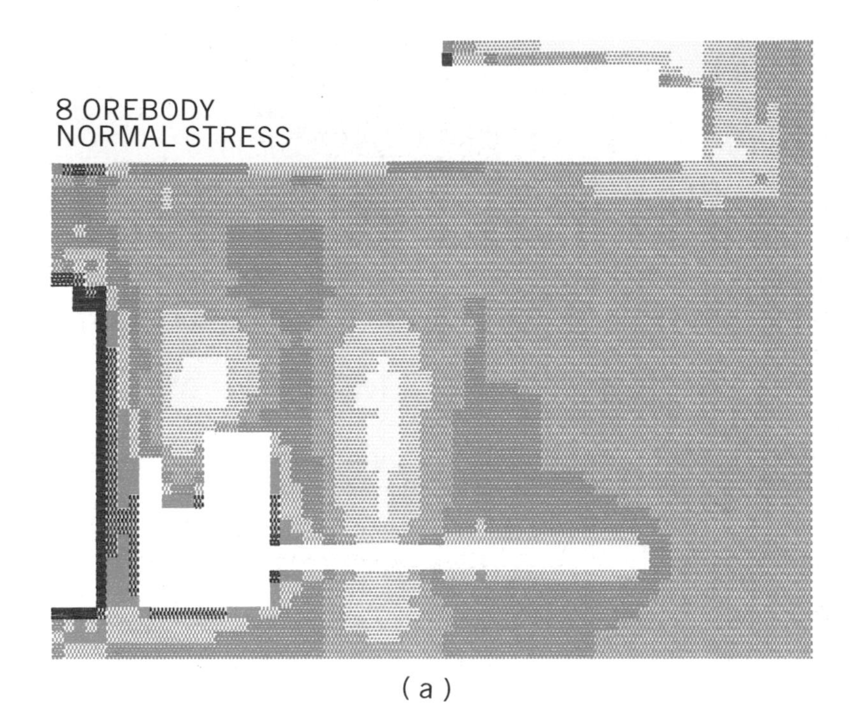


(a)

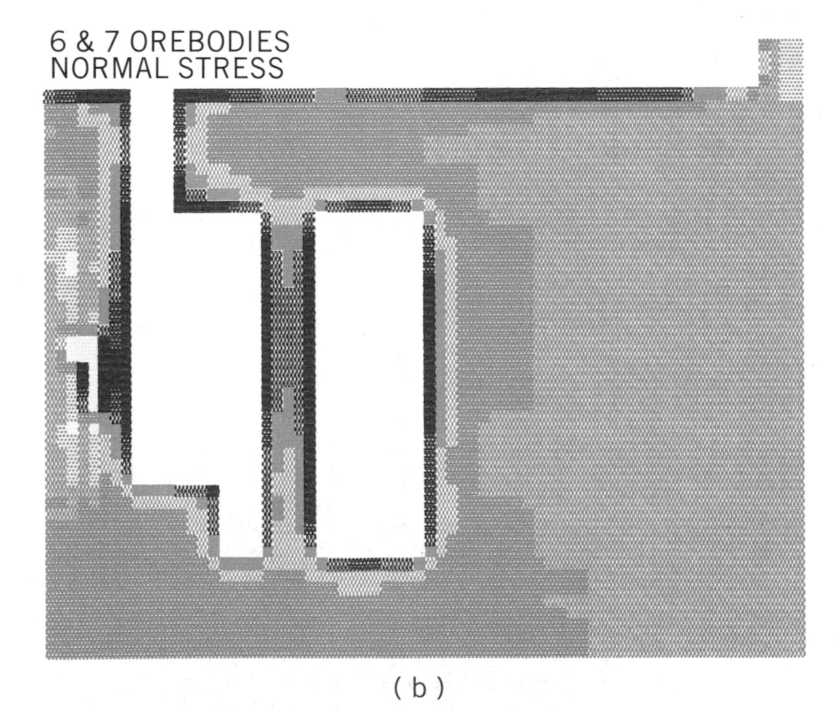


(b)

**Figure 10.10**
Contours of normal stress, $\underline{t}_{\underline{n}}$, across D.D. elements in the unexcavated (ore) region. (Courtesy Mount Isa Mines Ltd)

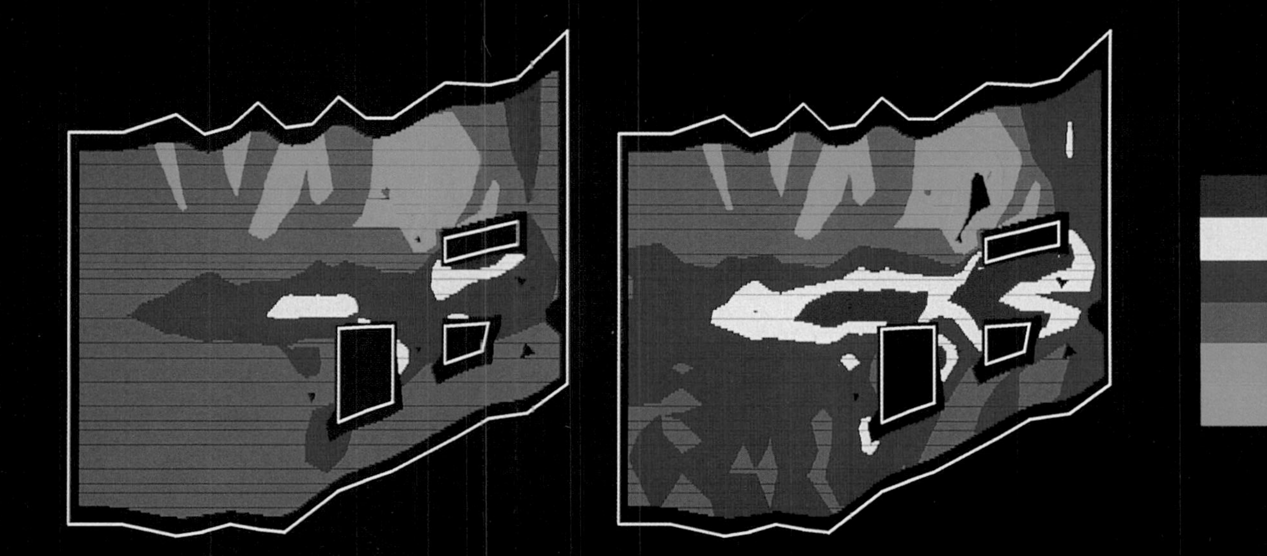


**Figure 10.19**
Contours of slip on the fault plane before and after G2 extension

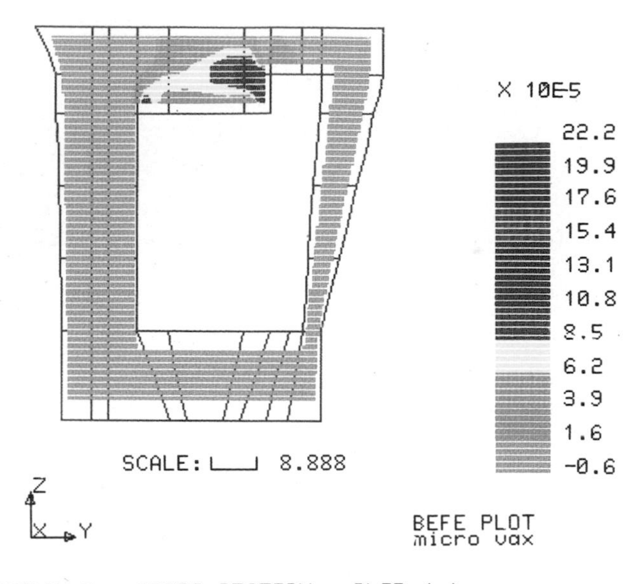


**Figure 10.17**
Slip on planes of weakness – section B-B

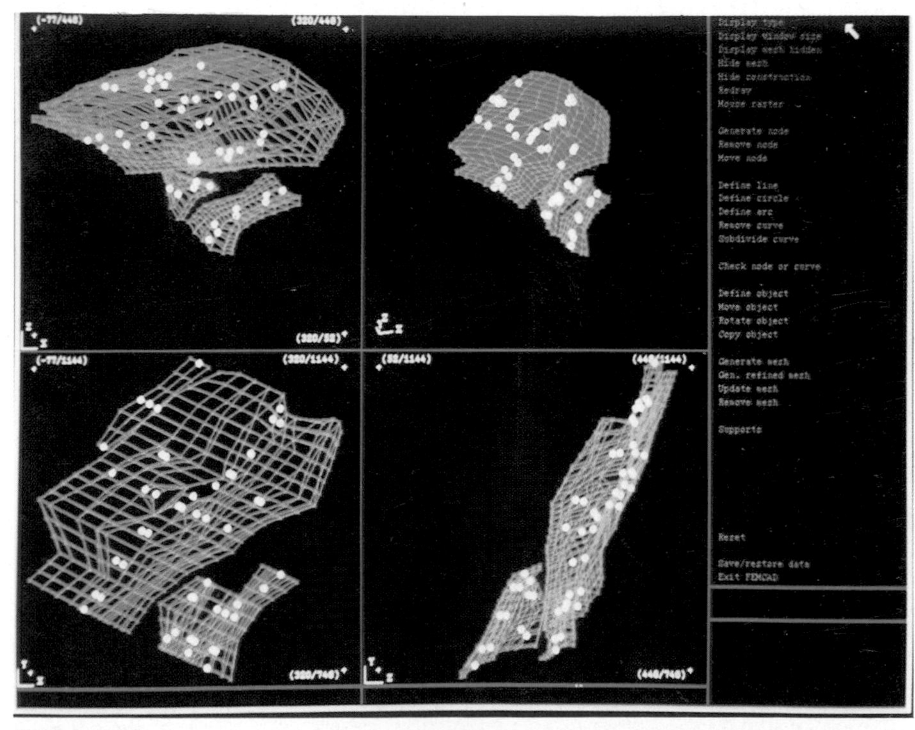


**Figure 10.20**
Perspective view of a three-dimensional boundary element mesh with approximately 600 elements generated using FEMCAD

unbounded problems in elastostatics.' *Int. j. numer. methods eng.*, **19**, 567–80.

Beer, G. (1985). 'An isoparametric joint element for finite element analysis.' *Int. j. numer. methods eng.*

Beer, G. (1986). 'Implementation of combined Boundary Element—Finite Element analysis with applications in geomechanics.' In: *Developments in Boundary Element Methods—4* (Banerjee and Watson, eds), Applied Science, London 191–226.

Beer, G. (1986). 'BEFE: Coupled Boundary Element—Finite Element Program.' In *Structural Analysis Systems—3* (Niku-Lari ed.), Oxford: Pergamon Press.

Beer, G. (1986). 'CYBER 205 Users: Simulating Rock Excavations.' *CSIRONET News*, No. 188, May.

Beer, G., and Mertz, W. (1989). 'A user friendly interface for numerical modelling in mining.' *Int. J. Rock Mech. Sci. and Geomech. Abstr.* (in press).

Beer, G., and Watson, J. O. (1989). 'Infinite Boundary Elements.' *Int. j. numer. methods eng.* (in press).

Crotty, J. H. (1982). User's Manual for Program BITEMJ—Two-dimensional Stress Analysis for Piecewise Homogeneous Solids with Structural Discontinuities, CSIRO, Division of Geomechanics. . . .

Crotty, J. M., and Wardle, L. J. (1985). 'Boundary integral analysis of piecewise homogeneous media with structural discontinuities.' *Int. J. Rock Mech. Min. Sci. and Geomech. Abstr.* **22**(6), 419–27.

Lachat, J. C., and Watson, J. O. (1975). 'A second generation boundary integral equation program for three-dimensional elastic analysis.' In T. A. Cruse and F. J. Rizzo (eds.), *Proc. ASME Conf. on Boundary Integral Equation Methods*, ASME, New York.

Sinha, K. P. (1979). 'Displacement discontinuity technique for analysis of stresses and displacements due to mining in seam deposits.' Ph.D. diss., University of Minesota.

Swoboda, G. (1986). FINAL—Finite element analysis linearer und Nichtlinearer Structuren Version 5.0. University of Innsbruck.

Swoboda, G. A., Mertz, W. G., and Beer, G. (1986). 'Application of coupled FEM–BEM analysis for three-dimensional tunnel analysis in *Boundary Elements*.' *Proc. Int. Conf.*, Beijing, China. Pergamon Press.

Tillmann, V. (1984). Ph.D. thesis, University of N.S.W.

Watson, J. O., and Cowling, R. (1986). 'Application of three-dimensional boundary element method to modelling of large mining excavations at depth.' *Proc. 5th Int. Conf. on Num. Meth. Geomech.*, Rotterdam: A. A. Balkema.

# 11 Discrete Element Method—Rigid Bodies

## 11.1 INTRODUCTION

The assumption of continuity is widely used as the basis for the idealization of many geomechanical materials including soils, rocks and ice. Powerful numerical techniques such as Finite Element, Finite Difference and Boundary Element are founded on the assumption of continuity and have proved extremely successful tools for the geotechnical engineer. However, there are many situations, especially in rock mechanics, when continuity cannot be assumed and other solution techniques must be utilized.

This chapter presents the basic ideas behind what is here termed the Discrete Element Method. While this term was originally applied in the 1960s to the Finite Element Method, it is used here to describe those techniques whose basic assumption is that of discontinuity between bodies and whose emphasis is on the solution of contact and impact between multiple bodies. Many variations of the Discrete Element Method exist within the area of rock mechanics, including rigid block method, Distinct Element Method and rigid block spring method and discontinuous deformation analysis. Other fields such as molecular dynamics, multibody dynamics and computer animation use methods which share much in common with discrete elements.

While the discrete element method has been presented in the literature as being distinct from the Finite Element Method it is shown here that the underlying basis can be cast into a familiar Finite Element form. Indeed, a standard Finite Element code can be embedded in a Discrete Element system (Barbosa and Ghaboussi, 1989) so that continuum deformations are handled using Finite Elements, while the contact and body interactions are handled using Discrete Element techniques.

In its most general form the Discrete Element Method is capable of analysing multiple interacting deformable continuous, discontinuous or fracturing bodies, undergoing large displacements and rotations. The method,

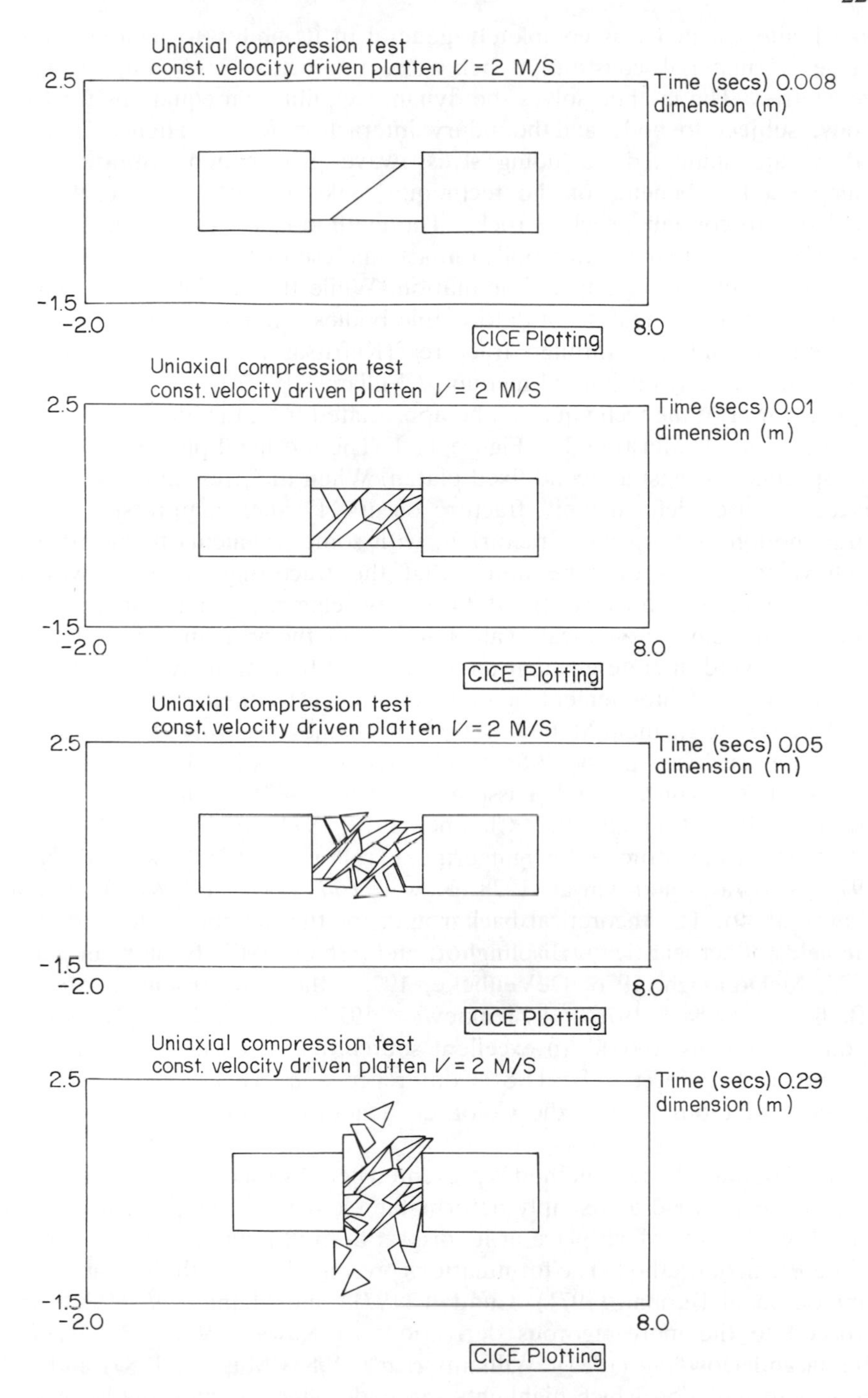


**Figure 11.1**
Illustration of fracturing specimen being crushed between two plattens

like Finite Elements, is completely general in its ability to handle a wide range of material constitutive behavior, interaction laws and arbitrary geometries. The method solves the dynamic equilibrium equations for each body, subject to body and boundary interaction forces. Highly dynamic effects are simulated, including stress wave propagation, vibration and damping. The benefits of the technique make it particularly useful for analysing discontinua, such as rocks. The method is also now being applied to such diverse areas as automobile impact, analysis of composites, machinery vibration, weapons effects and animation. While the emphasis here will be on the mechanical analysis of deformable bodies, the method has also been applied to fluid flow through fractures (Kafritsas *et al.*, 1985) and heat conduction through a jointed medium (Williams, 1985a).

The power of the technique can be appreciated from the idealized uniaxial compression test illustrated in Figure 11.1. The left-hand platen here drives the specimen against a second fixed platen. When the stress in the specimen exceeds a user defined limit, fracture occurs. Further compression causes large motion and repeated fracturing, with some fragments flying off with high velocity. It should be noted that the fracturing proceeds without user intervention; fracture orientation, new element generation and new interactions being automatically calculated within the program. This relatively straightforward discrete element analysis would be extremely difficult using a conventional finite element or finite difference technique.

The Discrete Element Method in rock mechanics has evolved from various disciplines such as, physics of particles (Ashley, 1967; Meirovitch, 1972; Likins, 1974; Huston and Passerells, 1979, 1980) from geomechanics (Burman, 1971; Cundall, 1971; Chappel, 1972, 1974; Byrne, 1974; Hocking, 1977) and from structural engineering (Trollope, 1967; Kawai, 1977a,b, 1979; Nakezawa and Kawai, 1978; Kawai *et al.*, 1978 and Watanabe and Kawai, 1980). The theoretical background for the method is derived from the fields of aeroelasticity (Bisplinghoff and Ashley, 1962; Bodley and Park, 1972; McDonough, 1976; DeVeubeke, 1976), the Finite Element Method (Bathe and Wilson, 1976 and Zienkiewicz, 1977), and the Finite Difference Method (Wilkins, 1969). An excellent source of references is contained in the proceedings of the 1st U.S. Conference and Workshop on Discrete Element Methods held at the Colorado School of Mines, October 17–19, 1989.

The Discrete Element method is presented here by following its progression from the early rigid and simply deformable elements to its present state of complex elements which place at its disosal the full range and power of the Finite Element Method. The formulations presented start with the engineering empiricism of Burman (1971), Cundall (1971) and Maini *et al.* (1978) and proceed to the more rigorous derivations of Kawai (1977), Belytschko, Plesha and Dowding (1984), Williams *et al.* (1985) Mustoe (1989) and the present formulation which highlights the underlying common mathematical basis of the Discrete Element and Finite Element Methods.

## 11.2 THE DISCRETE ELEMENT CONCEPT

In this section an overview of a typical Discrete Element program is given. Discrete Element techniques have been justified historically on an intuitive basis utilizing springs and masses. It is shown in Chapter 12 that many of these methods are equivalent to modal decomposition applied to a constant stress Finite Element. While modal decomposition has no significant advantages over a conventional Finite Element representation when implemented on a sequential computer, the modal approach does have advantages on the parallel architectures which have recently become available. For this and historical reasons, the solution of the dynamic equilibrium equations are presented here using a modal decomposition approach.

Only general forms of equations are presented in this present section and the reader is referred to later sections for detailed derivations.

In conventional analysis of a continuum using differential methods (Finite Elements and finite differences), a mesh of elements is constructed which are interconnected at the nodes and maintain displacement compatibility along the interelement boundaries. The system of equations is written for the entire assemblage of elements, inluding the constraints on the system. In the Discrete Element Method, each body communicates with surrounding bodies via boundary contacts which may change as a function of time, see Figure 11.2. (To emphasize the differences between discrete and finite elements, each body in the figure is composed of a single element. In practice a body can be composed of several Discrete Elements.) Figure 11.2(b) shows a typical Discrete Element mesh of a regularly jointed medium. In contrast to a Finite Element mesh there are no nodes common to more than one element. Thus, nodes 19, 24, 34 and 37, while possibly sharing the same spatial location, remain distinctly separate nodes within the solution algorithm. For clarity, the nodes are shown as spatially separate here. In a typical rock mechanics application the elements would represent the intact rock and the spaces between elements, the rock joints.

It should be noted that in Discrete Element analysis there is no restriction on where one element may make contact with another, and nodes may interact with nodes or nodes with element faces. The forces which are generated between contacting elements can be made to obey various interaction laws depending on the physical reality to be simulated. For example, interaction relationships for rock joints may include cohesion, dilation, damage to asperities, and stress dependent friction.

The connectivity or interaction of element to element is computed automatically in the algorithm, since the elements can rapidly change neighbouring elements upon motion or fracturing (Figure 11.2(a)).

The Discrete Element Method as described here can treat nonlinearities which may arise from large displacement, rotation, slip, separation and

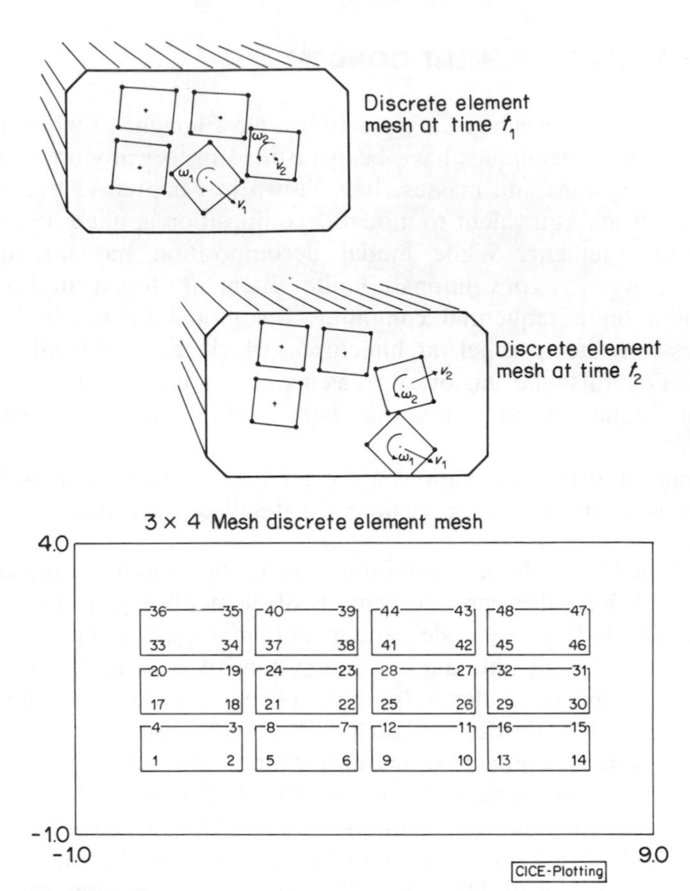


**Figure 11.2**
(a) illustration of possible Discrete Element connectivity with large motion between elements; (b) Discrete Element mesh of a continuum showing nodal points

material behaviour. The governing dynamic equilibrium equations for each discrete element can be written in the general form as:

$$[M]\{\ddot{u}\} + [C]\{\dot{u}\} + [K]\{u\} = \{f\} \tag{11.1}$$

in which $\{u\}$ is displacement, the superscript dots refer to differentiation with time, $[M]$ is the mass matrix, $[C]$ is the damping matrix, $[K]$ is the stiffness matrix, and $\{f\}$ is the applied load.

Consider the four-noded bilinear isoparametric finite element, illustrated in Figure 11.3. The motion and deformability of the element can be written as a superposition of the fundamental modes of the element. (This process is described in Chapter 12, Section 2.0.) In the case of small elastic incremental displacements, the zero frequency modes yield the rigid body motion (translation and rotation). The five higher-order modes provide the

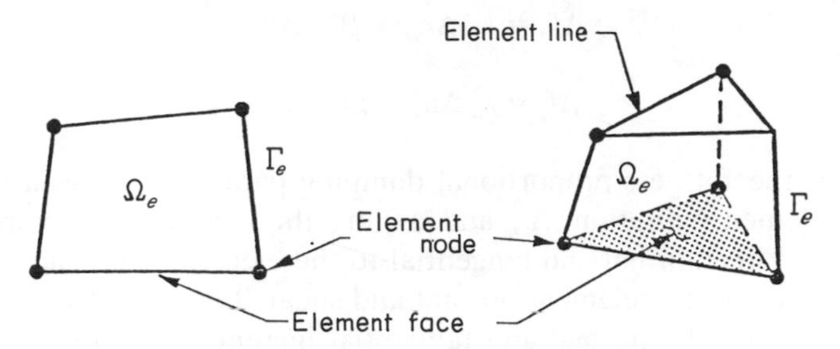


**Figure 11.3**
Two dimensional four-noded bi-linear isoparametric discrete element and six-noded three-dimensional wedge

deformational behaviour of the element. For higher-order elements, the required number of modes to describe the deformational behaviour of the element is increased. For example, in two dimensions the simple constant strain triangle requires three deformational modes while an eight noded parabolic isoparametric element requires 13 deformational modes to uniquely define its behaviour.

The orthogonality of the modes allows, in equation (11.1), the equations of motion and deformability of the element to be decoupled and written as separate equations. The equations, when integrated explicitly by the central difference method in time, yield equations of the general form

$$\{\ddot{u}\}^n = [M]^{-1}\left(\{f\}^n - [C]\{\dot{u}\}^{n-\frac{1}{2}} - [K]\{u\}^n\right) \tag{11.2}$$

$$\{\dot{u}\}^{n+\frac{1}{2}} = \{\dot{u}\}^{n-\frac{1}{2}} + \{\ddot{u}\}^n \Delta t \tag{11.3}$$

$$\{u\}^{n+1} = \{u\}^n + \{\dot{u}\}^{n+\frac{1}{2}} \Delta t \tag{11.4}$$

where $\Delta t$ is the time increment, the superscripts denote the time step number, and $[M]$, $[C]$ and $[K]$ are now the canonical forms of the mass, damping and stiffness matrices. The equations are derived in detail in Chapter 12, Section 2.0. Note that, when $[M]$ is diagonal, its inversion is trivial. The equation for the rigid body and the deformational modes of the Discrete Element are integrated separately in a staggered manner as described above. This staggered solution scheme is called an explicit–explicit partition, Bryne (1974). The stability limits for the time integration scheme are discussed in Belytschko *et al.* (1979) (see equation (11.36)).

The interaction of the Discrete Elements is computed from the relative motion of contacting elements which generate incremental interaction forces. For the simplest interaction law, employing a penalty function approach, the forces are given by

$$\Delta f_n = k_n \, \Delta u_n + \beta k_n \, \Delta \dot{u}_n \tag{11.5}$$

$$\Delta f_s = k_s \, \Delta u_s + \beta k_s \, \Delta \dot{u}_s \tag{11.6}$$

where $\beta$ is the stiffness proportional damping parameter necessary to damp element–element vibration, $\Delta f_n$ and $\Delta f_s$ are the incremental components of the contact force normal and tangential to the element–element interaction, $k_n$ and $k_s$ are the interelement normal and shear (tangential) stiffnesses, and $\Delta u_n$ and $\Delta u_s$ are the normal and tangential incremental components of the relative displacement between the discrete elements. The interelement stiffnesses may be real physical quantities as in the case of rock joints or artifacts used to enforce the correct interelement boundary constraints, as occurs in many impact problems. For physical rock joints the choice of stiffness $k_n$ and $k_s$ is governed by the measured joint stiffness (see Chapter 3 for detailed discussion). There is limited information in the available literature for damping on joints; however, data available indicates that this stiffness proportional damping is of the order of 1 to 2 percent prior to slippage on the joint. If the two elements are considered to be part of a continuum, then the stiffness is merely an artifact and must be computed to be sufficiently large to enforce compatibility between the Discrete Elements. In this case, the interaction stiffness relationships are analogous to a penalty function (Campbell, J. S., 1974) to enforce displacement compatibility in a weighted residual sense.

In a typical Discrete Element code the material properties and interaction relationships are input for each different material. Boundary conditions, applied loads, and body forces are required input as in any numerical analysis.

By stepping in time, the motion of each element is computed along with changes in loads due to gravity, buoyancy, drag, interaction, etc. The element moves and deforms under the action of these forces and moments. Incremental motion is computed with a relatively small time increment in the explicit–explicit scheme since dynamic wave propagation is computed within a body. This explicit–explicit scheme is utilized in fully dynamic problems while alternative implicit schemes (Belytschko *et al.*, 1984) are available for static analysis.

The incremental motion of the elements may give rise to interactions between elements and, thus, strain rates are generated within the elements. From the constitutive behaviour of the material, the incremental stresses are computed from the incremental strains. Depending on the constitutive behaviour of the material, the element will experience elastic or viscoplastic behaviour or possibly undergo fracturing.

Fracturing of a Discrete Element is relatively easy because all interactions with other elements occur through boundary forces. Thus, once the geometric bookkeeping of creating two elements from one has been accomplished, the standard Discrete Element algorithms take care of determining new contacts and forces.

While the structure for fracturing is relatively straight forward, the 'rules' for fracturing are not well developed, and at present detailed crack propagation with crack tip stress intensity factors has not been implemented.

## 11.3 RIGID BODY DYNAMICS

In this section the dynamic equation for rigid body motion are investigated as applied to a single Discrete Element acted upon by an arbitrary set of point forces. In general the forces depend on the motion of the body and lead to a set of highly nonlinear equations. The equations are linearized by adopting an incremental displacement approach using an updated Lagrangian basis.

The equations for rigid body motion are quoted here without reference. Interested readers are referred to Kibble (1966), Chorlton (1972), and Webster (1978).

### *Linear motion*

Newton's second law of motion can be written in the form

$$\frac{d\dot{u}_i}{dt} = \frac{\Sigma F_i}{m} \tag{11.7}$$

where $\dot{u}_i$ is the velocity in direction $i$, $F_i$ is a force component in direction $i$ and $m$ is the mass of the body.

Using a central difference time stepping scheme for the left-hand side, for time interval $t^{n-\frac{1}{2}}$ to $t^{n+\frac{1}{2}}$ gives

$$\left(\frac{d\dot{u}_i}{dt}\right)^n = \left(\frac{\dot{u}^{n+\frac{1}{2}} - u^{\cdot\, n-\frac{1}{2}}}{\Delta t}\right) \tag{11.8}$$

Rearranging equations (11.7) and (11.8)

$$\dot{u}_i^{n+\frac{1}{2}} = \dot{u}_i^{n-\frac{1}{2}} + \frac{\Sigma F_i^n}{m}\Delta t \tag{11.9}$$

Thus the velocity at time step $n + \frac{1}{2}$ is completely determined in terms of known quantities at time step $n$, and $n - \frac{1}{2}$.

The displacements $u_i$ at time step $n + 1$ may now be expressed in terms of previously calculated displacements as:

$$u_i^{n+1} = u_i^n + \dot{u}_i^{n+\frac{1}{2}}\Delta t \qquad (11.10)$$

Given the initial position, velocity and forces acting on a body at time step $t^n$, equations (11.9) and (11.10) allow us to calculate the acceleration, velocity and position (displacement) at time step $t^{n+1}$.

### *Angular rotation*

As equation similar to equation (11.7) can be written for angular rotation.

$$\frac{\mathrm{d}\dot{\theta}_i}{\mathrm{d}t} = \sum P_j\,(I_{ij})^{-1} \qquad (11.11)$$

where $\dot{\theta}_i$ is the angular velocity, $P_j$ is the applied moment in direction $j$ and $I_{ij}$ is a component of the moment of inertia tensor for the body.

Following the same process as for linear motion the following equations are derived

$$\dot{\theta}_i^{n+\frac{1}{2}} = \dot{\theta}_i^{n-\frac{1}{2}} + \sum P_j^n\,(I_{ij}^n)^{-1}\,\Delta t \qquad (11.12)$$

$$\theta_i^{n+1} = \theta_i^n + \dot{\theta}_i^{n+\frac{1}{2}}\,\Delta t \qquad (11.13)$$

where $\theta_i^{n+1}$ is the angular rotation at time step, $n + 1$.

At this point it is well to pause and notice the similarity between equation (11.7) and (11.11) and to introduce a terminology which will be used later in describing more general applications of the technique.

Both $u_i$ and $\theta_i$ are particular cases of generalized coordinates which define the state of the body; $F_i$ and $P_i$ are generalized forces, and $m$ and $I_{ij}$ are generalized masses which influence the body's acceleration with respect to the generalized coordinate. In Section 12.0 equations for the deformational modes of elements, are derived which closely resemble these rigid body equations.

### *Damping*

When mass proportional damping is introduced by setting $c = \alpha m$ into equation (11.7) the following equation results:

$$\ddot{u} + \alpha\dot{u} = F/m \qquad (11.14)$$

Using the central difference approximation it follows that

$$\dot{u}^{n+\frac{1}{2}} = \left(a\dot{u}^{n-\frac{1}{2}} + \frac{\Delta t}{m}\sum F^n\right)b \qquad (11.15)$$

where

$$a = \left(1 - \frac{\alpha\,\Delta t}{2}\right)$$

and

$$b = \left(1 + \frac{\alpha\,\Delta t}{2}\right)^{-1}$$

The equation below can be similarly derived for angular rotation

$$\dot{\theta}^{n+\frac{1}{2}} = \left(\mathrm{a}\dot{\theta}^{n-\frac{1}{2}} + \Delta t \sum P_j^n \,(I_{ij}^n)^{-1}\right) b \qquad (11.16)$$

An understanding of the effect of damping on a simple one-dimensional system whose equation is given by $m\ddot{u} + c\dot{u} + ku = 0$ can be derived as follows:

The general solution of a one dimensional damped oscillator is given by

$$u = A\,\mathrm{e}^{\lambda t}$$

where

$$\lambda = \frac{-c \pm \sqrt{c^2 - 4\mathrm{km}}}{2m} \qquad (11.17)$$

The behaviour of the oscillator is dependent on the relative values of $c$, $k$, and $m$ and is summarized below:

(i) oscillatory (underdamped) behaviour $c^2 < 4$ km
(ii) critically damped $c^2 = 4$ km
(iii) overdamped behaviour $c^2 > 4$ km

and

(iv) free vibration $c = 0$

*Note*: that free vibration occurs at an angular frequency

$$\omega_0 = \sqrt{(k/m)}, \quad \text{linear frequency}\, f_0 = \frac{\omega_0}{2\pi} \qquad (11.18)$$

For mass proportional damping where $c = \alpha m$, critical damping is given by $\alpha_0 = 2\omega_0 = 4\pi f_0$. It should be noted that most real engineering systems have many modes of vibration and that the choice of a single value for $\alpha$

will critically damp only one frequency, causing others to be underdamped or overdamped.

When seeking static solutions to linear problems dynamic relaxation techniques may be employed. Underwood (1980) presents a comprehensive review of the subject and details the theory behind such techniques as density scaling and automatic damping schemes (Papadrakakis, 1981).

In the explicit algorithm the equations of rigid body motion equations (11.11) and (11.14) are integrated separately in a staggered manner using an explicit central difference technique. The stability limit for a time increment $\Delta t$, employing an explicit–explicit partition can be shown to be (Belytschko, Yen, and Mullen, 1979)

$$\Delta t \leqslant 2\left\{(1+\mu^2)^{\frac{1}{2}}-\mu\right\}/\omega_{\max} \tag{11.19}$$

where $\omega_{\max}$ is the maximum frequency of the combined system (rigid body motion and element deformability), and $\mu$ is the fraction of critical damping at frequency $\omega_{\max}$. This stability limit has been proven only for linear systems; however, considerable experience indicates that it can also be applied to nonlinear systems.

### *Update of coordinates*

In two dimensions the rigid body unknowns are the element centroid displacements $u_x$ and $u_y$, and the angular rotation about the centroid $\theta$, along with their corresponding derivatives.

Unlike finite elements these unknowns are derived with respect to the element centroid. The nodes in the discrete element define the element geometry and in some cases (2D analysis) the contact points between elements where interaction forces are applied. However, an element can have any number of nodes defining the geometry without necessarily affecting the number of unknowns and the three dimensional discrete element program CICE (Hocking *et al.*, 1985a,b,c) and its successor DECICE has elements with as many as 120 nodes (Figure 11.3). Given the initial element geometry, the position of each node can be updated as the element ranslates and rotates as follows:

Compute the new element nodal velocities and coordinates via

$$v_i^{n+\frac{1}{2}} = V_i^{n+\frac{1}{2}} + e_{ijk}\omega_j^{n+\frac{1}{2}}\left(x_k^n - X_k^n\right) \tag{11.20}$$

$$x_i^{n+1} = x_i^n + v_i^{n+\frac{1}{2}}\,\Delta t \tag{11.21}$$

where

| | |
|---|---|
| $v_i^{n+\frac{1}{2}}$ | is the velocity of the element node at $t^{n+\frac{1}{2}}$, |
| $x_i^n$ | the global coordinates of the element node at $t^n$, |
| $X_i^n$ | the global coordinates of the element centroid at $t^n$, |
| $V_i^{n+\frac{1}{2}}$ | the velocity of the element centroid at $t^{n+\frac{1}{2}}$ |

and

| | |
|---|---|
| $e_{ijk}$ | the Levi–Civita permutation tensor density, defined by |

$$\begin{aligned} e_{ijk} &= 1 \text{ if}_{i,j,k} \text{permute cyclically} \\ &= +1 \text{ if}_{i,j,k} \text{permute anticyclically} \\ &= 0 \text{ if any indices}_{i,j,k} \text{ are equal.} \end{aligned}$$

## 11.4 CONTACT DETECTION GEOMETRIC MODELLING AND OBJECT REPRESENTATION

The essence of contact detection is the determination of the surfaces and volumes of intersection between two bodies. Penalty function and Lagrangian multiplier techniques can be used to enforce contact conditions, once the geometry of the contact has been determined. When only two or three bodies are being analysed the contact surfaces can be input by the analyst at problem setup, and only minimal checking is required by the program to isolate the exact parts of the surfaces in contact. (This is the present mode of operation of most FE codes.) With several hundred bodies present, it is impractical or impossible to determine all possible contact surfaces at problem setup. In this case the program itself must determine contacts.

It is relatively easy to construct a contact algorithm that checks every face of every element, with every face of every other element. Unfortunately, the number of checks required is proportional to $N^2$, where $N$ is the number of faces, and if each check is conducted in sufficient detail to determine the exact geometrical relationships of the bodies, the algorithm is extremely time consuming.

For this reason contact detection algorithms scale very badly with increasing numbers of bodies. One aim of present research is to find the algorithms which scale better, say as $N \log_2 N$ or possibly linearly with $N$.

An example of the complexity of the possible interactions of even simple geometric bodies is illustrated in Figure 11.4 which shows two cubes intersecting. Although the penetration of the bodies is exaggerated for clarity, the geometries are those that must be resolved by the contact algorithm.

The essential problem, therefore, in contact detection is to devise a robust algorithm applicable to a wide range of contact problems which executes

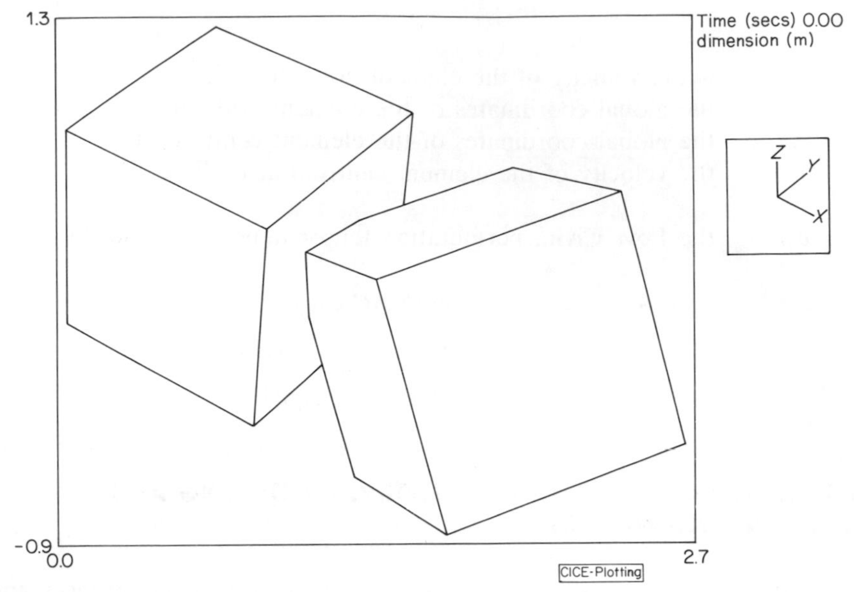


**Figure 11.4**
Examples of interaction geometries between two cubes

efficiently on today's computers. In order to achieve this it is necessary to consider the manner in which computer programs represent objects and utilize geometric data.

It is important to realize that the way in which we represent objects, in say Finite Element codes, is only one of many possible object representation schemes. Recent developments in object representation show that other methods, particularly those based on superquadrics (Williams and Pentland, 1988) may have significant advantages for contact detection.

The representation of physical objects, in standard numerical modelling procedures, consistss of discretizing the boundary of the object, and usually the interior, into a number of elements where the element geometry is defined by a number of nodal coordinate points. Element surfaces and the spatial relationship between the elements is defined by an ordered array of nodes called the element connectivity array. In the case of Finite Elements, the connectivity is used to determine how the elements are joined together so that the stiffness matrix can be constructed. The element connectivity is input by the user at the start of the analysis and is usually assumed to remain constant thereafter.

In Discrete Element analysis, while the initial connectivity must still be input, so that the elements belonging to each body are defined, the connectivity between bodies is assumed to change throughout problem execution. Thus, bodies in contact at the start of the analysis may loose old contacts and develop new ones. The Discrete Element program must be capable of continuously updating element contacts.

In general terms it is said that the program is capable of 'recognizing' the geometric environment. The algorithms required to achieve this are similar to those encountered in computer vision. For example, the program must recognize that two elements have made contact, in the same way that the software, controlling a robotic arm, must recognize when the hand is about to make contact with an object it is required to grasp.

Recognizing geometrical relationships requires that the program be provided with definite laws for decision making. These laws and algorithms for their implementation are a major stumbling block of present discrete element programs.

In the absence of a formal theory of geometric modelling, discrete element codes have relied on a wide range of empirical rules. These rules are discussed in the section on geometric rules of inference. At present the algorithms utilizing these rules are inefficient and extremely time consuming.

The following discussion details some of the basic data structures which can be utilized in a discrete element code for contact detection.

### *Graph based and cell decomposition models*

The familiar Finite Element representation of an object which uses elements and nodes is a particular instance of a 'graph-based model'. While most numerical modelling codes, reported to date, use this representation, other representations are possible. The cell decomposition method, based on quadtrees or hextrees in two dimensions and on octrees in three dimensions, provides an alternative method of describing an object. Since these methods are already being used successfully in some fields, including that of automatic mesh refinement, we shall take a brief diversion to outline their advantages.

*Cell decomposition for object representation.* Graph-based methods address objects using points relative to some coordinate system. Cartesian $(x, y)$ coordinates and latitude–longitude are both examples of systems for addressing points. In contrast the cell decomposition method assigns addresses to areas, rather than single points. These areas can be as large or as small as are required.

In two dimensions a hierarchical structure based on hexagons is particularly efficient, and gives rise to the name 'hextree'. The first level of a hextree is formed by taking a single hexagon and its six neighbours (Figure 11.5). A second level is formed by taking first level aggregate and its six first-level neighbours. This process continues until the required level of representation is reached. An example of the level of detail attainable is given by the fact that a 20-level system can represent the whole surface of the earth to within 10 cm resolution.

The address of any particular hexagon is composed of a series of digits. The lowest order digit identifies which of the seven cells (0–6) in the first level aggregate is being addressed. The second digit identifies in which of the first level cells the second level resides, and so on.

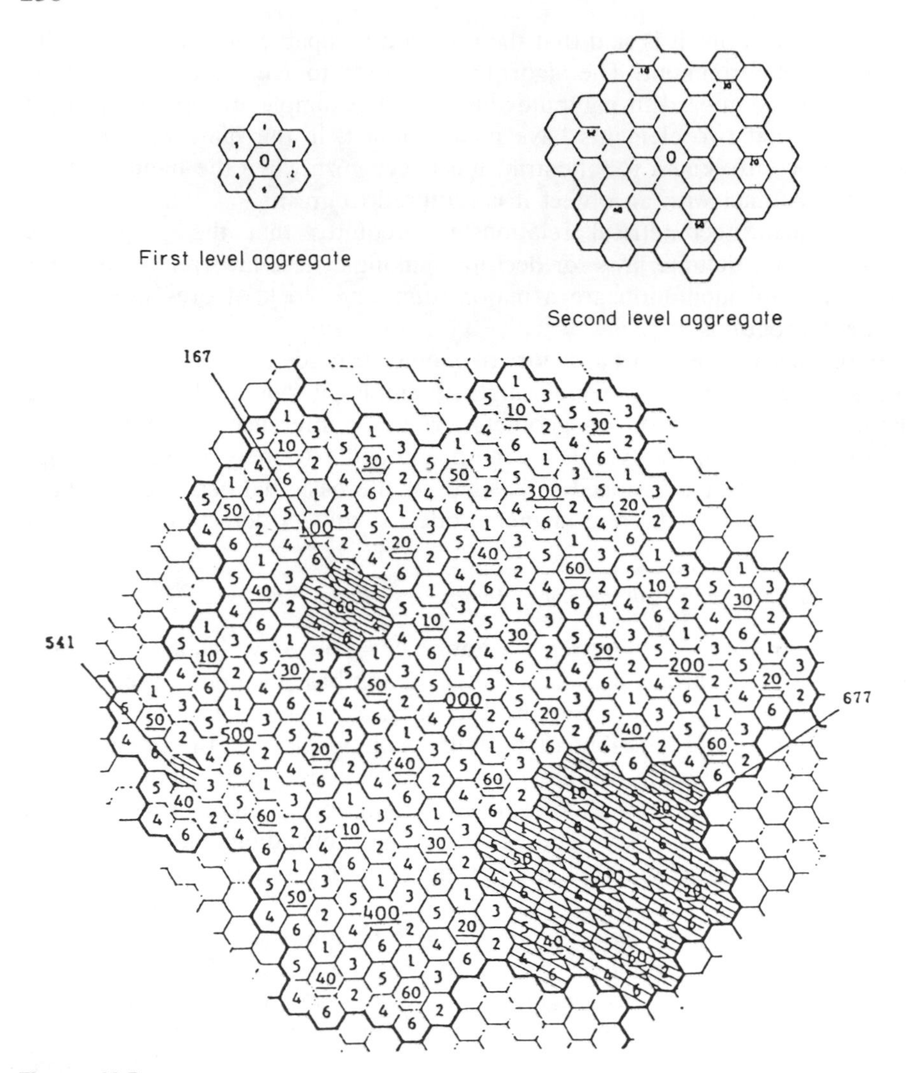


**Figure 11.5**
Three levels of hextree aggregate illustrating addressing scheme and addresses for 541, 167, 677. (Note a 7 indicates all cells at the level being addressed)

This system of representing spatial data has particular advantages. Firstly, all addressing involves integer arithmetic. Secondly, only those areas where objects exist need be represented in the tree. Both these features make retrieval and manipulation of spatial data extremely efficient.

*Graph-based data structures.* In the graph-based model, each discrete element is represented by arrays of nodal, edge, and face data, which contain spatial coordinate and topological relationships. The faces are bounded by specific edges and by other adjacent faces. The edges are bounded by nodes which may be shared with other edges. Two types of information are stored—the pointers defining the element's topology (usually ordered nodal arrays), and the coordinates of the nodal points. During program execution both the pointer information and the coordinate data must be updated.

Although all topological information can be derived from just the connectivity data stored in a Finite Element program, numerous calculations may be required to do so. If geometric data are accessed frequently it is more efficient to store the information in data structures than to recalculate it from element connectivity data. A typical geometrical modelling representation of a tetrahedron is shown in Figure 11.6. From the data stored, it is immediately known that node N1 is connected by edges to nodes N2, N3, and N4, is part of edges E1, E2, and E3 and is common to faces F1, F2, and F3. This representation contains redundant information that increases storage requirements but speeds up search algorithms.

Another example of storing geometrical data is given in Figure 11.7, which details the connectivity matrix for nodes and faces of a wedge shaped

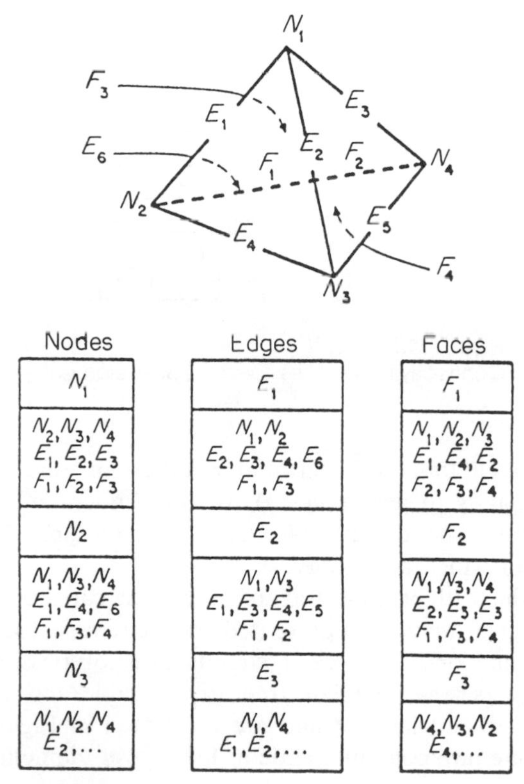

$N_1$
$F_3$
$E_1$
$E_3$
$E_6$
$E_2$
$F_2$
$N_4$
$F_1$
$N_2$
$E_5$
$E_4$
$F_4$
$N_3$
Nodes
$N_1$
$N_2, N_3, N_4$
$E_1, E_2, E_3$
$F_1, F_2, F_3$
$N_2$
$N_1, N_3, N_4$
$E_1, E_4, E_6$
$F_1, F_3, F_4$
$N_3$
$N_1, N_2, N_4$
$E_2, \ldots$
Edges
$E_1$
$N_1, N_2$
$E_2, E_3, E_4, E_6$
$F_1, F_3$
$E_2$
$N_1, N_3$
$E_1, E_3, E_4, E_5$
$F_1, F_2$
$E_3$
$N_1, N_4$
$E_1, E_2, \ldots$
Faces
$F_1$
$N_1, N_2, N_3$
$E_1, E_4, E_2$
$F_2, F_3, F_4$
$F_2$
$N_1, N_3, N_4$
$E_2, E_3, E_3$
$F_1, F_3, F_4$
$F_3$
$N_4, N_3, N_2$
$E_4, \ldots$


**Figure 11.6**
A graph based model of a tetrahedral element

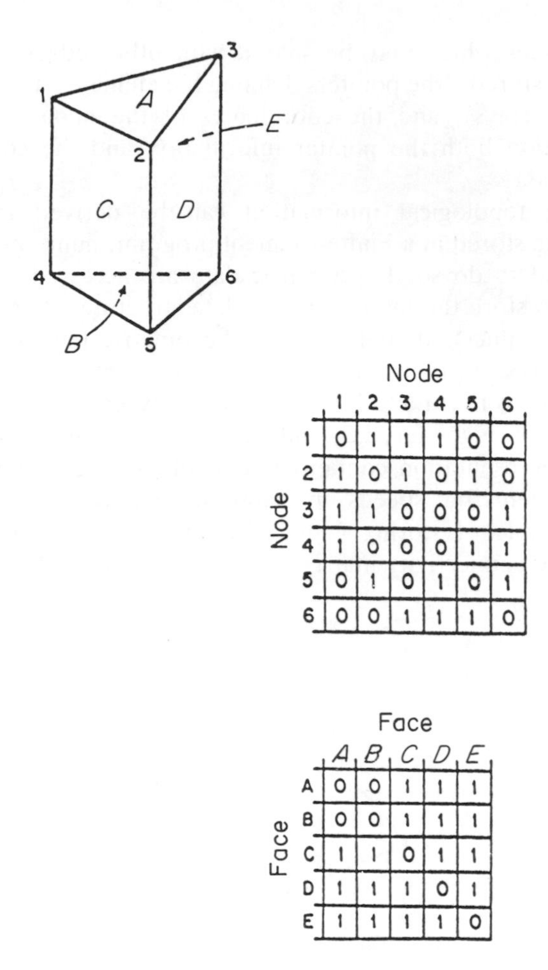


| Node | 1 | 2 | 3 | 4 | 5 | 6 |
|---|---|---|---|---|---|---|
| 1 | 0 | 1 | 1 | 1 | 0 | 0 |
| 2 | 1 | 0 | 1 | 0 | 1 | 0 |
| 3 | 1 | 1 | 0 | 0 | 0 | 1 |
| 4 | 1 | 0 | 0 | 0 | 1 | 1 |
| 5 | 0 | 1 | 0 | 1 | 0 | 1 |
| 6 | 0 | 0 | 1 | 1 | 1 | 0 |

| Face | A | B | C | D | E |
|---|---|---|---|---|---|
| A | 0 | 0 | 1 | 1 | 1 |
| B | 0 | 0 | 1 | 1 | 1 |
| C | 1 | 1 | 0 | 1 | 1 |
| D | 1 | 1 | 1 | 0 | 1 |
| E | 1 | 1 | 1 | 1 | 0 |

**Figure 11.7**
Connectivity matrices for a wedge element

element. The search time to determine whether an edge exists between nodes $i$ and $j$, that is to access $a(i, j)$, is independent of the number of vertices and edges. The disadvantage is that it requires $N * N$ storage locations, many of which are empty.

The way in which algorithms traverse the data can be represented by graphs. The traversal may be represented by a circuit if the start and end points are the same or by a tree if circuits are not possible. Systematic algorithms to traverse the complete tree are available and are classified as preorder, postorder and inorder and are illustrated in Figure 11.8. These algorithms translate into computer code as loops over elements, faces, edges and nodes.

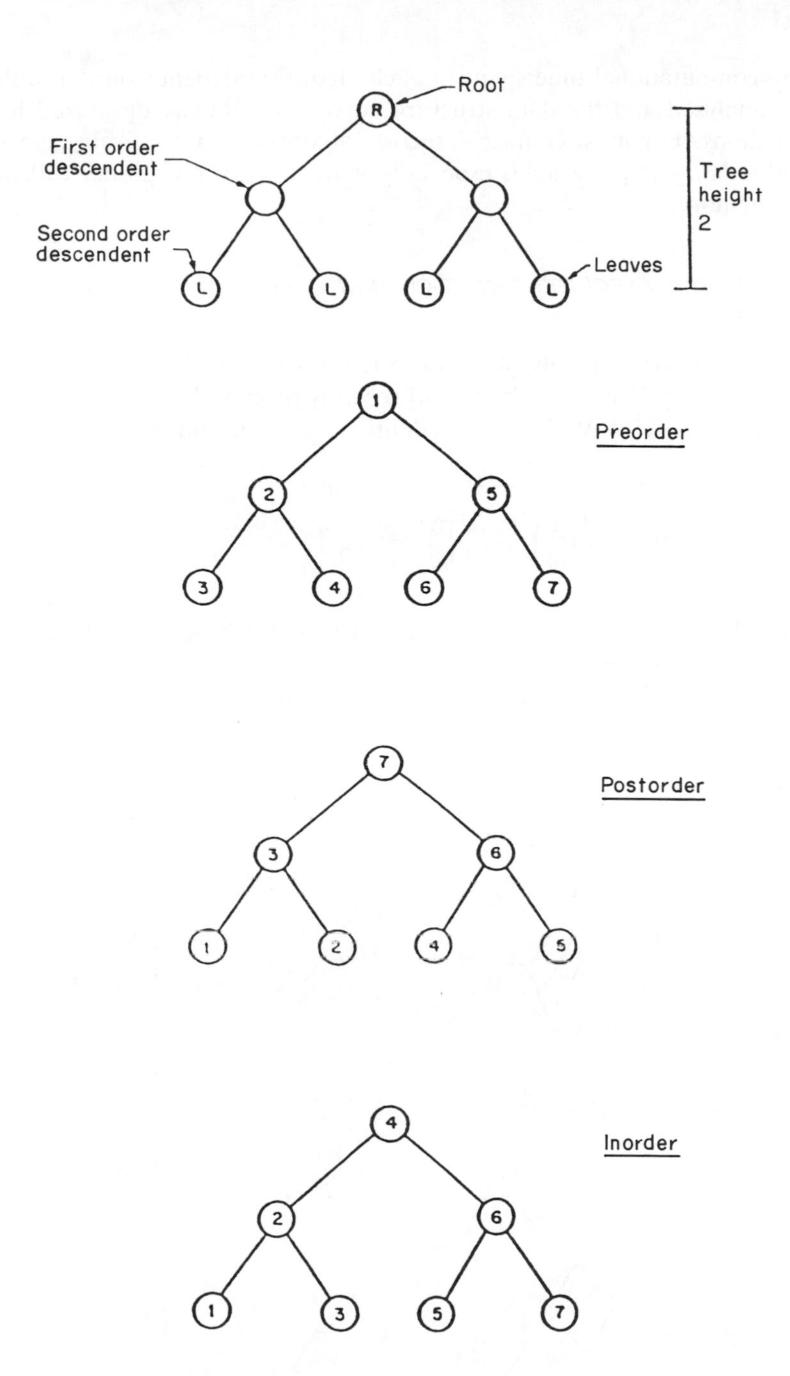


**Figure 11.8**
Methods of traversing a binary tree

The computational time spent in each algorithm depends on the problem being analysed and the data structure cannot be globally optimized for all applications. For most contact detection algorithms, a relatively high level of redundancy is preferable, especially with modern computers with large main memory.

### *Superquadric object representation and volumetric modelling primitives*

Superquadrics are a family of parametric functions, first investigated by the Danish designer Peit Hein (1965) and recently proposed for use in multibody dynamic analysis by Williams and Pentland (1988), and whose equation is given by:

$$\left\{\left(\frac{X}{a}\right)^{2/\beta} + \left(\frac{Y}{b}\right)^{2/\beta}\right\}^{\beta/\alpha} + \left(\frac{Z}{c}\right)^{2/\alpha} = 1 \tag{11.22}$$

Figure 11.9 illustrates the diverse shapes which can be generated by varying the $\alpha$ and $\beta$ parameters.

The superquadric family can also be represented parametrically by latitude and longitude parameters as:

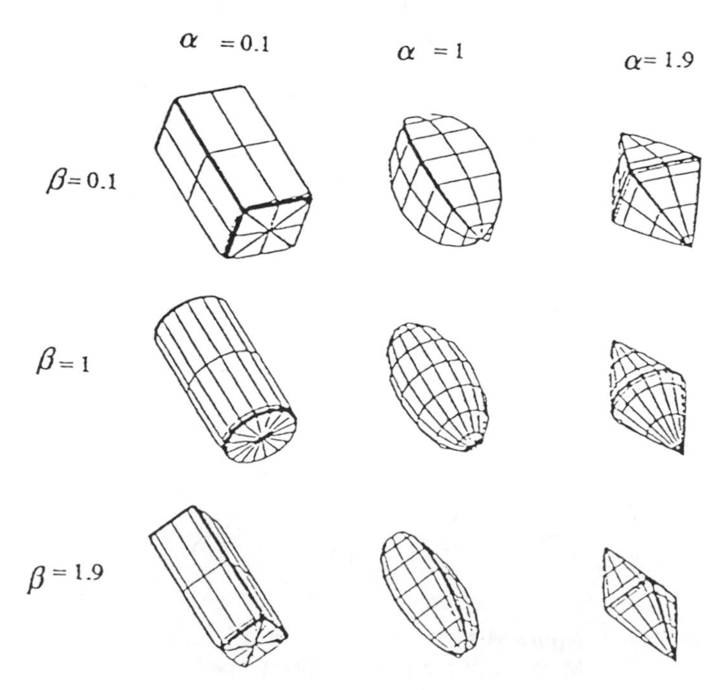


**Figure 11.9**
Superquadric functions for various parameter values

$$\mathbf{R}(\eta, \omega) = \begin{bmatrix} a \;\; \cos^{\beta}(\eta) \; \cos^{\alpha}(\omega) \\ b \;\; \cos^{\beta}(\eta) \; \sin^{\alpha}(\omega) \\ c \;\; \sin^{\beta}(\eta) \end{bmatrix} \tag{11.23}$$

Where **R** is a three-dimensional vector containing $x$, $y$, $z$, components of the surface point. The normal to the surface at point $(\eta, \omega)$ is given by

$$\mathbf{N}(\eta, \omega) = \begin{bmatrix} 1/a \;\; \cos^{2-\beta}(\eta) \; \cos^{2-\alpha}(\omega) \\ 1/b \;\; \cos^{2-\beta}(\eta) \; \sin^{2-\alpha}(\omega) \\ 1/c \;\; \sin^{2-\beta}(\eta) \end{bmatrix} \tag{11.24}$$

If equation (11.22) is rewritten as $F(x, y, z) = 0$, then F(x, y, z) provides a measure of the distance of the point $(x, y, z)$ from the surface of the superquadratic. For $\alpha = \beta = 2/n$, $F(X, Y, Z)$ gives the $L_n$ distance metric. This property provides an extremely useful check, sometimes called an 'inside–outside' check, on whether a point lies inside or outside the surface. For $F < 0$, the point lies inside, for $F > 0$, the point is outside, and for $F = 0$, the point lies exactly on the surface. This check is exactly that needed when determining if one body is in contact with another, i.e. an interference check.

The superquadric function provides a wealth of information about the geometry of the object. For example, it is straightforward to calculate the surface normal, the surface curvature, the volume, and various moments. These properties can be extremely difficult to claculate using representations based on surface patches (e.g. cubic splines) or cells.

The superquadric forms a basic building block, which is a superset of the standard constructive solid geometry (CSG) modelling primatives. It has been estimated that 80 per cent of all manufactured components can be represented by Boolean combinations of the CSG primitives.

The advantage of using superquadrics for contact detection is that the order of the scheme can be reduced by a factor of N, where N is the number of faces per body. Tests in the discrete element program Thingworld have resulted in speed-up of one to two orders of magnitude.

### *Fuzzy boundaries*

In the following discussion we assume a standard node and line boundary representation is used to represent bodies. The time step in any solution scheme is always finite and in one time step an element's position is updated so that at the next time step the element has 'jumped' to a new position (Figure 11.10). The quantum nature of this process means that intermediate positions cannot be considered in the contact algorithm. This introduces

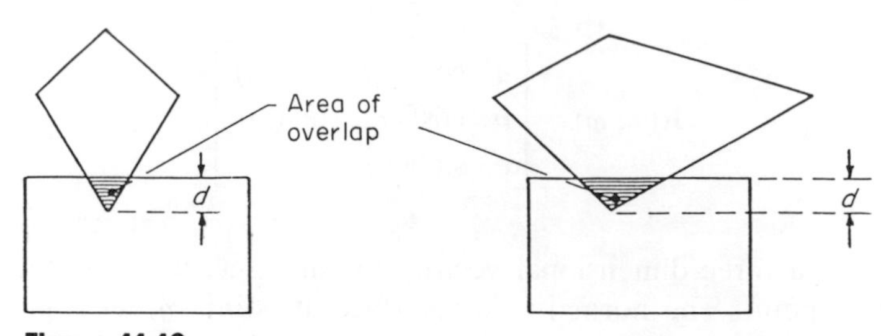


**Figure 11.10**
Two different contact cases with same penetration distance

problems, since a node can suddenly 'appear' in the interior of another element.

The solution is to define a boundary zone, both inside and outside the element's true boundary, where penetrating nodes can legally reside, and where interaction forces can be initiated. The zone must be sufficiently wide that no node can 'jump' over it in a single time step. The result of introducing this zone is to make the element's boundaries 'fuzzy'. All geometric checks concerning the element's boundaries must now contain tolerances. These tolerances and the fuzziness they imply, greatly complicate the generation of valid 'geometric rules'.

When using a penalty function scheme to impose contact constraints, the zone also acts like an arrestor wire on an aircraft carrier. The incoming node 'hooks up' on first being detected in the zone, and is slowed down over some finite distance by the contact forces. If the force generated by the penalty function is not sufficiently large, the node may penetrate so far into the element that the program becomes 'confused' (Figure 11.11). For example, the program may assume a penetrating node has entered through the face to which it is closest. If the node has actually entered through a more distant face, the program may generate forces in the wrong direction. Resolving such problems in three dimensions when many elements may possibly overlap can lead to excessively long computer execution times. The solution to the efficiency problem is to concentrate the computational effort only where troublesome contact problems exist. To accomplish this it is necessary to provide the program with a set of rules by which inferences can be made about what is or is not a difficult contact geometry.

## Geometric Rules of Inference

Given a particular geometric arrangement of bodies the human brain is adept at recognizing possible interactions. In particular it is able to remove irrelevant information from consideration. For example, given the distribution of bodies illustrated in Figure 11.12, the brain quickly determines that bodies 3 and 4 are unlikely to take part in any interactions for many time steps.

The idea that some information can be ignored, either completely, or for many time steps, and that certain checks can be conducted infrequently,

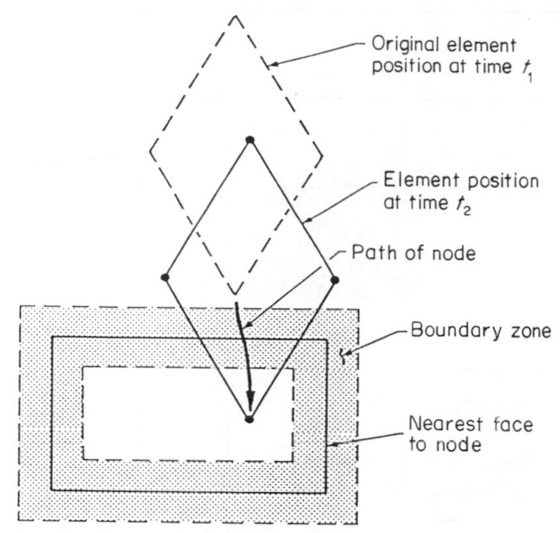


**Figure 11.11**
Example of over-penetration of node through boundary zone causing 'confusion' in force generation algorithm and generation in wrong direction

forms a basis for improving contact detection schemes. One particular scheme used in DECICE[21], called the Neighbour Sieve, utilizes this idea. It is described here in its two-dimensional form.

In two dimensions at least one node must take part in any contact between elements. (In three dimensions edge–edge interactions are also possible as illustrated in Figure 11.4).

In the ideal contact detection scheme, each node is visited in turn. Each face of every other element is then scanned to determine if the node lies on or inside it, i.e. interacts with it.

It is not necessary to check all element faces at every time step, since some faces will be so distant from the node that no contact can possibly occur for many time steps. Thus, a coarse (bounding box) check on the distance, between a node and the other elements will eliminate many faces as potential contacts. The faces, which are potential contacts, are flagged as neighbours of the node, and are stored in a 'neighbour array'. Thereafter, only faces in the neighbour array are considered in the check for contact. Finer tests utilizing more of the available geometric data can be constructed to eliminate still other faces. For example, a face whose outward normal $\mathbf{n}$ points in the same direction as a ray from the node to the face $\mathbf{n}$ ($\mathbf{n} \cdot \mathbf{p} > 0$), can be considered a 'hidden' face since it cannot possibly interact with the given node. In this manner a series of numerical sieves are constructed, the finest of which sifts only few faces for detailed contact (actual overlap), the coarsest sifts all faces, but both crudely and infrequently. The list of neighbours must be updated at intervals if all contacts are to be faithfully

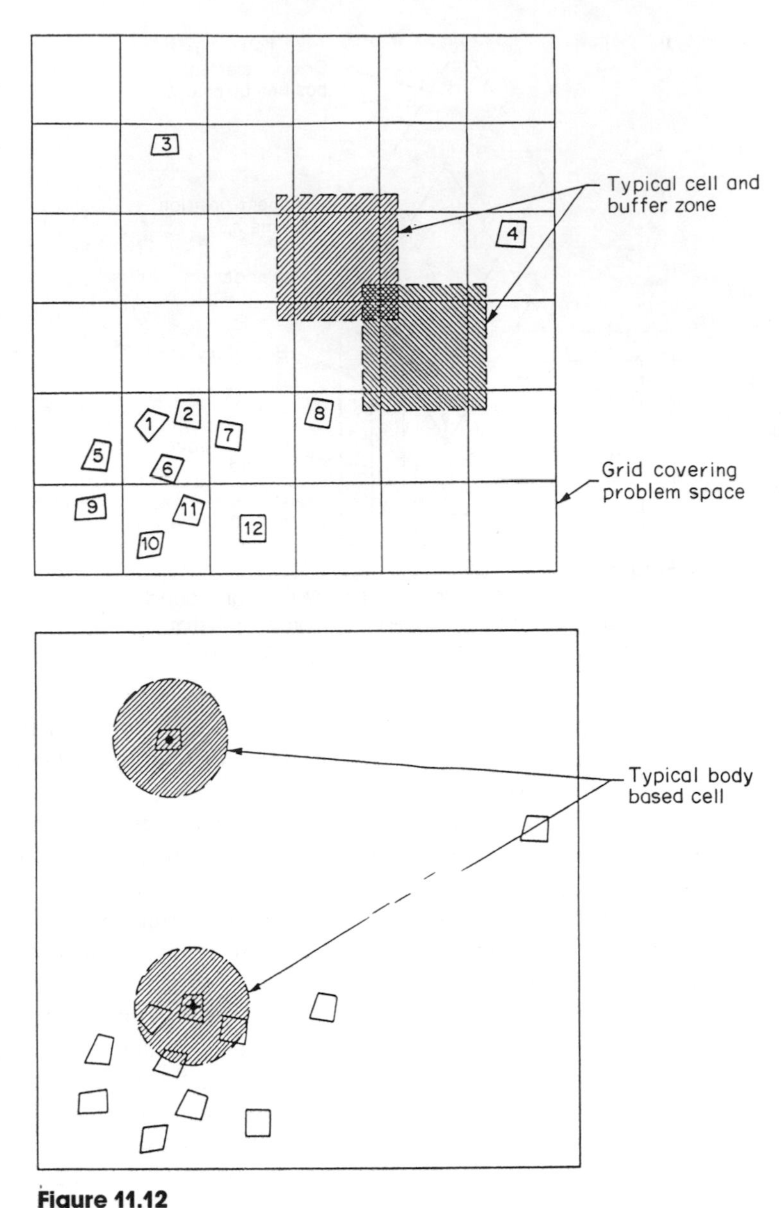


**Figure 11.12**
(a) Space-based grid dividing problem space into cell; (b) same problem with body based cells

detected. In the case of the distance check, a neighbour update must be executed if the relative motion between a node and a face exceeds a known distance (often taken as half the search radius for a body to be included in the neighbour list).

## *Spatial- and body-based contact search schemes*

*Space-based searches.* Consider the distribution of bodies (each consisting of one element) previously shown in Figure 11.12. If there are $N$ bodies, then $N(N - 1)/2$ checks are necessary to ensure all possible contacts are detected (each body must be checked against every other one). (In a program checks may actually be between nodes and faces but for simplicity we will describe the check here as between bodies.)

Now suppose the problem space is divided, by a grid, into $M$ cells. On average there will be $N/M$ bodies in each cell. To check for contacts in one cell $N/M(N/M - 1)/2$ checks are necessary, and to check all cells, $N(N/M - 1)/2$. An advantage has accrued because of the implicit assumption that bodies in one cell cannot interact with bodies in another cell. (The checks necessary to assign bodies to particular cells are neglected for the moment.) This scheme does not work well, as described, because bodies may span the grid between cells and the slightest motion may take a body from one cell to another, necessitating cell updates every time step.

This flaw can be overcome by assigning a 'buffer zone' to surround each cell. In this case any node within a cell has designated neighbours consisting of faces which lie in the cell plus buffer zone. A minor disadvantage is that some regions of the problem space are included in several cells, because of the overlapping buffer zones. However, once elements have been assigned to a cell, there is no need to update the cell assignments until a body has moved a distance equal to the buffer zone width. In this scheme it is relatively simple to monitor the changing geometry and determine when an update is necessary. This is an important consideration because it is possible to do more work calculating when an update is necessary, than is saved by not having to search all elements.

It is worth noting that a hexagonal grid is more efficient, even for this scheme, because the areas of overlap, caused by the buffer zones, is less, for a given buffer zone width.

A disadvantage of a scheme which divides up the problem space, is that no account is taken of the distribution of the bodies in that space. For example, the bodies could all be clustered in one cell, in which case no advantage accrues from this scheme.

An alternative is to use a cell decomposition method based, say, on a hextree. The resolution of the cells can be related to the density of the bodies, since cells (branches of the hextree) need only exist where bodies exist. Such schemes have been used in computer graphics, for ray tracing, with as many as 100 000 bodies.

*Body-based searches.* Another alternative is to associate the search cell with an element, so that each element has a cell surrounding it. It is as if each element has its own radar which scans out to a given distance. The element is only aware of other elements within its scan radius. The element completes a scan and memorizes all elements it detects in the so-called 'Neighbour List'. The program then proceeds to step in time. During each time step contact detection for any particular element uses only neighbours already selected during the previous scan. The procedure is valid so long as elements move less than the scan radius between updates, at which time another scan must be initiated.

In both space based cells and element based cells the extent of the buffer zone is critical to the efficiency of the scheme. If the relative motion of the bodies is small, a small buffer zone will be most efficient. The faster the motion of elements the larger the necessary buffer zone. A truly 'intelligent' program should be capable of adjusting the dimensions of the buffer zone and even the type of search scheme employed. Unfortunately, this is not easily achieved.

### *Moving contacts*

In many contact problems, the contact areas, established at problem initiation, are only modified by enlargement or translation. No new contacts are generated during the solution process. Typical problems where this assumption is valid is in the analysis of rolling gears, tyres, tank tracks, etc. Considerable simplifications to contact detection are possible when this assumption can be made.

Each contact region is visited in turn. From connectivity data, adjacent nodes, edges and faces can be checked, to determine if the region has enlarged. If new contacts are detected, then further adjacent elements are searched. This scheme is particularly effective because present contact information is used to infer where new contacts are possible. This scheme is successfully employed in the code UDEC. (Cundall et al., 1983).

## 11.5 CONTACT AND INTERACTION FORCES

The discussion to date has implicitly assumed that interaction forces are generated using a penalty function formulation. In this section various methods of imposing contact conditions between interacting bodies is discussed.

The numerical modelling of contact between bodies is a rather complex subject. Only a brief review is provided here and the reader is referred to the following papers in the Finite Element literature for a more detailed discussion of the various techniques available (Babiska, 1973; Brezzi, 1974; Hughes *et al.*, 1976, 1978; Oden *et al.*, 1980, 1982, 1983; Song *et al.*, 1980; Francavilla *et al.*, 1974; Sachdeva *et al.*, 1981; Okamoto *et al.*, 1979; Tseng *et al.*, 1981 and Hallquist *et al.*, 1985). It is noted, however, that any form of interaction law or numerical technique can be accommodated within the discrete element framework.

A most significant feature of contact problems involving friction is that the deformations, in general, depend on the loading history. It is necessary, therefore, that the loading process be followed in time.

Some exceptions are found for problems involving steady state conditions and the early investigations of contact concentrated on such steady-state solutions.

Here, we shall concentrate on methods applicable to dynamic impact problems where it is important to follow the correct loading path.

### *Penalty function techniques*

The penalty function technique of enforcing additional problem constraints has proved effective in many areas of numerical modelling and is discussed by Campbell (1974), Felippa (1975, 1986) and Zienkiewicz *et al*. (1984).

Consider a variational principle given by $\chi = 0$.

The usual Finite Element equations can be generated by taking variations with respect to the displacement unknowns as follows:

$$\frac{\partial \chi}{\partial u_i} = 0\,. \tag{11.25}$$

This leads to equations of the form

$$[K]\{u\} = \{f\} \tag{11.26}$$

Typically in contact problems we wish to force one node to move with the same displacement as another node. To be specific let us contain node $i$ to move with node $j$ such that $u_i - u_j = 0$

Since

$$u_i - u_j = 0 \tag{11.27}$$

then

$$k(u_i - u_j)^2 = 0$$

where $k$ is known as the penalty constant. We may now modify the variational function such that

$$\chi' = \chi + \frac{k}{2}(u_i - u_j)^2$$

and

$$\frac{\partial \chi'}{\partial u_k} = \frac{\partial \chi}{\partial u_k} \qquad k \neq i \quad \text{and} \quad k \neq j$$

$$\frac{\partial \chi'}{\partial u_i} = \frac{\partial \chi}{\partial u_i} + k(u_i - u_j)$$

$$\frac{\partial \chi'}{\partial u_j} = \frac{\partial \chi}{\partial u_j} - k(u_i - u_j) \tag{11.28}$$

The stiffness matrix $[K]$ is therefore modified as follows

$$\begin{bmatrix} K_{11} & \cdots\cdots\cdots\cdots\cdots\cdots\cdots\cdots\cdots\cdots\cdots \\ \vdots & \\ K_{i_1} & \cdots (K_{ii} + k) \cdots (K_{ij} - k) \cdots\cdots \\ \vdots & \\ K_{j_1} & \cdots (K_{ji} - k) \cdots (K_{jj} + k) \cdots\cdots \\ \vdots & \\ K_{n_1} & \cdots\cdots\cdots\cdots K_{nn} \cdots\cdots\cdots\cdots \end{bmatrix} \begin{bmatrix} u_1 \\ \vdots \\ u_i \\ \vdots \\ u_j \\ \vdots \\ u_n \end{bmatrix} \begin{bmatrix} f_1 \\ \vdots \\ f_i \\ \vdots \\ f_j \\ \vdots \\ f_n \end{bmatrix}$$

Thus $+k$ is added to diagonal terms $ii$ and $jj$ and $-k$ to off diagonal terms $ij$ and $ji$.

One problem with this method is in determining the value of the penalty constant $k$. If $k$ is too small the constraint will not be enforced effectively, while if $k$ is too large the constraint is enforced at the expense of the satisfaction of the remainder of the variational principle, i.e. the constraint numerically dominates all other factors in the solution process. Large values of $k$ also lead to ill conditioning of the equations making solution difficult.

The penalty constant can be interpreted as the stiffness of a spring between the nodes $i$ and $j$. When the spring is weak the nodes are only loosely connected while a stiff spring forces the nodes to move together.

If the contact is across a rock joint the spring stiffness may be related to the joint normal or shear stiffness depending on the joint orientation and the directions of the unknowns $u_i$ and $u_j$.

### *Lagrange multiplier formulation*

Contact constraints may also be enforced using a Lagrange multiplier procedure (Babushka, 1973). The procedure is first demonstrated here for two unknowns and is then generalized to more unknowns spanning an $n$ dimensional space.

Suppose the extrema of the function $X(x, y)$ are sought. If there is no auxiliary condition, the following equations are solved simultaneously:

$$\frac{\partial X}{\partial x} = 0, \qquad \frac{\partial X}{\partial y} = 0 \tag{11.29}$$

The solution $x_0$, $y_0$ gives the position of the extremum $f(x_0, y_0)$. Now suppose the extremum is sought subject to the condition $y = \bar{y}(x)$. This line will not generally pass through point $x_0$, $y_0$ and therefore the solution will not be the same as before. The new solution may be obtained by inserting the value for $y$ into $X$ so that $X$ now depends only on $x$. Differentiation with respect to $x$ now yields

A brief outline of the method when applied to a Finite Element procedure is now given.

Consider a variational principle $\chi(u_i) = 0$ subject to the additional constraint $u_i - u_j = 0$. Argument the functional as follows

$$\chi' = \chi + \lambda(\mathrm{u}_i - \mathrm{u}_j) = 0 \tag{11.30}$$

$$\frac{\partial \chi'}{\partial u_n} = \frac{\partial \chi}{\partial u_n} \qquad k \neq i, \quad k \neq j$$

$$\frac{\partial \chi'}{\partial u_i} = \frac{\partial \chi}{\partial u_i} + \lambda$$

$$\frac{\partial \chi'}{\partial u_j} = \frac{\partial \chi}{\partial u_j} - \lambda$$

$$\frac{\partial \chi'}{\partial \lambda} = \mathrm{u}_l - u_j$$

The normal Finite Element equation

$$[K]\{u\} = \{f\} \tag{11.31}$$

must therefore be modified as follows

$$\begin{bmatrix} & & & & & & 0 \\ & & & & & & \vdots \\ & & & K & & & 1 \\ & & & & & & \vdots \\ & & & & & & -1 \\ & & & & & & \vdots \\ 0 & \cdots & 1 & \cdots & \cdots & -1 & \cdots\ 0 \end{bmatrix} \begin{bmatrix} u_1 \\ \vdots \\ u_i \\ \vdots \\ u_j \\ \vdots \\ \lambda \end{bmatrix} = \begin{bmatrix} f_1 \\ \vdots \\ f_i \\ \vdots \\ f_j \\ \vdots \\ 0 \end{bmatrix} \tag{11.32}$$

These equations can be solved for the unknowns $u_i$ and $\lambda$. It is easy to show that the Lagrange multiplier $\lambda$, corresponds to a traction. In the case of contact problems the multiplier is the traction on the contact surface.

Oden *et al.* (1982) discuss the theory of Lagrange multipliers in detail and shows that the coerciveness condition necessary to ensure a unique solution may not generally hold, i.e. the solution may not be unique.

This can be overcome by introducing a perturbed Lagrangian (Wriggens *et al.*, 1985). The perturbed Lagrangian not only ensures a unique solution but also removes the numerically troublesome zeros on the diagonal of the modified stiffness matrix. The modified variational functional is taken as

$$\chi' = \chi + \lambda(u_i - u_j) + \frac{\lambda^2}{2\epsilon} \tag{11.33}$$

(This resemblance of the perturbation term to a penalty function should be noted.) This leaves all equations unaltered except for the variation with respect to $\lambda$ which becomes

$$\frac{\partial \chi'}{\partial \lambda} = (u_i - u_j) + \frac{\lambda}{\epsilon} \tag{11.34}$$

The modified stiffness matrix thus becomes

$$\left[\begin{array}{c|c} K & A \\ \hline A & \frac{1}{\epsilon} \end{array}\right] \left\{\begin{array}{c} u \\ \hline \lambda \end{array}\right\} = \left\{\begin{array}{c} f \\ \hline 0 \end{array}\right\} \tag{11.35}$$

Of the two methods of imposing constraints within a variational method the penalty method is the most attractive for a computational scheme (Oden *et al.*, 1982; Hallquist *et al.*, 1985). The method introduces no new variables (the Lagrange multipliers (tractions) can still be obtained by a direct calculation post facto) while it is ensured that the solution will converge to an unique value.

Problems with penalty terms and the necessity of using reduced forms of integration are discussed by Zienkiewicz *et al.* (1977), Hughes (1977) and Malkus *et al.* (1978).

Hallquist *et al.* (1985), provides an excellent discussion of the numerical details necessary to include the various methods of imposing constraints in a variety of numerical codes, including three-dimensional explicit finite element codes.

A simple but effective penalty function method used in most discrete element codes is now described.

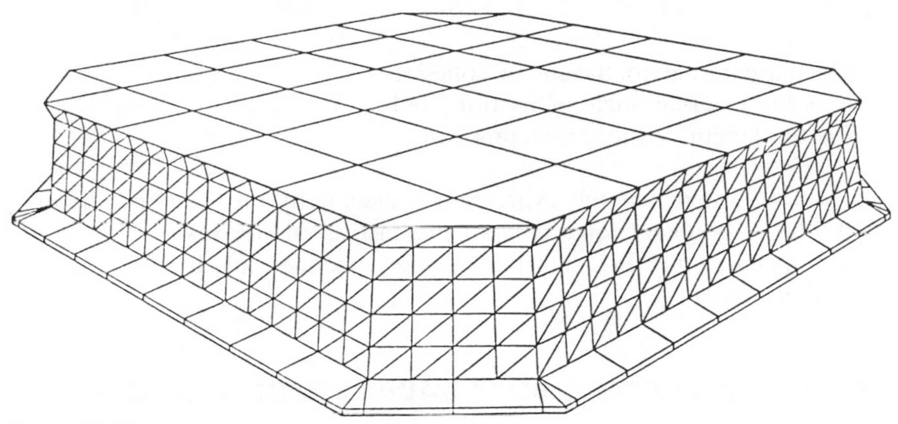

**Figure 11.13**
Complex Discrete Element with multiple faces

*The linear interaction model*

The linear interaction model for a node-face interaction between two elements is defined in two parts:

(i) The linear normal and shear interaction stiffness relationships

$$R_n = k_n u_n \quad \text{and} \quad R_s = k_s u_s, \tag{11.36}$$

where

the subscripts $n$, and $s$, denote the normal and shear directions of the interacting face
$R_n$ and $R_s$ are the generated normal and shear forces acting at the point of contact;
$u_n$ and $u_s$ are the normal and shear displacements of the node relative to the interacting face;
and $k_n$ and $k_s$ are constant normal and shear interaction stiffness coefficients.

and

(ii) the interface friction (or strength) criteria which limits the magnitude of the shear interaction force $R_s$, with the inequality

$$R_s \leqslant Cl + R_n \tan \Phi \tag{11.37}$$

where

$C$ is an external coefficient of cohesion,
$\Phi$ is the external angle of friction, and
$l$ is the length of the interacting face.

It should be noted that the expression is essentially the Coulomb friction law written in terms of force components instead of stresses.

The equations (11.34) and (11.35) are extended easily to nonlinear interaction models.

## 11.6 EXAMPLES OF RIGID BODY DISCRETE ELEMENT ANALYSIS

In many rock mechanics applications, especially concerning tunnel stability, the rock joint behaviour is the controlling parameter and the intact rock deformations may have little effect. In such cases the rigid body Discrete Element Method provides an excellent analysis tool.

The stability of tunnels in a jointed rock has been investigated by Belytschko *et al.* (1984) in which the intact rock is treated as a rigid element and the joint behaviour as a constraint imposed using a penalty function technique. Only static solutions are examined and an efficient implicit solution scheme is adopted using a secant iterative procedure. As a matter of interest it should be noted that the secant method has advantages over the more usual tangent stiffness method when a stress dependent Young's modulus is used and the material is unloading in some directions, as is the case during tunnel an excavation. The implicit scheme makes the method more efficient than an explicit dynamic relaxation scheme for static problems.

Investigations of rock fragmentation and cratering using rigid element codes BLOCK (Taylor, 1983), BUMP (Schamaun, 1984), CAROM (Gorgham–Bergeron, 1985) and DMC, (Taylor & Preece, 1989) have been conducted by Sandia National Laboratories. Figure 11.14 shows a snapshot of a cratering experiment simulation using BLOCK (Schamaun, 1984). The analysis simulates a single blasthole in a mine floor. The rock mass is represented by 480 blocks with symmetry imposed along the blasthole centreline. The block sizing was randomly distributed with a mean size decreasing towards the blasthole.

The method used in BLOCK, BUMP and CAROM is derived from the early BLOCK code of Cundall detailed in Maini *et al.* (1978). To facilitate more efficient code the link list scheme has been replaced. In the case of CAROM the elements have been restricted to discs which greatly simplifies interaction calculations and results in a very fast execution speed.

Three-dimensional rigid sphere codes DMC (Taylor & Preece, 1989) and a series of both two and three dimensional codes by Walton (Walton et al., 1988, Walton, 1982, 1983) have been used to analyze flows of granular

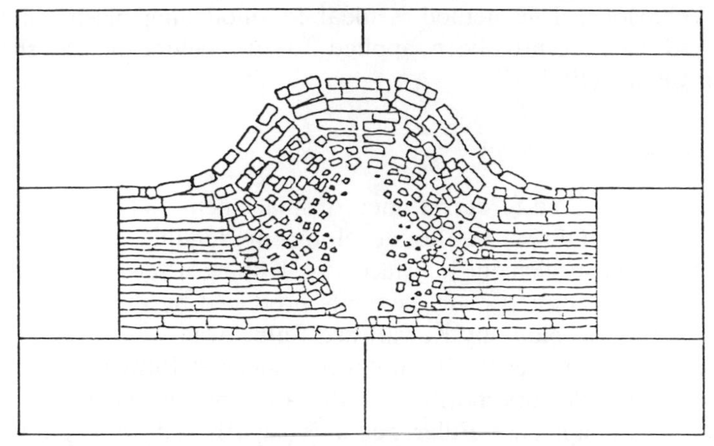

**Figure 11.14**
Cratering simulation using rigid block code BLOCK. (Schamaun (1984))

materials. Notable works on flows of granular materials include those by Thorton (1989) and Hakuro et al., 1988, 1989.

*Plates using rigid elements*

A novel Discrete Element Model for continuum plates called the Rigid Body Spring Model (RBSM) has been developed by Kawai (1977, 1979 and Kawai *et al.*, 1980), who shows that the basic equations can be derived from an energy functional. Although, at present the method cannot handle discontinuous bodies, its formulation contains most of the facets of the discrete element method. In Kawai's model the body is divided into rigid triangles as shown in Figure 11.15. The triangles are connected together by springs along the edges. These springs are adjusted to give the correct

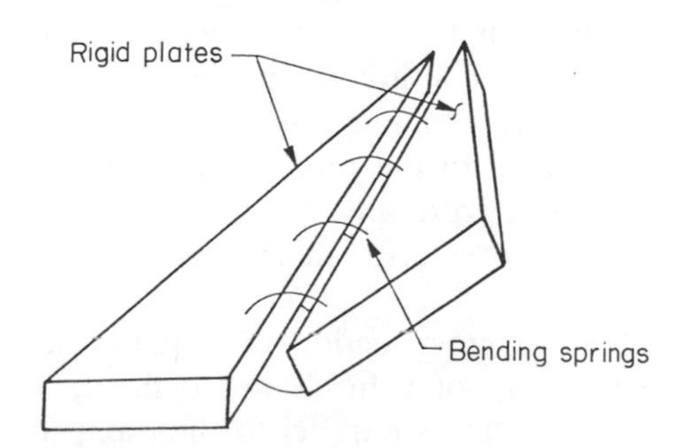


**Figure 11.15**
Rigid plate elements of Kawai, 1980

bending behaviour. The method is ideal for modelling plastic hinge-type failures and has recently been applied to the failure of ice sheets by Yoshimura *et al.* (1985).

### *Plate elements*

Another approach used to derive linear displacement elements is described by Mustoe *et al.* (1987). In this development bending deformations are confined to adjacent element interfaces by defining a set of finite interface stiffnesses. The beam and plate elements produced by this method (which is similar to the technique by Kawai) are Euler beams and Kirchoff plates respectively. It is interesting to note the shear deformation effects are included by a simple superposition of the appropriate linear displacement element. For example, an Euler beam is superposed with a four-noded plane stress quadrilateral, and a Kirchoff plate with an eight-noded hexahedral element.

The plate element is developed using the Kirchoff hypothesis (which is analogous to the simple theory for the bending of beams) in which the bending moment-radius of curvature equations are given by

$$M_l = D\left(\frac{1}{r_l} + \frac{\nu}{r_m}\right), \quad M_m = D\left(\frac{1}{r_m} + \frac{\nu}{r_l}\right), \quad M_{lm} = D(1-\nu)\frac{1}{r_{lm}} \tag{11.38}$$

where $l$, $m$ and $n$ are the local plate axis directions,

$$D = Et^3\{12(1-\nu^2)\}, \quad 1/r_l = \frac{\partial^2 w}{\partial l^2}, \quad 1/r_{lm} = \frac{\partial^2 w}{\partial l\, \partial m}, \quad \frac{1}{r_m} = \frac{\partial^2 w}{\partial m^2}$$

and the moments $M_l$, $M_m$ and $M_{lm}$ are defined per unit length.

The Discrete Element plate is essentially four Discrete Elements connected together with a series of normal and shear springs which form four hinge lines as shown above in Figure 11.16(a). In order to compute the normal and shear stiffness coefficients the constant curvature modes produced by a constant moment within a continuous plate (see Figure 11.16(b)) are equivalenced to the discrete element system described above. Calculation of normal stiffnesses parallel to the $l$-axis is performed with the following procedure. Suppose $M_l$ is nonzero and $M_m = M_{lm} = 0$, which results in the deformation shown in Figure 11.17(a) then from above equations $M_l = D(1 - \nu^2)/r_l$.

Figure 11.17(b) shows a cross-section of a plate element at a vertical boundary where two springs of value $k$ are defined. Therefore the total number of normal spring stiffness parallel to the $l$-axis for the four-element arrangement is eight.

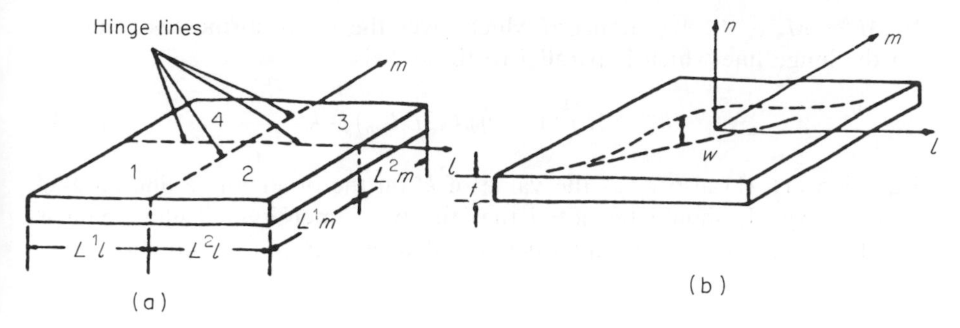


**Figure 11.16**
(a) Discrete Element plate showing hinge lines; (b) a continuous plate representing a typical finite element configuration

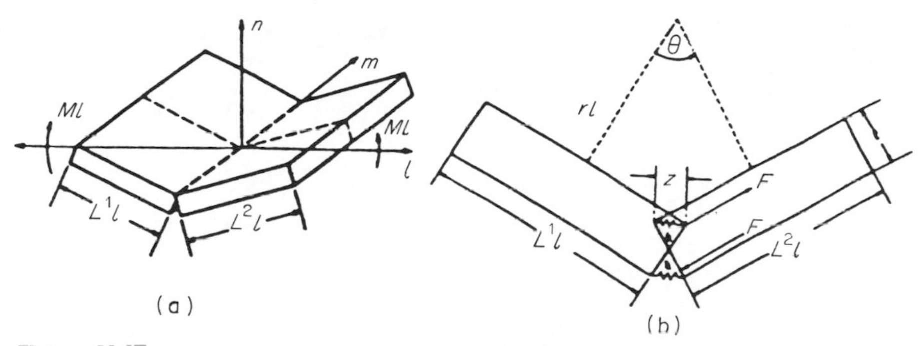


**Figure 11.17**
(a) Illustration of Discrete Element plate undergoing bending in ℓ-direction; (b) illustration of Discrete Element plate undergoing bending of interspace normal stiffnesses

Upon equating the moment curvature relations for the continuous plate and the equivalent Discrete Element assemblage the value of the normal stiffness coefficient parallel to the $l$-axis, $k^l{}_n$, is given by

$$k_n^l = (EtL_m)/(12L_l) \tag{11.39}$$

A similar argument using $M_l = Mlm = 0$ and a nonzero $M_m$ gives the normal stiffness coefficients in the $m$-direction a value of

$$k_n^m = (EtL_l)/(12L_m) \tag{11.40}$$

The derivation for the shear stiffnesses (which act at the nodal locations as the normal stiffnesses illustrated in Figure 17(b) is performed as for the normal stiffness calculation except that the constant twist mode is provided by $M_l = M_m = 0$, $M_{lm}$ nonzero which gives the shear stiffnesses value $k_s^m$ on the hinge line which is parallel to the $m$-axis

$$k_s^m = (Et^3(1 + \nu))/(12L_lL_m) = k_s^l \tag{11.41}$$

Equation (11.41) also gives the value of $k_s^l$ on the other hinge line parallel to the $l$-axis. It should be noted that the two-dimensional Euler beam is simply derived by considering a bending deformation in one direction only.

# 12 The Discrete Element Method—Deformable Bodies

## 12.1 DYNAMICS OF DEFORMABLE BODIES

Before launching into a mathematical exposition of deformable elements let us familiarize ourselves with the problem.

Consider a single deformable element (Figure 12.1) acted upon by a constant force $F$. Given the geometry, density, and elastic modulus of the element, the problem is to determine its behaviour. Using Newton's Laws it has already been shown how to determine the motion of the centroid and any rotation about the centroid. The problem not yet solved is to determine the deformation of the element about the centroid.

The first problem that must be addressed is that the centroid of the element is in motion and a set of axes fixed in the element at the centroid is both translating and rotating. In order to drive the deformational equations of motion with respect to these axes it is necessary to transform the dynamic equilibrium equations, usually formulated with respect to an inertial frame

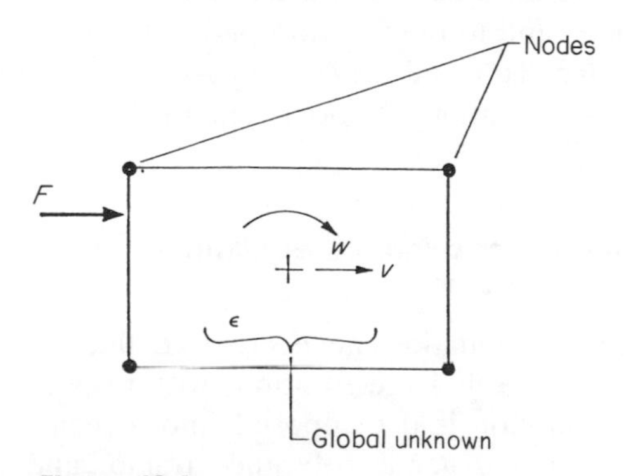


**Figure 12.1**
Deformation of a single Discrete Element under force loading

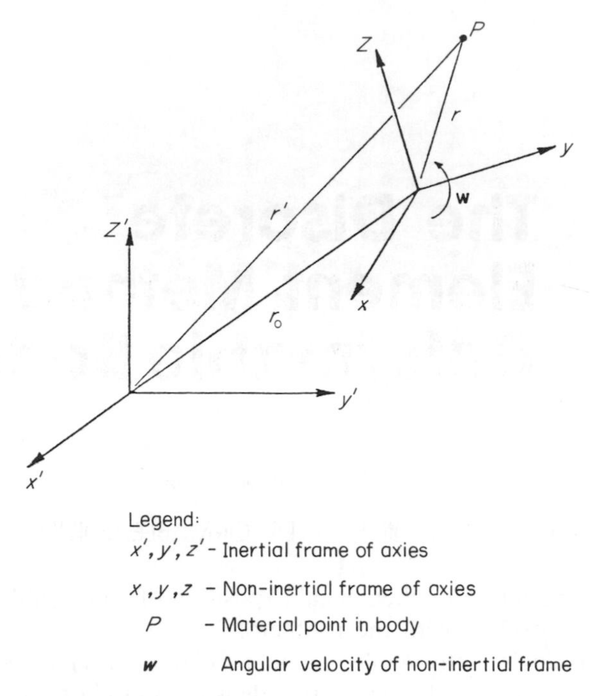


**Figure 12.2**
Relationship between the inertial and noninertial system of axes

of reference, to one coincident with the principal axes of the element which are both translating and rotating (Figure 12.2). With respect to these axes (Tisserand axes) the element has zero relative linear and angular momentum and the deformation displacements can be represented exactly by an expansion in the natural elastic modes of the element. The modal approach can be quite generally applied to large rotation and large strain problems. In the latter case we must pay strict attention to choosing appropriate stress and strain measures and to the rotation path followed between undeformed and deformed states. To avoid these complexities we restrict ourselves here to a small strain formulation. A more general formulation is contained in Williams and Pentland (1988).

### 12.1.1 Transformation of dynamic equilibrium equations to a noninertial reference frame

This section details the equations necessary to define the motion of a flexible body undergoing arbitrarily large rotations with respect to an inertial frame of reference. The motion is decomposed into a mean rigid body motion, defining a non inertial dynamic reference frame, and a relative motion, defining the deformations with respect to that frame.

Following Fraeijs de Veubeke (1976), the dynamic reference frame is taken as satisfying the Tisserand conditions of zero relative linear and angular momentum. By minimization of the mean square of relative displacements rather than minimizing the relative kinetic energy, which generates nonlinearities, a linearized form of the zero angular momentum condition is generated. This allows the relative displacements to be represented exactly by an expansion in the natural elastic modes of the body.

Using Hamilton's principle and the concept of generalized coordinates, all the rigid body and deformational equations are derived. These equations are simplified by the assumption of small strain within a time step. The equations are not limited to small deformations, and large distortions are treated correctly.

### *Fundamental axioms of mechanics*

As stated by Eringen (1980), the fundamental axioms of mechanics, written with respect to an inertial frame, are as follows:

### *Axiom 1 Principle of conservation of mass*

The total mass of a body remains unaltered during deformation

$$\frac{d}{dt}\int \rho \, dV = 0 \tag{12.1}$$

where

$\rho$ is the mass density at a material point in the primed system integration is over the volume of the body

and

$\frac{d}{dt}$ denotes the material time derivative.

### *Axiom 2 Principle of balance of momentum*

The time rate of change of momentum is equal to the resultant force $F$ acting on the body.

$$\frac{d\mathbf{P}}{dt} = \mathbf{F} \quad \text{or} \quad \frac{d}{dt}\int \rho v_k \, dV = F_k \tag{12.2}$$

where

**P** is the momentum
**F** is the force

$v_k$ is the velocity in direction $k$
$F_k$ is the force in direction $k$
$\rho$ is the density

*Axiom 3 Principle of the balance of the moment of momentum*

The time rate of change of moment of momentum about a fixed point O, is equal to the resultant moment about O.

$$\frac{d\mathbf{H}}{dt} = \mathbf{L} \quad \text{or} \quad \frac{d}{dt}\int \rho e_{kim} x_i v_m \, dV = L_k \tag{12.3}$$

where

$\mathbf{H}$ is the moment of momentum about O
$\mathbf{L}$ is the resultant moment about O
$\rho$ is the mass density
$e_{kim}$ is the Levi–Civita permutation tensor density
$x_i$ is the moment arm about O in direction $i$
$v_m$ is the velocity in direction $m$
$L_k$ is the moment about O in direction $k$
$V$ is the volume

*Axiom 4 Principle of conservation of energy*

The time rate of change of the kinetic energy is equal to the sum of the rate of work $W$, of the external forces plus all other energies that enter or leave the body per unit time.

$$\frac{d}{dt}(K + E) = W + \Sigma U_\alpha \tag{12.4}$$

where

$K$ is the kinetic energy
$E$ is the internal energy
$W$ is the external work done
$U_\alpha$ is the $\alpha$th kind of energy per unit time (e.g., heat, electrical energy)

From these basic axioms the following governing equations with respect to an intertial frame (denoted by prime superscript) are derived:

$$\frac{d\mathbf{P}'}{dt} = \mathbf{F}' \tag{12.5}$$

$$\frac{d\mathbf{H}'}{dt} = \mathbf{L}' \tag{12.6}$$

where

$$
\begin{aligned}
\mathbf{F}' &= \int \mathbf{t} \, \mathrm{d}S + \int \mathbf{R} \, \mathrm{d}V \\
\mathbf{L}' &= \int \mathbf{r}' \times \mathbf{t} \, \mathrm{d}S + \int \mathbf{r}' \times \mathbf{R} \, \mathrm{d}V \\
\mathbf{P}' &= \int \rho \, \dot{\mathbf{r}}' \, \mathrm{d}V \\
\mathbf{H}' &= \int \rho (\mathbf{r}' \times \dot{\mathbf{r}}') \, \mathrm{d}V
\end{aligned}
$$

and where

$S$ and $V$ follow the deformation of the body
$\mathbf{t}$ is the boundary traction
$\mathbf{R}$ is the body force

No distinction is made between vector and tensor quantities, such as **t** and **R**, which are indifferent to reference frame by definition because they are defined in terms of surface forces and surface orientation.

The surface tractions **t**, and body forces **R**, contain all external influences on the body including buoyancy, drag, and gravity. In the case of buoyancy and water drag, the forces are nonzero only over the wetted surface.

Equations (12.5) and (12.6) can be rewritten in the familiar form as Euler's equations.

$$\frac{\mathrm{d}}{\mathrm{d}t} \int \rho \frac{\mathrm{d}\mathbf{r}}{\mathrm{d}t} \mathrm{d}V = \int \mathbf{t} \, \mathrm{d}S + \int \mathbf{R} \, \mathrm{d}V \tag{12.7}$$

$$\frac{\mathrm{d}}{\mathrm{d}t} \int \rho \left( \mathbf{r}' \times \frac{\mathrm{d}\mathbf{r}'}{\mathrm{d}t} \right) \mathrm{d}V = \int \mathbf{r}' \times \mathbf{t} \, \mathrm{d}S + \int \mathbf{r}' \times \mathbf{R} \, \mathrm{d}V \tag{12.8}$$

Applying Green's Theorem to each of the above equations leads to Cauchy's laws of motion which, expressed in tensor notation, are:

$$\rho \ddot{r}_i = \sigma_{ij,j} + R_i \tag{12.9}$$

and

$$\sigma_{ij} = \sigma_{ji} \tag{12.10}$$

To derive an expression for kinetic energy, Cauchy's first law is multiplied by $\dot{\mathbf{r}}'$ and integrated over the body and then, using Green's Theorem, the following is derived:

$$K = \int \mathbf{t} \cdot \dot{\mathbf{r}}' \, \mathrm{d}S + \int \mathbf{R} \cdot \dot{\mathbf{r}}' \, \mathrm{d}V - \int T : \nabla \dot{\mathbf{r}}' \, \mathrm{d}V \tag{12.11}$$

where

$$K = \tfrac{1}{2} \int \rho \, \dot{\mathbf{r}}^{2\prime} \, \mathrm{dV}$$

and

$T$ is the stress dyadic.

The last term on the right-hand side is just the rate of change of strain energy of the body and can also be written in tensor notation as

$$\int \sigma_{ij} \dot{\epsilon}_{ij} \, \mathrm{d}V \tag{12.12}$$

The stress dyadic can be written as

$$\begin{aligned} T = {} & \sigma_{xx}\mathbf{ii} + \sigma_{yx}\,\mathbf{ji} + \sigma_{zx}\mathbf{ki} \\ & + \sigma_{xy}\mathbf{ij} + \sigma_{yy}\,\mathbf{jj} + \sigma_{zy}\mathbf{kj} \\ & + \sigma_{xz}\mathbf{ik} + \sigma_{yz}\,\mathbf{jk} + \sigma_{zz}\mathbf{kk} \end{aligned} \tag{12.13}$$

*Transformation of equations to an arbitrary reference frame*

It is desired to transform the basic equations, derived in an inertial frame, to an arbitrary noninertial frame, which is both translating and rotating. Here, all quantities referred to the inertial frame are primed; and all quantities in the dynamic, noninertial, frame are unprimed (Figure 12.2).

The relationship between position vectors for a specific point and the material derivatives in the two reference frames are:

$$\mathbf{r}' = \mathbf{r}_0 + \mathbf{r} \tag{12.14}$$

$$\frac{\mathrm{d}\mathbf{r}'}{\mathrm{d}t} = \boldsymbol{\omega} \times \mathbf{r}' + \frac{\partial \mathbf{r}'}{\partial t} = \boldsymbol{\omega} \times \mathbf{r}' + \dot{\mathbf{r}}' \tag{12.15}$$

where $\boldsymbol{\omega}$ is the angular velocity in the inertial frame.

These equations lead to the following:

$$\frac{\mathrm{d}\mathbf{r}'}{\mathrm{d}\mathbf{t}} = \frac{\mathrm{d}\mathbf{r}_0}{\mathrm{d}t} + \frac{\mathrm{d}\mathbf{r}}{\mathrm{d}t} = \frac{d\mathbf{r}_0}{\mathrm{d}t} + \boldsymbol{\omega} \times \mathbf{r} + \dot{\mathbf{r}} \tag{12.16}$$

$$\begin{aligned} \frac{\mathrm{d}^2\mathbf{r}'}{\mathrm{d}t^2} = {} & \frac{\mathrm{d}^2\mathbf{r}_0}{\mathrm{d}t^2} + \dot{\boldsymbol{\omega}} \times \mathbf{r} + \boldsymbol{\omega} \times (\boldsymbol{\omega} \times \mathbf{r}) \\ & + 2\boldsymbol{\omega} \times \dot{\mathbf{r}} + \ddot{\mathbf{r}} \end{aligned}$$

Substituting these relationships into equations (12.5), (12.6) and (12.11), the following are derived:

$$\mathbf{F} = m\ddot{\mathbf{r}}_0 + \dot{\boldsymbol{\omega}} \times \int \rho \mathbf{r} \, \mathrm{d}V$$

$$+ \boldsymbol{\omega} \times \left\{ \boldsymbol{\omega} \times \int \rho \mathbf{r} \, \mathrm{d}V + 2\mathbf{P} \right\} + \dot{\mathbf{P}} \tag{12.17}$$

$$\begin{aligned} \mathbf{L} = & -\ddot{\mathbf{r}}_0 \times \int \rho \mathbf{r} \, \mathrm{d}V + \dot{\boldsymbol{\omega}} \cdot I \\ & + \boldsymbol{\omega} \times (\boldsymbol{\omega} \cdot I) + \boldsymbol{\omega} \cdot \dot{I} \\ & + \boldsymbol{\omega} \times \mathbf{H} + \dot{\mathbf{H}} \end{aligned} \tag{12.18}$$

where

**I** is the moment of inertia dyadic. Also we derive

$$\begin{aligned} \int \dot{\mathbf{r}} \cdot \mathbf{R} \, \mathrm{d}V + \int \dot{\mathbf{r}} \cdot \mathbf{t} \, \mathrm{d}S = & \; \ddot{\mathbf{r}}_0 \cdot \mathbf{P} + \dot{\boldsymbol{\omega}} \cdot \mathbf{H} - \tfrac{1}{2} \boldsymbol{\omega} \cdot \dot{I} \cdot \boldsymbol{\omega} \\ & + \dot{\mathbf{K}} + \int \mathrm{T} : \nabla \dot{\mathbf{r}} \, \mathrm{d}V \end{aligned} \tag{12.19}$$

These three equations govern the motion of the body in terms of quantities measured in a noninertial reference frame. The first two equations represent momentum conservation and the third represents energy conservation.

### *Choosing a noninertial reference frame*

Let us choose the origin of the noninertial frame as the centre of mass of the body. This is equivalent to the following:

$$\dot{\mathbf{r}}_0 = \frac{1}{m} \int \rho \dot{\mathbf{r}} \, \mathrm{d}V \tag{12.20}$$

where

$m$ is the mass of the body

Using this, it is easily shown that

$$\int \mathrm{P} \dot{\mathbf{r}} \, \mathrm{d}V = \mathbf{P} = 0 \tag{12.21}$$

and

$$\dot{\mathbf{P}} = 0 \tag{12.22}$$

Let us also choose the axes so that the angular momentum of the body as measured by an observer in the noninertial frame is zero. This may be expressed as follows:

$$\boldsymbol{\omega} \cdot I = \mathbf{H}' - \mathbf{r}_0 \times \mathbf{P}' \tag{12.23}$$

It follows that

$$\mathbf{H} = \dot{\mathbf{H}} = 0 \tag{12.24}$$

The expressions for momenta and kinetic energy in the inertial frame reduce to the following:

$$\mathbf{P}' = m\dot{\mathbf{r}}_0 \tag{12.25}$$

$$\mathbf{H}' = m(\mathbf{r}_0 \times \dot{\mathbf{r}}_0) + \boldsymbol{\omega} \cdot I \tag{12.26}$$

$$K' = \tfrac{1}{2}m\dot{\mathbf{r}}_0^2 + \tfrac{1}{2}\boldsymbol{\omega} \cdot I \cdot \boldsymbol{\omega} + K \tag{12.27}$$

Relative to this specific reference frame, the equations of motion equations (12.17) through (12.19) reduce to the following:

$$\mathbf{F} = m\ddot{\mathbf{r}}_0 \tag{12.28}$$

$$\mathbf{L} = \dot{\boldsymbol{\omega}} \cdot I + \boldsymbol{\omega} \times (\boldsymbol{\omega} \cdot \mathbf{I}) + \boldsymbol{\omega} \cdot \dot{\mathbf{I}} \tag{12.29}$$

$$\int \dot{\mathbf{r}} \cdot \mathbf{R} \, \mathrm{d}V + \int \dot{\mathbf{r}} \cdot \mathbf{t} \, \mathrm{d}S = -\tfrac{1}{2}\boldsymbol{\omega} \cdot \dot{I} \cdot \boldsymbol{\omega} + \dot{K} + \int T : \nabla\dot{\mathbf{r}} \, \mathrm{d}V \tag{12.30}$$

where $K$ is the kinetic energy with respect to the noninertial reference frame.

The left-hand side of the last equation is the rate of work done by external forces on the body. The first two terms on the right-hand side are the kinetic energy with respect to the dynamic frame and the last term is the rate of change of internal strain energy in the body.

Equations (12.28) through (12.30) define the equations of motion relative to the dynamic axes fixed so that at all times the origin is at the centre of mass and the axes are orientated so that the angular momentum with respect to the frame is zero.

The translational motion of the centre of mass is given by equation (12.28).

The rotational motion of the body about the centre of mass is given by equation (12.29). For small angular velocities and small strains, the equation reduces to the familiar:

$$\mathbf{L} = \dot{\boldsymbol{\omega}} \cdot \mathbf{I}$$

The internal deformation of the body is defined by equation (12.30). We shall now show that by writing the deformation in terms of the natural vibration modes, decoupled equations can be derived for each mode.

## 12.2 MODAL DECOMPOSITION

With respect to the Tisserand axes the position of a point within the body is $\mathbf{r}$ (Figure 12.2).

We may write

$$\mathbf{r} = \bar{\mathbf{r}} + \mathbf{g} \tag{12.31}$$

where $\bar{\mathbf{r}}$ is the position vector to the undisturbed point and $\mathbf{g}$ is the elastic deformation vector.

The expression is valid for large amplitudes, but the rotations then give rise to strains because of nonlinear terms in the strain measure. The elastic deformation vector $\mathbf{g}$ may be expanded in terms of a finite number of vibration modes, such that

$$\mathbf{g}(x, y, z, t) = \sum_i \boldsymbol{\phi}_i (x, y, z)\xi_i(t) \tag{12.32}$$

where

$\xi_i$ are generalized coordinates and
$\boldsymbol{\phi}_i$ is the vector form of the natural modes.

It should be noted that $i$ represents the $i$th mode, and summation is over all modes.

The natural modes possess the following orthogonality property

$$\int \mathrm{P}\boldsymbol{\phi}_i \boldsymbol{\phi}_j \,\mathrm{d}V = 0 \quad \text{if } i \neq j$$

$$= m_i \quad \text{if } i = j \tag{12.33}$$

where

$m_i$ is the generalized mass (also sometimes called the effective mass) associated with mode $i$, and is defined by this equation.

The modes must also satisfy the linear and angular momentum equations, which for free vibration lead to

$$\int \boldsymbol{\phi}_i \,\mathrm{d}V = 0 \tag{12.34}$$

$$\int \mathbf{r} \times \boldsymbol{\phi}_i \,\mathrm{d}V = 0 \tag{12.35}$$

It is shown by Fraeijs de Veubeke (1976), that if the expansion of relative displacements **g**, is limited to natural modes, then the principle of minimum square average displacement is automatically satisfied and the equations of mean motion (rigid body motion), equations (12.28) and (12.29), are just those associated with the motion of the dynamic axes.

*Hamilton's principle applied to generalized modes*

By applying various principles such as weighted residual, Lagrange's principle, Hamilton's principle, or virtual work (d'Alembert's principle), the equations for each mode can be decoupled. Here we present the derivation based on Hamilton's principle which can be stated as follows.

$$\delta\int_{t_0}^{t_1}(K-U)\,\mathrm{d}t=\int_{t_0}^{t_1}\delta W\,\mathrm{d}t \tag{12.36}$$

where

| | |
|---|---|
| $K$ | is the kinetic energy |
| $U$ | is the internal energy |
| $\delta W$ | is the virtual work |

*Kinetic energy*

The expression for kinetic energy, leads to the following:

$$K'=\tfrac{1}{2}m\dot{\mathbf{r}}_0^2+\tfrac{1}{2}\boldsymbol{\omega}\cdot I\cdot\boldsymbol{\omega}+\tfrac{1}{2}m_i\dot{\xi}_i^2 \tag{12.37}$$

where

$$m_i=\int\rho\phi_i\phi_i\,\mathrm{d}V$$

It should be noted that the orthogonality of the natural modes ensures no cross terms in the expression for generalized mass $m_i$.

*Internal strain energy/elastic restoring terms*

The first two equations for free vibration are given in equations (12.34) and (12.35). The third equation is

$$\Gamma(\boldsymbol{\phi})=\rho n^2\boldsymbol{\phi} \tag{12.38}$$

where

$n$ are the angular frequencies of the vibration modes (the eigenvalues)

and

$\boldsymbol{\Gamma}$ is an operator depending on the deformation properties of the body.

### *Internal strain energy*

Bisplinghoff and Ashley (1975), show that for small strains, the internal strain energy can be expressed in the form

$$U = \tfrac{1}{2}\int \mathbf{g} \cdot \boldsymbol{\Gamma}(\mathbf{g})\,\mathrm{d}V \tag{12.39}$$

Using

$\mathbf{g} = \sum_{1} \boldsymbol{\phi}_i \xi_i$ and the orthogonality conditions yields

$$U = \tfrac{1}{2}\sum_{i} \int \boldsymbol{\phi}_i \cdot \Gamma(\boldsymbol{\phi}_i)\,\mathrm{d}V\,\xi_i \xi_i \tag{12.40}$$

Using

$$\Gamma(\phi_i) = \rho n_i^2 \phi_i$$

gives

$$U = \tfrac{1}{2}\sum_{i} n_i^2 \int \boldsymbol{\phi}_i \cdot \boldsymbol{\phi}_i\,\rho\,\mathrm{d}V\,\xi_i \xi_i \tag{12.41}$$

but

$$\int \boldsymbol{\phi}_i \cdot \boldsymbol{\phi}_i\,\rho\,\mathrm{d}V = m_i$$

Thus

$$U = \tfrac{1}{2}\sum_{i} m_i n_i^2 \xi_i^2 \tag{12.42}$$

*Work done by external forces*

If

$$\mathbf{P} = \int \mathbf{t}\, \mathrm{d}S + \int \mathbf{R}\, \mathrm{d}V \tag{12.43}$$

where

$\mathbf{t}$ are the surface tractions per unit area

and

$\mathbf{R}$ are the body forces per unit volume.

The increment of work done during the generalized displacement $\mathbf{dg}$ is given by

$$\delta W = \mathbf{P} \cdot \mathbf{dg} = \int \mathbf{t} \cdot \mathbf{dg}\, \mathrm{d}S = \int \mathbf{R} \cdot \mathbf{dg}\, \mathrm{d}\mathbf{V}$$

$$= \sum_i \left\{ \int \mathbf{t} \cdot \boldsymbol{\phi}_i\, \mathrm{d}\xi_i \cdot \mathrm{d}S + \int \mathbf{R} \cdot \boldsymbol{\phi}_i \right\} \mathrm{d}\xi_i\, \mathrm{d}\mathbf{V} \tag{12.44}$$

Substituting for kinetic energy, internal energy and work done into Hamilton's equation (12.36) and taking the variation with respect to the generalized coordinates $\xi_i$, gives

$$\int \left\{ m_j\, \ddot{\xi}_j + m_j n_j^2\, \xi_j - \int (\mathbf{t} \cdot \boldsymbol{\phi}_j)\, \mathrm{d}S - (\mathbf{R} \cdot \boldsymbol{\phi}_j)\, \mathrm{d}V \right\} \mathrm{d}\xi_j\, \mathrm{d}t = 0 \tag{12.45}$$

Since the $\mathrm{d}\xi_j$ are perfectly arbitrary we deduce

$$m_j \ddot{\xi}_j + m_j n_j^2 \xi_j = \int (\mathbf{t} \cdot \boldsymbol{\phi}_j)\, \mathrm{d}S + \int (\mathbf{R} \cdot \boldsymbol{\phi}_j)\, \mathrm{d}V \tag{12.46}$$

This equation determines each deformation mode of the body.

(Using a suitable definition of $\boldsymbol{\phi}_j$, this equation can also be used to represent rigid body motion.)

The first term on the left-hand side is the generalized acceleration, e.g., the linear acceleration, the angular acceleration, or the deformation mode acceleration.

The second term on the left-hand side is the restoring force due to the generalized displacement $\xi_j$. For rigid body motion it is zero, since the eigenvalues are zero. For deformation modes it depends on the deformational moduli of the body and the geometry.

The terms on the right-hand side are the generalized forces. For linear motion the $\phi$ are unity, and the terms are summed to the resultant force. For angular rotation the $\phi$ are terms such as $x$ and $y$, and the terms represent the resultant couple. For other modes, the $\phi$ may be various functions of the coordinates.

It should be noted that damping has been excluded in this derivation. In its most general form it will alter the eigenvectors and eigenvalues of the body. However, as detailed by Bathe and Wilson (1976), if a prudent choice is made and the damping is small, the damping matrix will not significantly alter the eigenvalues or eigenvectors, e.g. mass, stiffness or Raleigh damping.

## 12.3 MODAL DECOMPOSITION IN FINITE ELEMENTS

In the solution of the problem

$$M\ddot{\mathbf{u}} + K\mathbf{u} = \mathbf{f} \tag{12.47}$$

where **M** is the matrix, **K** is the symmetric, positive definite stiffness matrix, **u** is the vector of unknown nodal displacements, and **f** is the given load vector, we can set

$$\mathbf{u} = \boldsymbol{\phi}\alpha(t) \tag{12.48}$$

where $\phi$ is the matrix of eigenvectors of the standard eigenproblem

$$K\alpha\boldsymbol{\phi} = \lambda M\alpha\boldsymbol{\phi} \tag{12.49}$$

and $\alpha(t)$ is the vector of modal participation factors. Equation (12.41) is then rewritten as

$$M\boldsymbol{\phi}\ddot{\alpha} + K\boldsymbol{\phi}\alpha = \mathbf{f} \tag{12.50}$$

and after premultiplying with $\boldsymbol{\phi}^T$ we arrive at

$$\boldsymbol{\phi}^T M\boldsymbol{\phi}\ddot{\alpha} + \boldsymbol{\phi}^{\mathbf{T}}\mathbf{K}\boldsymbol{\phi}\alpha = \boldsymbol{\phi}^T \cdot \mathbf{f} \tag{12.51}$$

As $\boldsymbol{\phi}$ contains the eigenvectors of matrix **K**, $\boldsymbol{\phi}^T\mathbf{K}\boldsymbol{\phi}$ is in fact the stiffness matrix in its canonical form, and $\boldsymbol{\phi}^T M\boldsymbol{\phi}$ the mass matrix, so that

$$\boldsymbol{\phi}^T K\boldsymbol{\phi} = \Lambda \qquad \boldsymbol{\phi}^T M\boldsymbol{\phi} = I \tag{12.52}$$

where $\Lambda$ is the diagonal matrix, with the eigenvalues $\lambda_i$ on the diagonal, and $I$ is the identity matrix. The system of $n$ linear equations can now be decomposed into $n$ independent equations

$$\ddot{\alpha}_i + \lambda_i \alpha_i = \boldsymbol{\phi}_i^T \cdot \mathbf{f} \qquad i = 1, n \tag{12.53}$$

from which we can solve for $\alpha_i$ directly.

Equations (12.51) and (12.53) are once again the now familiar equations for determining modal response.

## 12.4 DETAILED ONE-DIMENSIONAL FORMULATION

### *Modal description for one-dimensional contact problems*

The principals of modal decomposition are best illustrated by a simple one dimensional example.

Consider the one-dimensional linear finite element shown in Figure (12.3) with shape functions $N_1$ and $N_2$ given by:

$$N_1 = \tfrac{1}{2}(1 - \xi), \qquad N_2 = \tfrac{1}{2}(1 + \xi) \tag{12.54}$$

For simplicity the element is taken as extending from $x = -L/2$ to $x = +L/2$. Let the Young's modulus of the element be denoted by $E$, the density by $\rho$, and the cross sectional area by $A$.

If the element is acted on by forces $F_1$ and $F_2$ at nodes 1 and 2 respectively and no damping is considered, the motion and deformation of the element is given by

$$[M]\{\ddot{u}\} + [K]\{u\} = \{F\} \tag{12.55}$$

where the mass matrix $[M]$ and stiffness matrix $[K]$ are determined as follows:
Noting

$$\frac{\partial N_1}{\partial x} = -\frac{1}{L}, \qquad \frac{\partial N_2}{\partial x} = +\frac{1}{L}$$

and

$$u(x) = \sum_{i=1}^{2} N_i u_i$$

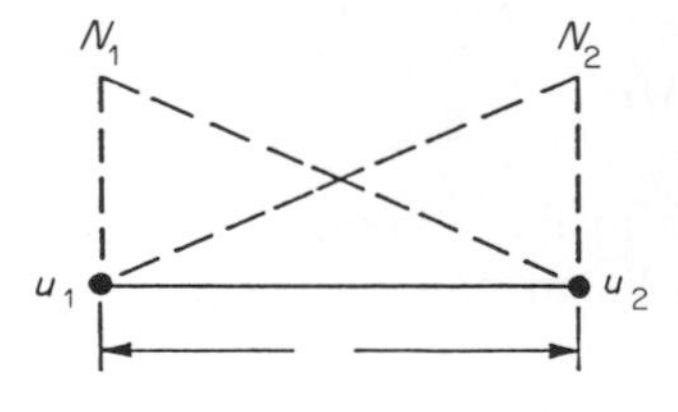


**Figure 12.3**
A linear one-dimensional element with shape functions $N_1$ and $N_2$

It follows that

$$\epsilon_{xx} = \frac{\partial u}{\partial x} = \sum_{i=1}^{2} \frac{\partial N_i}{\partial x} u_i$$

$$= \left(-\frac{1}{L}\right) u_1 + \left(\frac{1}{L}\right) u_2 \qquad (12.56)$$

This is usually expressed in typical Finite Element notation as

$$\epsilon_{xx} = [B]\{u\} \quad \text{where } [B] = \left[-\frac{1}{L}, \frac{1}{L}\right] \qquad (12.57)$$

The stiffness matrix is given by

$$[B]^T[D][B]\,\mathrm{d}V$$

In this example, this becomes

$$[K] = \int_{\xi=-1}^{+1} A \left\{-\frac{1}{L}, \frac{1}{L}\right\}^T [E] \left\{-\frac{1}{L}, \frac{1}{L}\right\} \frac{L}{2}\,\mathrm{d}\xi$$

$$= \frac{EA}{L}\begin{bmatrix} 1 & -1 \\ -1 & 1 \end{bmatrix} \qquad (12.58)$$

Choosing a lumped mass matrix (it is left as an exercise to repeat the calculations with a consistent mass matrix derived using

$$m_{ij} = \frac{AL\rho^{+1}}{2} \int_{\xi=-1} N_i N_j \,\mathrm{d}\xi)$$

the following expression results

$$[\mathrm{M}] = \frac{\mathrm{AL}\rho}{2}\begin{bmatrix} 1 & 0 \\ 0 & 1 \end{bmatrix} \qquad (12.59)$$

Equation (12.55) now can be written

$$\frac{\mathrm{AL}\rho}{2}\begin{bmatrix} 1 & 0 \\ 0 & 1 \end{bmatrix}\begin{Bmatrix} \ddot{u}_1 \\ \ddot{u}_2 \end{Bmatrix} + \frac{AE}{L}\begin{bmatrix} 1 & -1 \\ -1 & 1 \end{bmatrix}\begin{Bmatrix} u_1 \\ u_2 \end{Bmatrix} = \begin{Bmatrix} F_1 \\ F_2 \end{Bmatrix} \qquad (12.60)$$

Now let

$$u_i = [\phi]\alpha_i(t) \quad \text{and} \quad \alpha_i(t) = \alpha_i\, \mathrm{e}^{i\omega t}$$

The eigenvalue equation becomes

$$\left\{-\rho\omega^2 \frac{AL}{2}\begin{bmatrix}1 & 0\\0 & 1\end{bmatrix} + \frac{AE}{L}\begin{bmatrix}1 & -1\\-1 & 1\end{bmatrix}\right\}\{\phi\} = 0 \tag{12.61}$$

It follows that

$$\begin{bmatrix}\left(\dfrac{E}{L} - \rho\omega^2\dfrac{L}{2}\right) & -\dfrac{E}{L}\\ -\dfrac{E}{L} & \left(\dfrac{E}{L} - \rho\omega^2\dfrac{L}{2}\right)\end{bmatrix}\begin{Bmatrix}\phi_1\\\phi_2\end{Bmatrix} = 0 \tag{12.62}$$

Which gives

$$\left(1 - \frac{\rho\omega^2L^2}{2E}\right)^2 - 1 = 0$$

or

$$\left[\left(1 - \frac{\rho\omega^2L^2}{2E}\right) - 1\right]\left[\left(1 - \frac{\rho\omega^2L^2}{2E}\right) + 1\right] = 0 \tag{12.63}$$

The two roots are given by

$$\omega = 0 \quad \text{and} \quad \omega = \frac{2}{L}\sqrt{\frac{E}{\rho}}$$

Rewriting equation (12.62) gives

$$\begin{bmatrix}\left(1 - \dfrac{\rho\omega^2L^2}{2E}\right) & -1\\ -1 & \left(1 - \dfrac{\rho\omega^2L^2}{2E}\right)\end{bmatrix}\begin{Bmatrix}\phi_1\\\phi_2\end{Bmatrix} = 0 \tag{12.64}$$

For

$$\omega = 0, \qquad \phi_1 - \phi_2 = 0 \quad \text{or } \phi_1 = \phi_2$$

For

$$\omega = \frac{2}{L}\sqrt{\frac{E}{\rho}}, \qquad \phi_1 + \phi_2 = 0 \quad \text{or } \phi_1 = -\phi_2$$

Thus the two eigenvectors are given by $\begin{Bmatrix} 1 \\ 1 \end{Bmatrix}$ and $\begin{Bmatrix} 1 \\ -1 \end{Bmatrix}$

These can be made orthonormal to $[M]$ so that:

$$\{\phi\}^T[M]\{\phi\} = \delta_{ij} \tag{12.65}$$

by multiplying each vector by '$a$' where

$$a = \frac{1}{\sqrt{\rho LA}}$$

The $[M]$ orthogonal eigenvectors are

$$\frac{1}{\sqrt{\rho LA}}\begin{Bmatrix} 1 \\ 1 \end{Bmatrix} \text{ and } \frac{1}{\sqrt{\rho LA}}\begin{Bmatrix} 1 \\ -1 \end{Bmatrix}$$

Equation (12.55) can now be transformed as follows

$$[\phi]^T[M][\phi]\{\ddot{\alpha}\} + [\phi]^T[K][\phi]\{\alpha\} = [\phi]^T\{F\} \tag{12.66}$$

where

$$[\phi]^T[M][\phi] = \begin{bmatrix} 1 & 0 \\ 0 & 1 \end{bmatrix}$$

The canonical form of the stiffness matrix is given by

$$\begin{aligned}[\phi]^T[K][\phi] &= \frac{1}{\rho LA}\begin{bmatrix} 1 & 1 \\ 1 & -1 \end{bmatrix}\frac{AE}{L}\begin{bmatrix} 1 & -1 \\ -1 & 1 \end{bmatrix}\begin{bmatrix} 1 & 1 \\ 1 & -1 \end{bmatrix} \\ &= \frac{4E}{\rho L^2}\begin{bmatrix} 0 & 0 \\ 0 & 1 \end{bmatrix}\end{aligned} \tag{12.67}$$

The generalized forces are given by

$$[\phi]^T\{F\} = \frac{1}{\sqrt{\rho LA}}\begin{Bmatrix} (F_1 + F_2) \\ (F_1 - F_2) \end{Bmatrix}$$

The transformed equation is therefore

$$\begin{bmatrix} 1 & 0 \\ 0 & 1 \end{bmatrix}\begin{Bmatrix} \ddot{\alpha}_1 \\ \ddot{\alpha}_2 \end{Bmatrix} + \frac{4E}{\rho L^2}\begin{bmatrix} 0 & 0 \\ 0 & 1 \end{bmatrix}\begin{Bmatrix} \alpha_1 \\ \alpha_i \end{Bmatrix} = \frac{1}{\sqrt{\rho LA}}\begin{Bmatrix} (F_1 + F_2) \\ (F_1 - F_2) \end{Bmatrix} \tag{12.68}$$

Thus

$$\ddot{\alpha}_1 = \frac{1}{\sqrt{\rho LA}}(F_1 + F_2) \tag{12.69}$$

and

$$\ddot{\alpha}_2 + \frac{4E}{\rho L^2}\alpha_2 = \frac{1}{\sqrt{\rho LA}}(F_1 - F_2)$$

or

$$\ddot{\alpha}_2 + \omega^2\alpha_2 = \frac{1}{\sqrt{\rho LA}}(F_1 - F_2) \tag{12.70}$$

Remembering

$$\{u\} = [\phi]\{\alpha\} = \frac{1}{\sqrt{\rho LA}}\begin{bmatrix} 1 & 1 \\ 1 & -1 \end{bmatrix}\begin{Bmatrix} \alpha_1 \\ \alpha_2 \end{Bmatrix} \tag{12.71}$$

we may write

$$\begin{Bmatrix} \alpha_1 \\ \alpha_2 \end{Bmatrix} = \frac{\sqrt{\rho LA}}{2}\begin{bmatrix} 1 & 1 \\ 1 & -1 \end{bmatrix}\begin{Bmatrix} u_1 \\ u_2 \end{Bmatrix} \tag{12.72}$$

Substituting this into equations (12.63) and (12.70) the following are derived

$$\rho LA\left(\frac{\ddot{u}_1 + \ddot{u}_2}{2}\right) = F_1 + F_2 \tag{12.73}$$

$$\frac{\rho L^2 A}{L}\left(\frac{\ddot{u}^1 - \ddot{u}^2}{L}\right) + \frac{2EA}{L}\frac{(u^1 - u^2)}{L} = F_1 - F_2 \tag{12.74}$$

Noting that $(u_1 + u_2)/2$ is the displacement of the centre of mass the first equation determines the motion of the centre of mass and can be rewritten

$$M\ddot{u}_g = \sum F \tag{12.75}$$

where $u_c$ is the displacement of the centre of mass and

$$M = \rho AL$$

Noting that $(u_1 - u_2)/L$ is the strain in the element (the length of the element being $L$) the second equation can be rewritten

$$\frac{ML}{2}\left(\frac{\ddot{u}_1 - \ddot{u}_2}{L}\right) + 2E\left(\frac{u_1 - u_2}{L}\right)A = (F_1 - F_2)$$

or

$$M_g\ddot{\varepsilon} + 2\sigma A = (F_1 - F_2) \tag{12.76}$$

where $\sigma$ is the internal stress in the element and $M_g$ is called the generalized (or effective) mass.

### *Finite element description for one-dimensional contact problems*

An equivalent formulation can be derived using a standard Finite Element approach. The method described below should be compared with the modal description.

The dynamic equilibrium equation for a single linear one-dimensional finite element is given by equation (12.60)

$$\frac{AL\rho}{2}\begin{bmatrix}1 & 0\\0 & 1\end{bmatrix}\begin{Bmatrix}\ddot{u}_1\\\ddot{u}_2\end{Bmatrix} + \frac{AE}{L}\begin{bmatrix}1 & -1\\-1 & 1\end{bmatrix}\begin{Bmatrix}u_1\\u_2\end{Bmatrix} = \begin{Bmatrix}f_1\\f_2\end{Bmatrix}$$

Consider now the two elements shown in Figure 12.4(a) with connectivity enforced at node 2.

For node 1

$$m_1\ddot{u}_1 = f_1 - \frac{AE_1}{L_1}(u_1 - u_2) \tag{12.77}$$

where $E_1$ is the modulus for element 1 and $m_1$ is the nodal mass contribution from element 1.

For node 2, which spans two elements, the following equations are derived for each element singly (the external force $f$, is arbitrarily shared between the two elements and the total mass at node 2, given by $m_2$, is assumed to derive from each element equally).

**Figure 12.4**
(a) Two one-dimensional Finite Elements with common node 2; (b) two separate one-dimensional finite elements with possible interaction between Nodes 2 and 3

$$m_2\ddot{u}_2/2 = \frac{f_2}{2} - \frac{E_1}{L_1}(-u_1 + u_2) - R \quad \text{for element 1} \tag{12.78}$$

$$m_2\ddot{u}_2/2 = \frac{f_2}{2} - \frac{E_2}{L_2}(u_2 - u_3) + R \quad \text{for element 2} \tag{12.79}$$

where $R$ is the reactions of element 1 on element 2.

The resulting equation for the combined elements is

$$m_2\ddot{u}_2 = f_2 + A\left\{\frac{E_1}{L_1}u_1 - \left(\frac{E_1}{L_1} + \frac{E_2}{L_2}\right)u_2 + \frac{E_2}{L_2}u_3\right\} \tag{12.80}$$

Now consider the two elements as separated and interacting with an interaction force $F$, as illustrated in Figure 12.4(b) where $F = k_{23}\,(u_3 - u_2)$ and $k_{23}$ denotes the interaction stiffness between nodes 2 and 3.

In this case the equation for $u_2$ is

$$m_2\ddot{u}_2 = f_2 + A\left\{\frac{E_1}{L_1}u_1 - \left(\frac{E_1}{L_1} + \frac{k_{23}}{A}\right)u_2 + \frac{k_{23}}{A}u_3\right\} \tag{12.81}$$

Comparing equations (12.80) and (12.81) it is seen that the effect of replacing element connectivity by an interaction stiffness is to replace $AE^2/L_2$ by $k_{23}$.

Similarly for node 3 the following is derived

$$m_3\ddot{u}_3 = f_3 + A\left\{\frac{k_{23}}{A}u_2 - \left(\frac{k_{23}}{A} + \frac{E_2}{L_2}\right)u_3 + \frac{E_2}{L_2}u_4\right\} \tag{12.82}$$

Consider now the problem of a body consisting of two elements interacting with a body consisting of a single element. The following table illustrates the quantities associated with each element and node.

| | | | | | | | | | | | |
|---|---|---|---|---|---|---|---|---|---|---|---|
| displacement | | $u_1$ | | $u_2$ | | $u_3$ | | $u_4$ | | $u_5$ | |
| | / | | \ / | | \ / | | \ / | | \ / | | \ |
| modulus $E$ | 0 | | $E_1$ | | $E_2$ | | 0 | | $E_3$ | | 0 |
| stiffness $k$ | 0 | | 0 | | 0 | | $k_{34}$ | | 0 | | 0 |
| geometry | | O | —— | O | —— | O | | O | —— | O | |
| | | $u_1$ | | $u_2$ | | $u_3$ | | $u_4$ | | $u_5$ | |

It is noted that the correct equation for each node can be assembled if the element moduli or stiffnesses to each side of the node are known, and the correct nodal mass is constructed.

For example, consider node 3. Noting that either the element modulus or interaction stiffness will occur but not both, the equation can be written as

$$m_3 \ddot{u}_3 = f_3 + (P_{23} u_2 - (P_{23} + P_{34}) u_3 + P_{34} u_4) \tag{12.83}$$

where $m$ is the total nodal mass and $P_{23}$ denotes either $AE_2/L_2$ or the stiffness, $k_{23}$, between nodes 2 and 3.

Introducing mass proportional damping so that $c = \alpha m$ and stiffness proportional damping so that $c = \beta k$ the following general equation for node $i$ is derived

$$\begin{aligned}
\dot{u}_i^{n+\frac{1}{2}} = {} & R_1 R_2 \dot{u}_i^{n-\frac{1}{2}} + \frac{R_1 \Delta t}{m_i} f_i^n \\
& + \frac{R_1 \Delta t}{m_i} (P_{i-1,i} u_{i-1}^n - (P_{i-1,i} + P_{i,i+1}) u_i^n \\
& + P_{i,i+1} u_{i+1}^n) \\
& + \frac{R_1 \Delta t}{m_i} (\beta_{i-1,i} u_{i-1}^{n-\frac{1}{2}} - (\beta_{i-1,i} + \beta_{i,i+1})\, u_i^{n-\frac{1}{2}} \\
& + \beta_{i,i+1}\, \dot{u}^{n-\frac{1}{2}}{}_{i+1})
\end{aligned} \tag{12.84}$$

where

$$R_1 = \left(1 + \frac{\alpha \Delta t}{2}\right)^{-1}$$

and

$$R_2 = \left(1 - \frac{\alpha \Delta t}{2}\right)$$

## 12.5 GENERALIZED MODAL METHODS FOR THE ANALYSIS OF DISCRETE SYSTEMS

### 12.5.1 Introduction to simply deformable elements

The Discrete Element Method solves the dynamic equilibrium equations, given below, for individual elements,

$$\rho \ddot{u}_i = R_i + \frac{\partial \sigma_{ij}}{\partial x_j} \quad \text{in region } \Omega \tag{12.85}$$

subject to boundary conditions $u_i = u_0$ on $\Gamma_1$ and $F_i = F_0$ on $\Gamma_2$ where $\rho$ is the density, $\ddot{u}_i$ the acceleration in direction $i$, $R_i$ the body force, $\sigma_{ij}$ the stress tensor (compression positive), and $x_j$ the Cartesian coordinate of point $p$ in the body at which all quantities are evaluated.

For rigid bodies equations (12.85) reduce to the standard equations of motion

$$\sum F_i = m\ddot{u}_i \text{ (translation of centre on mass in } i \text{ direction)} \tag{12.86}$$

$$\sum C = I\ddot{\theta} \text{ (rotation about centre of mass)} \tag{12.87}$$

where $F_i$ is the resultant force in the $i$th direction, $C$ is the couple about the centre of mass, $m$ is the element mass, and $I$ is the moment of inertia about the centre of mass.

The rigid block method was extended by Cundall *et al.* (1978) so that each element could also deform about its centroid. These elements were called Simply Deformable Elements. Although the original formulation in terms of element strains has a number of problems, which are identified below, the method can be adjusted to be correct by using other element modes. We shall follow the historical approach here and deal first with the strain mode approach.

It was proposed that equations for the element strain can be written independent of rigid body modes as

$$m^k \ddot{\epsilon}^k = \sigma_A^k - \sigma_I^k \tag{12.88}$$

where

- $m^k$ is the generalized (effective) mass of the element corresponding to strain mode $k$
- $\epsilon^k$ is the internal strain velocity rate of the strain mode $k$
- $\sigma_A^k$ is a generalized force (applied stress) on the element corresponding to the strain mode $k$, e.g. $\epsilon_{xx}$, $\epsilon_{yy}$, $\epsilon_{xy}$
- $\sigma_I^k$ is the internal stress on the element corresponding to strain mode $k$.

These so called 'strain equations' have been used in several discrete (distinct) element formulations in the literature. To date their justification has been based on the assumption that the element deformation can be written as a superposition of the rigid body motion and the strains, such that the displacement at any point p in the element is given by:

$$u_i = \bar{u}_i + \omega_{ij}\bar{x}_j + \epsilon_{ij}\bar{x}_j \tag{12.89}$$

where

- $\bar{x}_i$ is the coordinate of point $p$ relative to the element centroid
- $u_i$ is the displacement of any point $p$ in the element
- $\bar{u}_i$ is the displacement of the centroid in direction $i$

$\bar{x}_j$ is the coordinate of point $p$ relative to the centroid
$\omega_{ij}$ is the rotation tensor defined by $\omega_{ij} = e_{ijk}\omega_k$ where $e_{ijk}$ is the permutation tensor and $\omega_k$ is the rotation vector
$\epsilon_{ij}$ is the strain tensor

and summation over repeated indices is implied.

Equation (12.89) is a special case of the more general expansion for the displacement at any point in the element in terms of the modes $\phi$, such that

$$u_i = \sum_N \phi_i^N \alpha^N \tag{12.90}$$

where $\phi^N$ is the $N$th modal vector and $\alpha^N$ is the corresponding time dependent participation factor. It is also assumed that the modes $\phi$ are chosen to be orthogonal so that $\boldsymbol{\phi}^M \cdot \boldsymbol{\phi}^N = 0$ if $M \neq N$.

Substituting equation (12.90) into equation (12.85) weighting each with $\phi^L{}_i$ and integrating over the volume gives

$$\sum_M \int \rho\phi_i^L \phi_i^M \, \mathrm{d}V \, \ddot{\alpha}^M = \int \phi_i^L \, R_i \, \mathrm{d}V + \int \phi_i^L \frac{\partial \sigma_{ij}}{\partial x_j} \, \mathrm{d}V \tag{12.91}$$

Using the orthogonality of modes, the equation for the participation of mode $\phi^L$ is derived

$$\int \rho\phi_i^L \, \phi_i^L \, \mathrm{d}V \, \ddot{\alpha}^L = \int \phi_i^L \, R_i \, \mathrm{d}V + \int \phi_i^L \frac{\partial \sigma_{ij}}{\partial x_j} \, \mathrm{d}V \tag{12.92}$$

This can be combined with the force boundary condition

$$\phi_i^L (F_i - F_0) \, \mathrm{d}\Gamma_2 = 0$$

to give

$$m^L \ddot{\alpha}^L = S_A^L - S_I^L \tag{12.93}$$

where

$$m^L = \int \rho\phi^L{}_i \, \phi^L{}_i \, \mathrm{d}V$$

$$S_A^L = \int \phi^L{}_i \, R_i \, \mathrm{d}V + \int \phi^L{}_i \, F_0 \, \mathrm{d}\Gamma_2$$

$$S_I^L = \int \phi^L{}_i \, \sigma_{ij,j} \, \mathrm{d}V$$

The close correspondence of equation (12.93) to that of equation (12.88) is noted. The strains in equations (12.87) and (12.88) are particular cases of participation factors $\alpha$ and the coordinates $\bar{x}_j$ give the form of the mode $\phi$. Now for the specific choice of modes given by equation (12.88) it remains to be shown that the orthogonally conditions are satisfied.

### 12.5.2 Orthogonality conditions

Consider the deformation modes given in equation (12.88). To be specific let us examine them for a linear four-noded quadrilateral (Figure 12.5(a) such that the nodal displacements are given by $(u)^T = \{u_1, u_2, u_3, u_4, u_5, u_6, u_7, u_8\}$ where $u_1 = u_x$ for node 1 and $u_2 = u_y$ for node 1, etc. The rigid body modes can be written as follows

$$\{\phi^1\}^T = \{1, 0, 1, 0, 1, 0, 1, 0\} \qquad x \text{ translation}$$

$$\{\phi^2\}^T = \{0, 1, 0, 1, 0, 1, 0, 1\} \qquad y \text{ translation}$$

$$\{\phi^3\}^T = \{-y_1, x_1, -y_2, x_2, -y_3, x_3, -y_4, x_4\} \qquad \text{rotation}$$

where $x_i$ and $y_i$ are the coordinates of the nodes with respect to the centre of mass. The strain modes are given by

$$\{\phi^4\}^T = \{x_1, 0, x_2, 0, x_3, 0, x_4, 0\} \qquad \text{strain } x$$

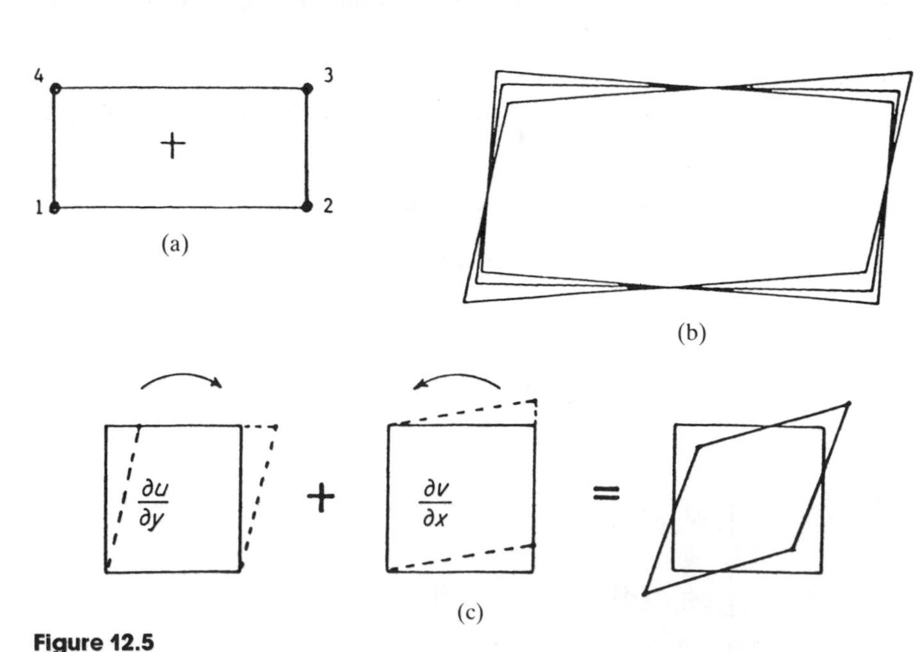


**Figure 12.5**
(a) Linear four-node quadrilateral; (b) illustration of shear giving rise to rigid body rotation; (c) illustration of rotation cancellation in shear strain of square

$\{\phi^5\}^T = \{0, y_1, 0, y_2, 0, y_3, 0, y_4\}$ strain $y$

$\{\phi^6\}^T = \{y_1, x_1, y_2, x_2, y_3, x_3, y_4, x_4\}$ shear strain

Two other modes are necessary to span the space, but they were not originally considered, and for the present they will be specified only as $\{\phi^7\}$ and $\{\phi^8\}$ and will be assumed orthogonal to the other modes. (It is easily shown that these modes correspond to the so called hourglass modes.)

Now if the principle of modal superposition is to be applied the strain deformation modes must be orthogonal to the rigid body translation and rotation modes (Bisplinghoff *et al.*, 1962; de Veubeke, 1972). These conditions can be written as

$$\int \rho \boldsymbol{\phi}^N \, dV = 0 \quad \text{(for all modes } N\text{, excluding translation)} \tag{12.94}$$

$$\int \rho(\mathbf{r} \times \boldsymbol{\phi}^N) \, dV = 0 \quad \text{(for all modes } N\text{, excluding rotation)} \tag{12.95}$$

For a rectangular element with masses lumped at the nodes, equations (12.94) and (12.95) imply the conditions detailed in Table 12.1. (Similar conditions can be deduced for distributed mass idealizations.)

**Table 12.1**
Conditions that the element must satisfy if strains are modes

| Condition \ Deformation mode | $\epsilon_{xx}$ | $\epsilon_{yy}$ | $\epsilon_{xy}$ |
|---|---|---|---|
| $\int \rho\phi^N dV=0$ | $\Sigma m_i x_i=0$ | $\Sigma m_i y_i=0$ | $\Sigma m_i(x_i+y_i)=0$ |
| $\int \rho(r\times\phi^N)dV=0$ | $\Sigma m_i x_i y_i=0$ | $\Sigma m_i x_i y_i=0$ | $\Sigma m_i((x_i)^2-(y_i)^2)=0$ |

where

$m_i$ is the mass node $i$

$x_i$, $y_i$ are the coordinates of node $i$.

There are four independent conditions and these must be satisfied if each strain deformation mode is to be orthogonal to the rigid body modes.

The conditions $\Sigma m_i x_i = 0$ and $\Sigma m_i y_i = 0$ are satisfied if the deformations are measured with respect to local axes with origin at the centre of mass of the element. The condition $\Sigma m_i x_i y_i = 0$ can be satisfied by particular choice of the orientation of the local axes. These axes are called the principal axes of the element. The final condition $\Sigma m_i (x_i^2 - y_i^2)$ cannot in general be satisfied. The particular case when $\Sigma m_i(x_i^2 - y_i^2) = 0$ is satisfied occurs when the principal moments of inertia are equal. It is therefore concluded that for general shaped elements the shear strain deformation produces a rotation of the element and cannot be used in a formulation dependent on superposition. Figure 12.5(b) illustrates how shear strain and rotation are coupled for a rectangular element.

### 12.5.3 Deducing a pseudo shear mode orthogonal to rotation

In order to deduce a 'shear type' mode which is orthogonal to the other modes it is necessary to understand what causes the shear to give rise to a rotation.

The shear strain $\epsilon_{xy}$ is defined as $\frac{1}{2}(\partial v/\partial x + \partial u/\partial y)$ or as $\frac{1}{2}(\epsilon_{yx} + \epsilon_{xy})$ and can be viewed as the sum of two separate deformations $\epsilon_{yx}$ and $\epsilon_{xy}$. For a square element the rotation inherent in the one deformation is countered by the opposite rotation in the other as illustrated in Figure 12.5(c). When the element considered is not a square the rotations do not cancel each other and a resultant rotation accrues.

To find a mode with no rotation let us consider a 'pseudo' shear mode given by $\bar{\epsilon}_{xy} = a\epsilon_{yx} + b\epsilon_{xy}$ where $a$ and $b$ are constants for a given element geometry. The corresponding $\phi$ is

$$[\bar{\phi}_6]^T = \{ay_1, bx_1, ay_2, bx_2, ay_3, bx_3, ay_4, bx_4\}$$

The condition that this is orthogonal to the rotation mode $[\phi_3]$ is, for a lumped mass approach

$$[\phi^3]^T[\bar{\phi}_6] = 0 \tag{12.96}$$

or

$$a\sum y_i^2 = b\sum x_i^2, \qquad a:b = \sum x_i^2 : \sum y_i^2$$

It is easily shown that in general the ratio $a : b$ is actually the ratio of the moments of inertia of the element $I_y : I_x$, about the principal local axes of the element.

It should be noted that the actual coordinate values of $x_1$, $y_1$ etc. of the nodes should be substituted in checking orthogonality. It is then verified that $\{\bar{\phi}_6\}$ is orthogonal to all other modes $\{\phi\}$.

Using similar logic as above it can be demonstrated that, for a generally shaped two dimensional quadrilateral, it is possible to choose modes which are orthogonal to each other and which are closely related to the strains of the element. In general these modes are *not* eigenmodes of the element. (Appendix IV shows that it is relatively straightforward for a rectangular element to choose the actual eigenmodes.)

The modified strain modes can be used as the basis of an explicit scheme similar to that proposed by Cundall *et al.* (1978). However, we must ask ourselves what, if anything, has been gained by our mental gymnastics. To answer this question let us explore in more detail alternative formulations.

### 12.5.4 Alternative formulations

Consider the dynamic equilibrium equation (12.85) applied to a linear finite element. It is shown in standard Finite Element texts, Bathe and Wilson (1976), that the equations can be written in matrix form as

$$[M]\{\ddot{u}\} = \{f\} - [K]\{u\} \tag{12.97}$$

where $\{\mathbf{u}\}$ is the vector of nodal displacements, $[M]$ is the mass matrix, $[K]$ the stiffness matrix, and $\{f\}$ the vector of nodal forces. Here all quantities are evaluated at time step $n$, and except for $\{\ddot{u}\}$ are assumed to be known at this time.

In general the stiffness matrix $[K]$ is nondiagonal. If the matrix $[M]$ is made diagonal, then a standard explicit scheme results, and each nodal acceleration $\ddot{u}_i$ can be calculated independently from $\{\ddot{u}\} = [M]^{-1} (\{f\} - [K]\{u\})$.

For an explicit scheme to be possible the critical point to note is that only the mass matrix need be diagonal. Coupling between known quantities $u_i$ on the RHS does not inhibit the use of an explicit scheme.

Consider now deformation modes of the element specified by

$$[\phi]\{\alpha\} = \{u\}$$

where it is assumed that each mode is orthogonal to all others, but is not necessarily an eigenmode of the element. Premultiply both sides by $[\phi]^T$ and we derive

$$[\phi]^T[\phi]\{\alpha\} = [\phi]^T\{u\}$$

Now $[\phi]^T[\phi]$ is always diagonal and if we choose the modes to be orthonormal in addition to being orthogonal we may write

$$\{\alpha\} = [\phi]^T\{u\}$$

Substitute for $\{u\}$ and $\{\ddot{u}\}$ in equation (12.97) and premultiply by $[\phi]^T$ to give

$$[\phi]^T[M][\phi]\{\ddot{\alpha}\} = [\phi]^T\{f\} - [\phi]^T[K][\phi]\{\alpha\} \tag{12.98}$$

Now depending on the choice of $[\phi]$ various formulations are possible.

Choosing $[\phi] = [I]$, the identity matrix, returns the original explicit Finite Element form with $\{\alpha\} = \{u\}$. It should be noted that the nodal displacements still form an orthogonal set of modes, e.g.

$$\{\phi^1\}^T = \{1, 0, 0, 0, 0, 0, 0, 0\}$$

$$\{\phi^2\}^T = \{0, 1, 0, 0, 0, 0, 0, 0\}$$

such that $\{\phi_i\}^T\{\phi_j\} = 0$ if $i \neq j$.

The second possibility is to choose $[\phi]$ as the matrix of eigenmodes of the element, such that $[\phi]$ is the solution of the eigenproblem

$$[K]\alpha_i\{\phi_i\} = \lambda[M]\alpha_i\{\phi_i\}$$

As $[\phi]$ contains the eigenvectors of matrix $[K]$, $[\phi]^T[K][\phi]$ is in fact the stiffness matrix in its canonical form, and $[\phi]^T[M][\phi]$ is the diagonalized mass matrix, so that

$$[\phi]^T[K][\phi] = [\Lambda] \quad \text{and} \quad [\phi]^T[M][\phi] = [I]$$

where $[\Lambda]$ is the diagonal matrix, with the eigenvalues $\lambda_i$ on the diagonal, and $[I]$ is the identity matrix. The system of $n$ coupled linear equations can now be decomposed into $n$ independent equations

$$\ddot{\alpha}_i = \{\phi_i\}^T\{f\} - \lambda_i\alpha_i \qquad i = 1, n \tag{12.99}$$

from which we can solve for $\alpha_i$ directly.

The third possibility is to choose $[\phi]$ as a set of orthogonal modes, which are *not* eigenmodes, but which still diagonalize the mass matrix $[M]$. The modes may closely correspond to the strains of the element, but as shown in Section 12.5.2 *cannot* be the actual strains.

If either these modes or the eigenmodes of the element are chosen, it should be noted that the modes must be specified at all times with respect to local axes which rotate with the element. When modal equations are written in global coordinates the mass matrix becomes nondiagonal when the element rotates.

### 12.5.5 Rotation of modal equations

Let us suppose decoupled modal equations have been derived for some element with respect to fixed global axes. To be specific let the element be a rectangle aligned with the global axes.

Now consider the modal equations for the element in the original $x$, $y$ coordinate system

$$[M]\{\ddot{\alpha}\} + [K]\{\alpha\} = \{\bar{f}\} \tag{12.100}$$

where $[M]$ is diagonal by definition and $[K]$ is diagonal if the eigenmodes are chosen.

Now let the element rotate through angle $\theta$ about its centroid.

It is readily shown that the matrices $[M']$ and $[K']$ for the rotated element can be written with respect to the original global system as

$$[M'] = [R]^T[M][R], \qquad [K'] = [R]^T[K][R] \tag{12.101}$$

where $[R]$ is the antisymmetric rotation matrix. Thus equation (12.100) becomes

$$[R]^T[\bar{M}][R]\{\ddot{\alpha}\} + [R]^T[\bar{K}][R]\{\alpha\} = [R]^T\{\bar{f}\} \tag{12.102}$$

and are the new equations for the element with respect to the global axes.

It is easily shown that $[M']$ and $[K']$ are no longer diagonal and that equation (12.102) is now coupled.

To find the axes in which the equations are still decoupled, premultiply equation (12.102) by $[R]$. Noting $[R][R]^T = [I]$, the identity matrix

$$[M][R]\{\ddot{\alpha}\} + [K][R]\{\alpha\} = \{\bar{f}\} \tag{12.103}$$

Thus the new vectors for the rotated element are $[R]\{\alpha\}$, i.e. they are the old vectors rotated by the same amount as the element.

The point to note, however, is that the modes are orthogonal only with respect to axes which rotate with the element, and thus the equations remain decoupled only when written with respect to those same rotating axes. For elements undergoing large strain defining the path taken by axes rotating with the body in a non trivial exercise.

A special case exists when the modes are chosen as the nodal displacements such that $[\phi] = [I]$. In this case rotation leaves the identity matrix unchanged and the equations can be written with respect to global coordinates.

The main conclusion to be drawn from third discussion is that a standard finite element approach can be used for discrete elements, and a modal approach is only advantageous for special purposes, such as a reduced basis method.

## 12.6 CONSTITUTIVE RELATIONSHIPS

### 12.6.1 Constitutive relationships within elements

As shown in the previous sections a discrete element can be constructed to perform exactly as a Finite Element. This is also true of the range of material models available, so that any constitutive model available in Finite Elements is also available in Discrete Elements. For example, the Discrete Element behaviour can be elastoviscoplastic with any form of yield surface or plastic potential, and constitutive models described in Chapters 4 and 7 are, in principle, applicable.

In addition to standard plastic failure, Discrete Elements can also undergo brittle failure. When the stress state in an element exceeds some user defined limit, brittle fracture takes place and the element splits into two. Generation of new elements takes place automatically and since no global stiffness matrix need be constructed no problems arise in restructuring data. (Elements can also be removed with ease.) Figure 12.6 shows the fracturing which occurs when a tunnel is impacted by a missile.

In CICE and in DECICE the Mohr–Coulomb with tension cutoff brittle material model is used. Three modes of brittle failure can occur when using this material model: compressive, tensile, and flexural. The failure criteria for these cracking modes are described below.

### *Compressive failure criteria*

The Coulomb criteria for a compressive shear failure along a plane is

$$|\tau| \geqslant S_0 + \mu\sigma \tag{12.104}$$

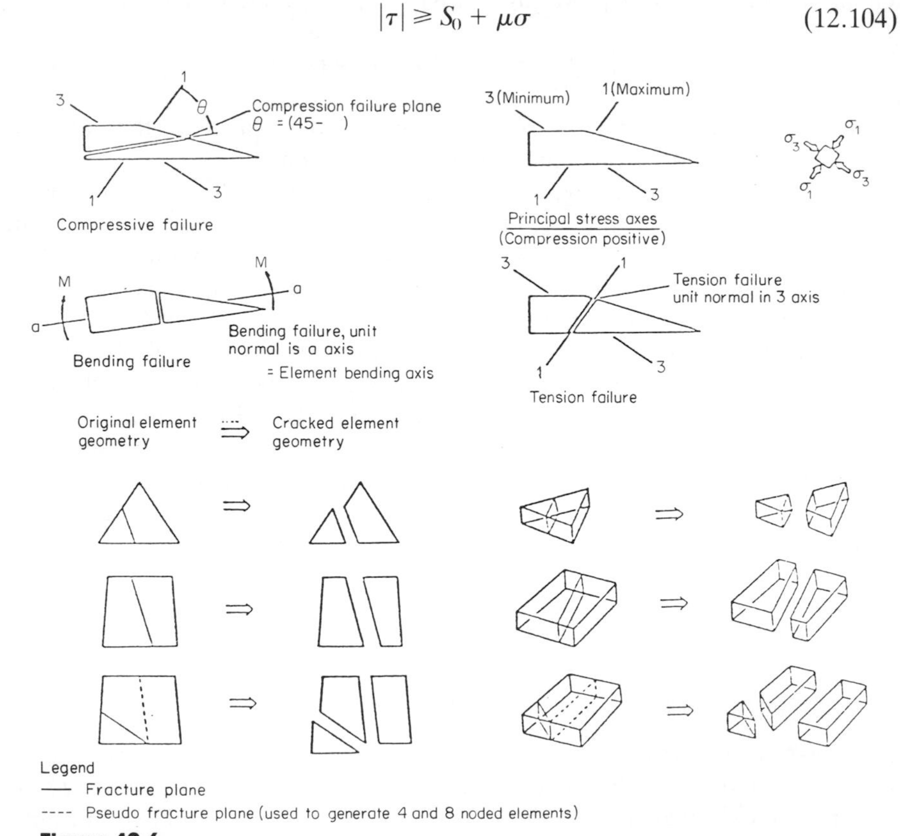


**Figure 12.6**
Two- and three-dimensional failure geometry of a single Discrete Element

where

| | |
|---|---|
| $\sigma$ and $\tau$ | are the normal and shear stresses across the plane |
| $S_0$ | is the shear strength (or internal cohesion) of the material |

and

| | |
|---|---|
| $\mu$ | is the coefficient of internal friction of the material. |

This failure criterion is applied along 'locked' interelement mesh lines (or faces) and determines whether or not the mesh line (or face) will fail. The shear strength, $S_0$ used in equation (12.104), is either specified directly or is computed as a function of the coefficient of internal friction and the unconfined compressive strength via

$$S_0 = C_0\left\{(\mu^2 + 1)^{\frac{1}{2}} - \mu\right\}/2 \tag{12.105}$$

In order to apply the failure criterion (12.106) to check whether an element will fracture through its centroid, it is necessary to recast it in terms of the maximum and minimum principal stresses denoted by $\sigma_1$ and $\sigma_3$, respectively, as

$$\sigma_1 \geqslant C_0 + \sigma_3 \tan^2(\pi/4 + \phi/2) \tag{12.106}$$

where

$\phi$ is the internal angle of friction defined by $\mu = \tan\phi$,

and

$C_0$ is the unconfined compressive strength.

It should be noted that the unconfined compressive strength is related to the shear strength and internal coefficient of friction via the equation

$$C_0 = 2S_0\left\{(\theta^2 + 1)^{\frac{1}{2}} + \mu\right\} \tag{12.107}$$

For this failure mode there are two possible directions of fracture inclined at equal angles $(\pi/4 - \phi/2)$ to the direction of the maximum principal stress on either side of it. These are called the conjugate directions and, in two dimensions, are defined by

$$\theta = \tan^{-1}\left\{(\sigma_1 - \sigma_x)/\tau_{xy}\right\} \pm (\pi/4 - \phi/2) \tag{12.108}$$

where $\theta$ is the angle between the fracture plane and the positive global $x$-axis (see Figure 12.6), $\sigma_x$ is the $x$-component in the stress tensor, and $\tau_{xy}$ is the $xy$ shear component of the stress tensor.

The sign of the second term on the right-hand of equation (12.108) may be determined by using a random number generator. In this manner a relatively random crack pattern is produced as a series of elements fracture.

In three dimensions the two conjugate fracture planes are defined by two planes which are perpendicular to the $\sigma_1\sigma_3$ plane and subtend an angle of $\pm(\pi/4 - \phi/2)$ with the positive $\sigma_1$ principal stress direction.

### *Tensile failure criteria*

The tensile failure criteria are based on the minimum principal stress $\sigma_3$, exceeding the tensile strength of the material $T_0$. The failure criterion in two dimensions is written as

$$\sigma_3 \geqslant T_0 \tag{12.109}$$

and the direction of the fracture plane is given by

$$\theta = \tan^{-1}\left\{(\sigma_1 - \sigma_x/\tau_{xy}\right\} \tag{12.110}$$

where $\theta$ is the angle between the fracture plane and the positive global $x$-axis, and $\sigma_3$ and $T_0$ are positive values representing the magnitude of tensile stresses. It should be noted that, in this case, the $\theta$-direction is aligned with the maximum principal stress direction. A similar definition holds for three dimensional situations.

### *Flexural failure criteria*

The flexural failure criteria in two dimensions for a beam element is given as

$$\sigma_f \geqslant B_0 \tag{12.111}$$

where $B_0$ is the flexural strength of the material and of is the maximum tensile fibre stress on a two-dimensional beam. It should be noted that $\sigma_f$ is comprised of two parts and can be written as

$$\sigma_f = \sigma_b + \sigma_l \tag{12.112}$$

where

$\sigma_b$ is the beam-bending stress,

and

$\sigma_l$ is the tensile component of the direct stresses $\sigma_{ij}$ along the axis of the beam.

The failure plane is perpendicular to the local beam axis of the element and is shown in Figure 12.6.

The flexural failure for a three-dimensional plate is of the same form as equation (12.112); however, in this case, $\sigma_f$ is defined as the maximum tensile fibre stress which acts on the lower or upper surface of the plate surface. It should be noted that determination of $\sigma_f$ requires the combination of the plate bending stresses ($\sigma^b_l$, $\sigma^b_m$, $\tau^b_{lm}$) and the three-dimensional direct stresses ($\sigma_x$, $\sigma_y$, $\sigma_z$, $\tau_{xy}$, $\tau_{yz}$, $\tau_{zx}$). From the combined stress state, the maximum and minimum principal stresses in the lower and upper surfaces of plate may be found and, hence, the maximum tensile fibre stress $\sigma_f$.

*Fractured element geometries*

In two dimensions three- and four-noded elements may crack through their centroids at any angle and generate further three- and four-node elements. Figure 12.6 illustrates some of the two-dimensional fractured element geometries.

In three dimensions all the two-dimensional fractured element geometries are allowable so long as the elements cracks through opposite faces. This means that any three-dimensional cuboid or wedge elements may crack into further cuboid and wedge elements. Some examples of three-dimensional cracked element geometries are given in Figure 12.6.

## 12.7 EXAMPLES OF DEFORMABLE BODY ANALYSIS

### 12.7.1 Two-dimensional applications

*Key block failure*

It has long been recognized that the stability of many underground openings is controlled by one or more blocks of rock, called keyblocks. If rock support is installed to stabilize these blocks the mechanics is such that all other blocks are also stable. Detailed calculation of forces required to stabilize keyblocks have been presented by Goodman *et al.* (1984) along with a general analysis method called 'Discontinuous Deformation Analysis' (Gen-hua Shi, 1984, 1988, 1989).

Figure 12.7(a) shows an underground tunnel which is intersected by a number of major joints. The orientation of the joints is such that stability of the roof depends on the stability of a single keyblock. The objective of the analysis is to determine the force required to stabilize the keyblock. Figure 12.7(b) shows the failure mechanism when no support is present. A support element was then inserted as shown in Figure 12.7(c). The element's centroid was fixed but the element was allowed to deform. The modulus of

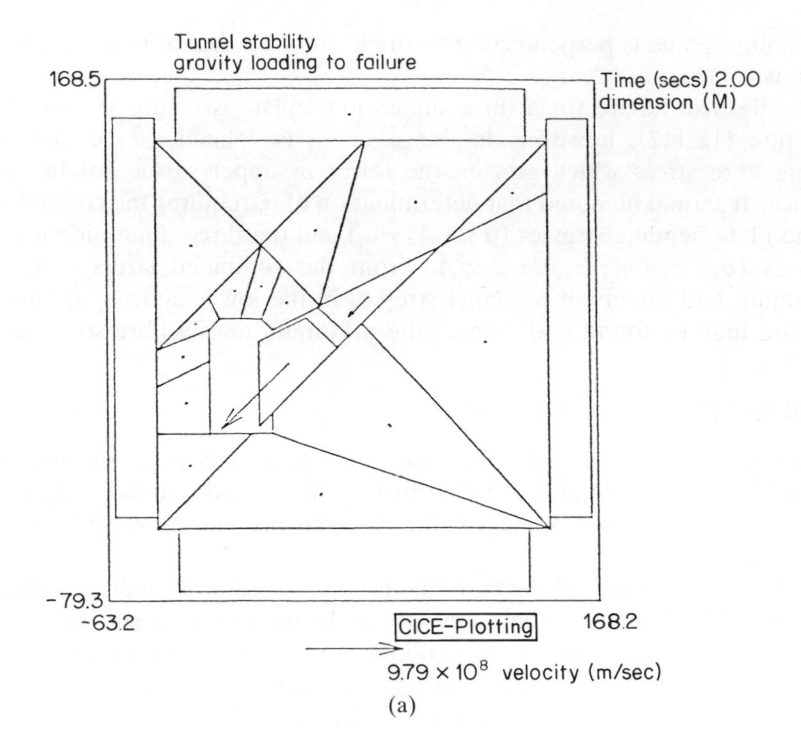
Tunnel stability
gravity loading to failure
168.5
Time (secs) 2.00
dimension (M)
-79.3
-63.2
CICE-Plotting
168.2
9.79 × 10^8 velocity (m/sec)

(a)

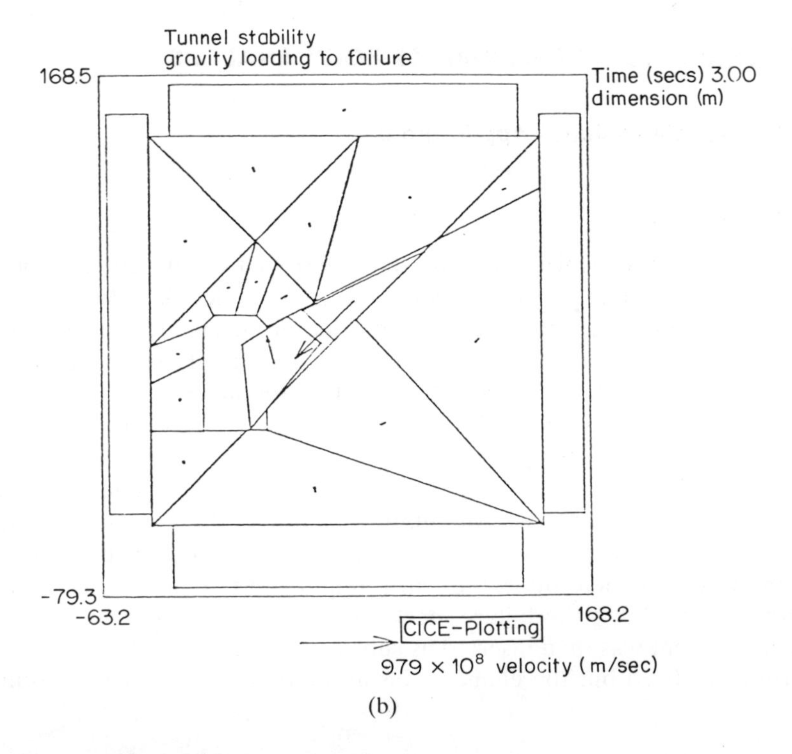
Tunnel stability
gravity loading to failure
168.5
Time (secs) 3.00
dimension (m)
-79.3
-63.2
CICE-Plotting
168.2
9.79 × 10^8 velocity (m/sec)

(b)

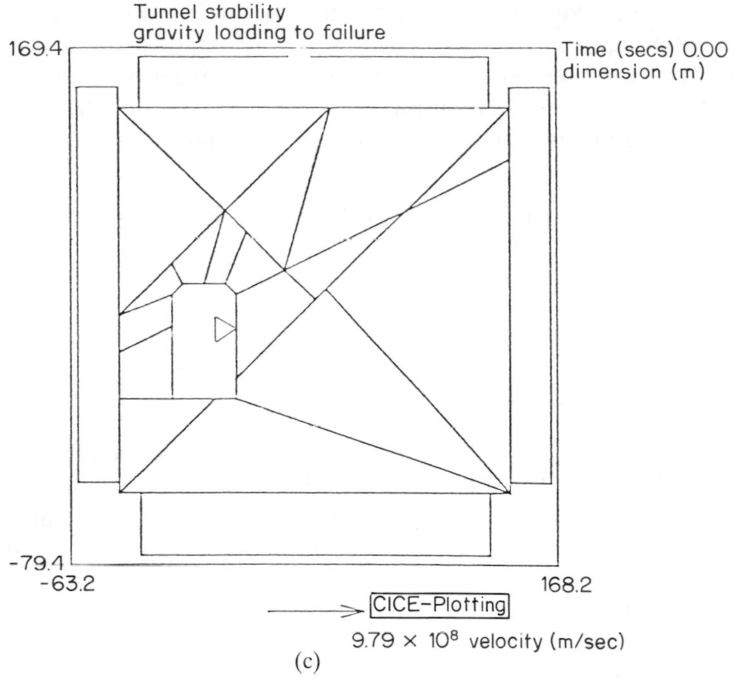


(c)

**Figure 12.7**
(a) Discrete Element mesh of underground tunnel intersected by a number of major joints; (b) Illustration of dynamic failure of the keyblock; (c) The minimum support force is calculated by placing a fixed element to support the keyblock and monitoring the force exerted on it

the element was taken to be large compared to the rock modulus so that its deformation was negligible. By monitoring the force on this element the minimum support force necessary to stabilize the keyblock was determined.

*Analysis of mine entry/pillar entry*

Analyses were performed using DECICE to investigate the mechanical behaviour of a typical mine entry/pillar system with prescribed initial vertical and horizontal stresses prior to the excavation of the mine entry. Two computations were made in which the jointed coal seam was modelled as (i) a frictional jointed discontinuum, and (ii) an elastic continuum. The results illustrate the effect on floor heave and pillar penetration. The initial stress loading case considered was equivalent to an overburden pressure generated by a vertical depth of 1000 m and a horizontal stress ratio of 0.33. A uniform density of 2500 kg/m$^3$ was assumed.

The problem considered was a cross-section through two entry systems. The calculations utilized the geometrical symmetry by examining one entry only, 7.5 meters wide with pillars on either side, founded on a 3 m seam of coal with joints sloping at 60° to the horizontal (see Figure 12.8). The coefficient of friction of this coal stratum was assumed to be in the range 10° to 30°. The analyses presented here assume an angle of friction of 30°.

An elastic constitutive relationship within the pillars and the coal were assumed in these analyses with the following material properties adopted,

| Rock type | Young's modulus (Pa) | Poisson's ratio |
|---|---|---|
| Coal | $1.38 \times 10^9$ | 0.3 |
| Sandstone | $2.76 \times 10^9$ | 0.17 |

The results obtained using a discontinuum representation of the coal seam are given in Figures 12.9 and 12.10. Figure 12.9 illustrates the principal stress components of the initial stress field within the coal stratum and Figure 12.10 the final stress distribution. Figure 12.11 shows the results of the analysis assuming the coal straum to be a continuum as in the Finite Element Method.

It can be seen that in the discrete discontinuum analysis, because of slippage, less horizontal stress is carried in the upper entry floor compared to the continuum analysis. The stress increases with depth however in order to adequately carry the horizontal thrust. A further analysis with a discontinuum model for the coal stratum with equal initial vertical and horizontal stress equivalent to 1000 m of overburden clearly shows the increase in floor heave, Figure 12.12, compared with the previous case, Figure 12.11. The results obtained whilst nonspecific, show good agreement

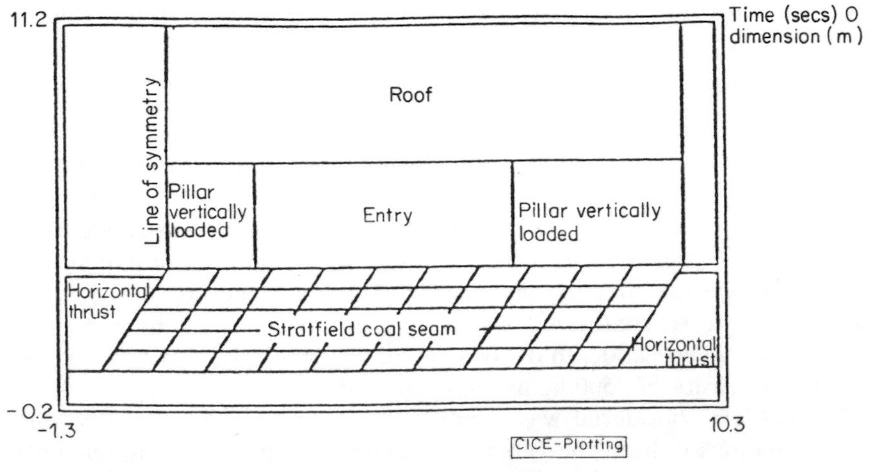


**Figure 12.8**
Discrete Element idealization of mine entry way

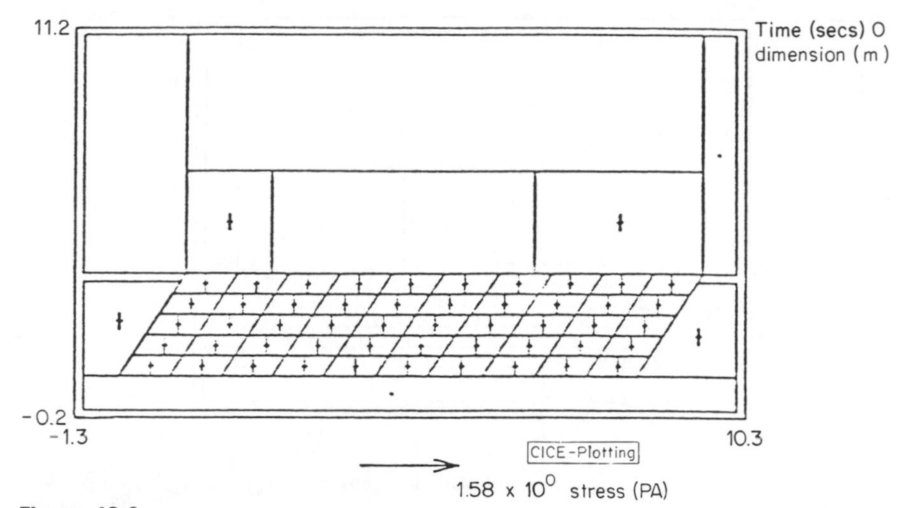


**Figure 12.9**
Mine entry discontinuum analysis—Initial stress state

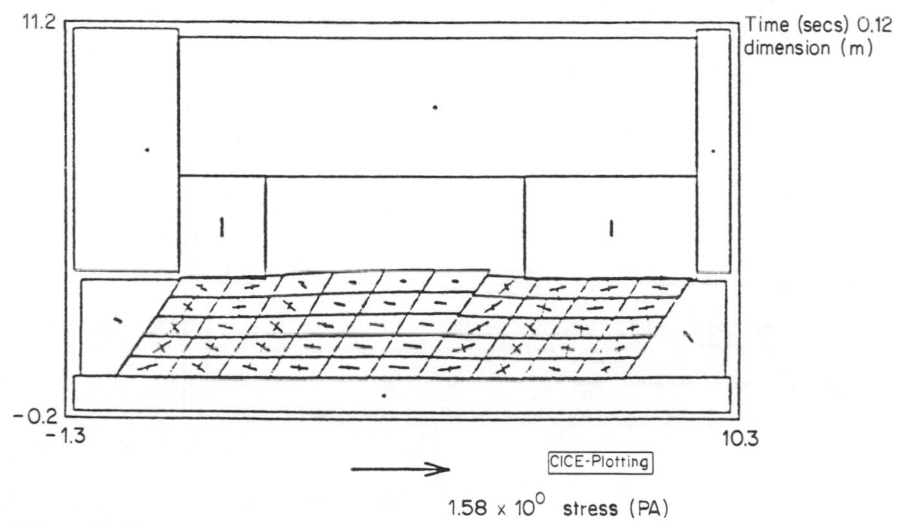


**Figure 12.10**
Mine entry discontinuum analysis—Final stress state

with observed behaviour in mine environments (for example, the Beckley coalbed, Aggson (1978)) and illustrate the ability of discrete element to model both stress distributions and finite joint displacements typically found in problematic mine entry situations.

### *Application of discrete elements to ice mechanics*

Recently there has been much interest in the behaviour of ice in the Arctic both for offshore oil exploration and with respect to submarine and missile penetration.

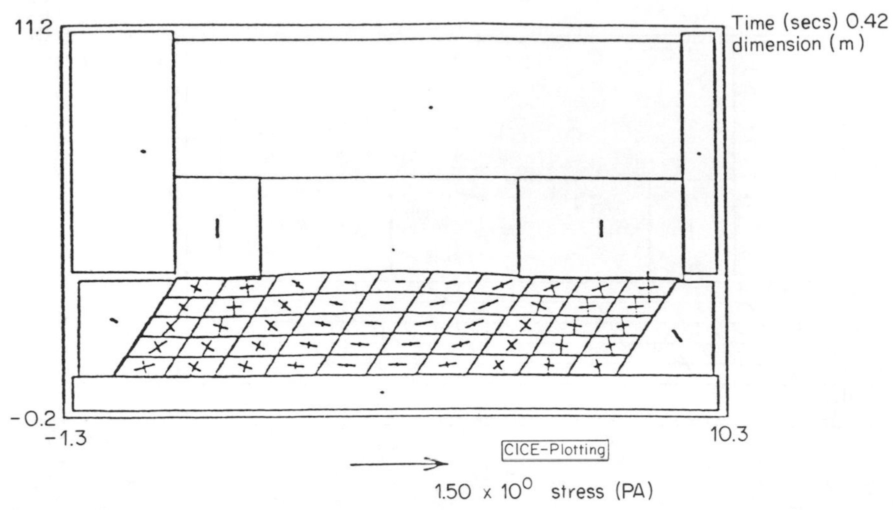


**Figure 12.11**
Mine entry continuum analysis—Final stress state

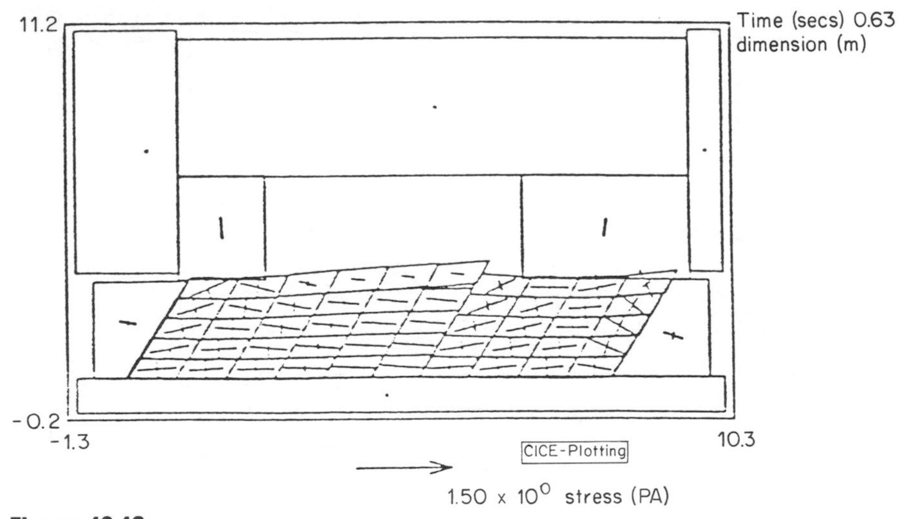


**Figure 12.12**
Mine entry discontinuum analysis—Increased overburden final stress state

One novel concept for offshore oil exploration required the construction of an artificial island from spray ice. (Spray ice is formed by pumping sea water into the frigid air through high-pressure nozzles, in much the same way as artificial snow is formed on ski slopes.)

A major consideration in the design procedure for an artificial ice island used for exploratory oil-drilling operations is the accurate prediction of the environmental ice-loading conditions to which the ice-island structure may be subjected during its operational life (Mustoe *et al.*, 1986; Williams *et al.*, 1986).

A typical design problem involves the analysis of the interaction between the ice island structure and the oncoming first-year ice sheet (see Figure 12.13(a). Figure 12.13(b) gives a schematic illustration of the local ice ridging failure mechanisms which occur as the ice sheet moves toward the island.

A Discrete Element idealization for this type of problem is shown in Figure 12.14. This plane strain discretization contains two-dimensional membrane elements which model the man-made ice material and soil, and beam elements for the floating ice sheet. It should be noted that the planc strain idealization for a circular island geometry will lead to a conservative estimate for the prediction of ice loads. The model consists of the following three different types of material:

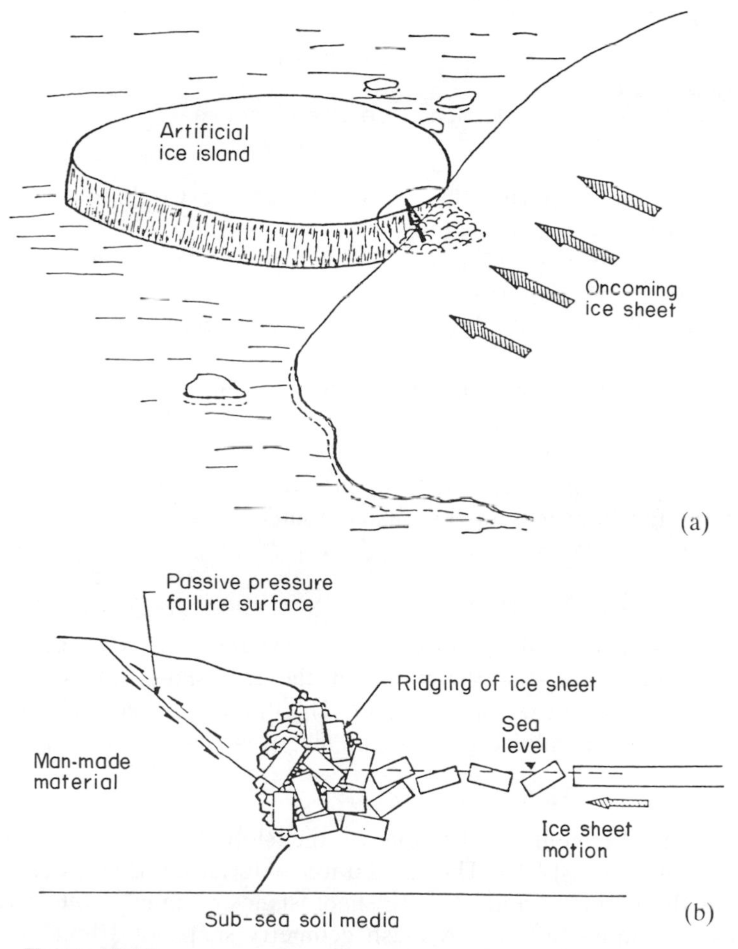


**Figure 12.13**
Arctic artifical ice island drilling structure. (a) Schematic of island—ice sheet interaction process; (b) illustration of local ice ridging mechanisms

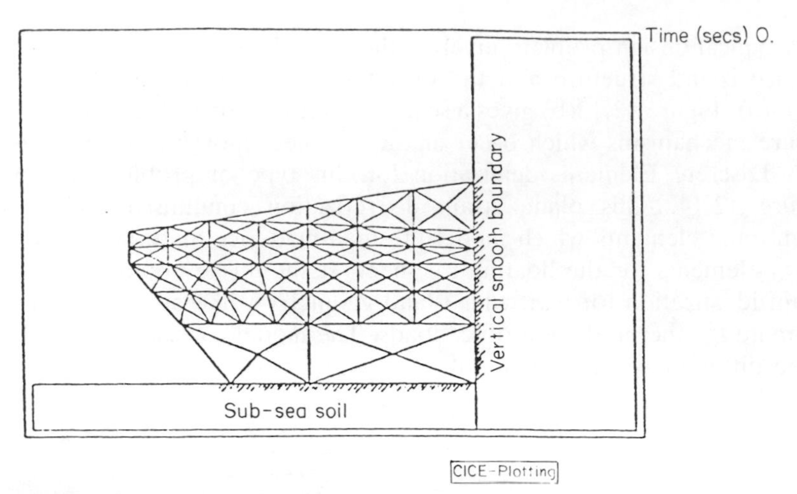


**Figure 12.14**
Discrete Element idealization of ice island problem

(a) A floating first-year ice sheet (2 m thick)—a Mohr–Coulomb elasto–brittle material with a tension cutoff.
(b) Man-made ice material above (and below) waterline—an elastoplastic Mohr–Coulomb material.
(c) Seabed soil—an elastoplastic Mohr–Coulomb material.

The Discrete Element calculation of the ice island—ice-sheet interaction was conducted to two stages:

(i) The isostatic equilibrium state of the ice island under the influence of gravity and buoyancy forces was computed

and

(ii) The floating ice sheet was introduced and the ice-sheet element furthermost from the perimeter of the ice island was prescribed a constant velocity toward the island. The initial ice-sheet velocity, which was applied to all ice-sheet elements, was defined as 1 m/s in this analysis.

The Discrete Element idealization of the sloped ice island problem is illustrated in Figure 12.14. The seabed-ice material interface is assumed to be perfectly bonded, and the ice-sheet-island perimeter interface as a frictional unbonded surface. A mesh geometry snapshot after 7.5 seconds (see Figure 12.15) shows the development of a local passive failure at the perimeter of the ice island due to the incoming ice sheet. After another 2.5 seconds sufficient horizontal asymmetry is generated in the ice sheet via the

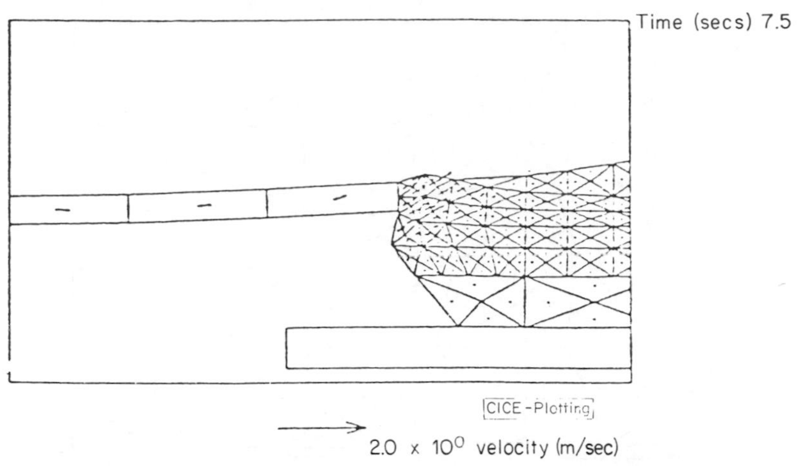


**Figure 12.15**
Ice island problem—Initial local passive failure of island perimeter

passive resistance of the island to initiate a series of flexural cracks as shown in Figure 12.16.

Figure 12.17 shows the distribution of principal plastic strains within the ice material at the island at 10.0 seconds.

### 12.7.2 Three-dimensional applications

A three-dimensional application of the Discrete Element Method is presented for a structural mechanics problem, involving brittle fracture of sea ice impacting an arctic offshore platform.

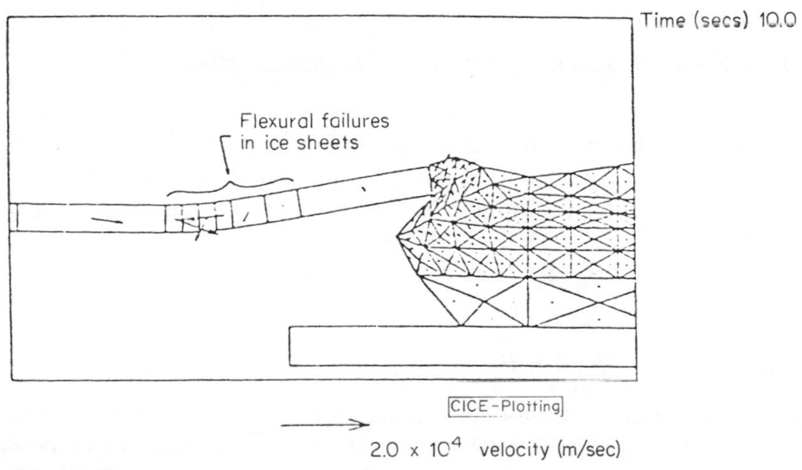


**Figure 12.16**
Ice island problem—Flexural failure within oncoming ice sheet

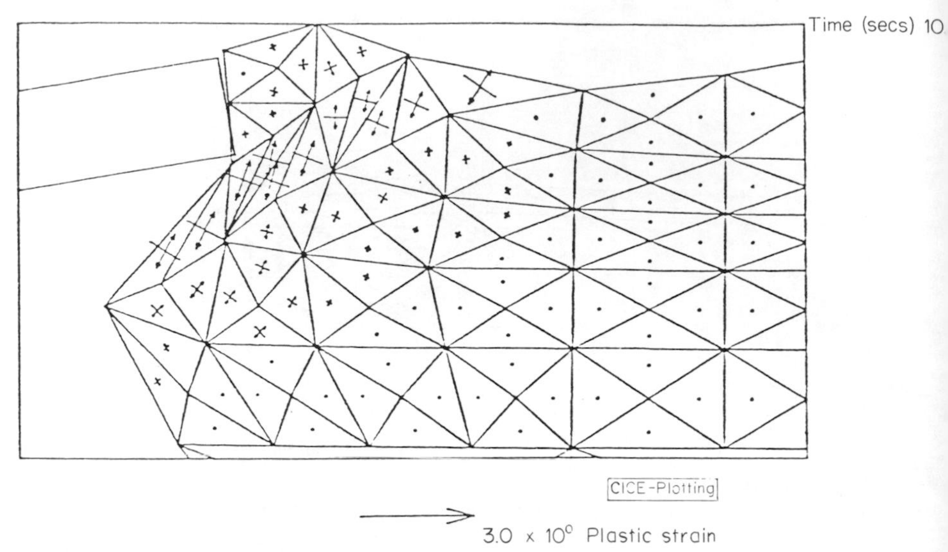


**Figure 12.17**
Ice island problem—Plastic strain distribution within island perimeter

### *Offshore platform—sea ice interaction*

The simulation of sea ice impacting a large bottom founded offshore conical platform is illustrated in Figure 12.18 taken from Hocking *et al.* (1987). The ice sheet is initially flaw free and is driven against the platform by environmental driving forces. The sheet initially fractures radially followed by circumferential cracks. Upon further motion the fragmented ice sheet begins to clear around the structure. Note the override and submergence of many of the fragmented ice pieces.

### *Underwater craft surfacing through a brittle ice sheet*

In this analysis the conning tower of the craft is modelled with a rigid multifaceted element with a constant driving vertical velocity of 1.5 m/s. The elastic-brittle ice sheet is idealized with a discretization of plates with a Young's modulus and strength parameters internal angle friction of 30°, flexural strength of 1.0 MPa and tensile cutoff of 1.0 MPa. A series of snapshots of the surfacing craft are shown in Figure 12.19.

### *Representation of joint sets*

The determination of three-dimensional rock block geometries from surface mappings of joint planes is a difficult task without the aid of computer graphics. The graphics capabilities of a Discrete Element or Finite Element code can often provide a useful tool for generating a detailed visual representation of the rock mass.

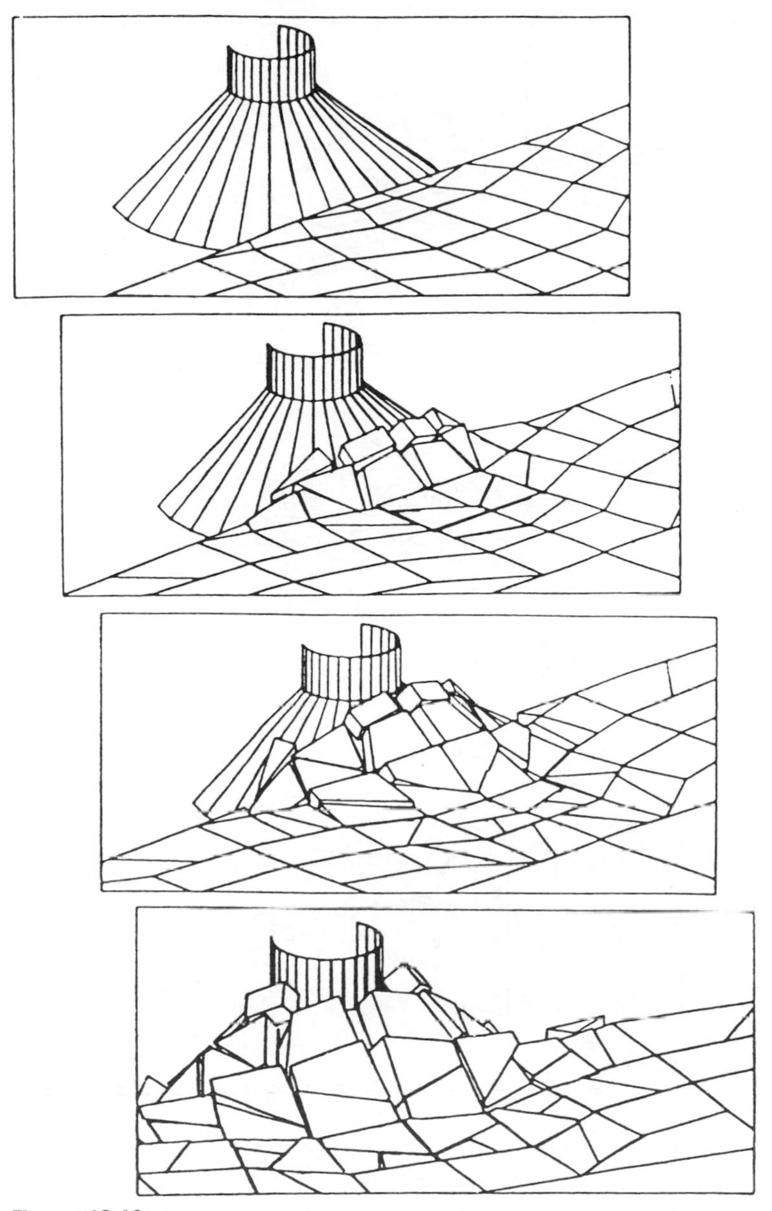

**Figure 12.18**
Discrete Element simulation of a sea ice sheet impacting an offshore platform

Figures 12.20 and 12.21 show a block of rock cut by a number of joint sets. In Figure 12.20 the joint sets are parallel but are unevenly spaced while in Figure 12.21 the orientation and spacing is random. The preprocessor RANJ03D (Williams *et al.*, 1985c) was used to generate the Discrete Elements shown in the figures by specifying the joint plane positions and

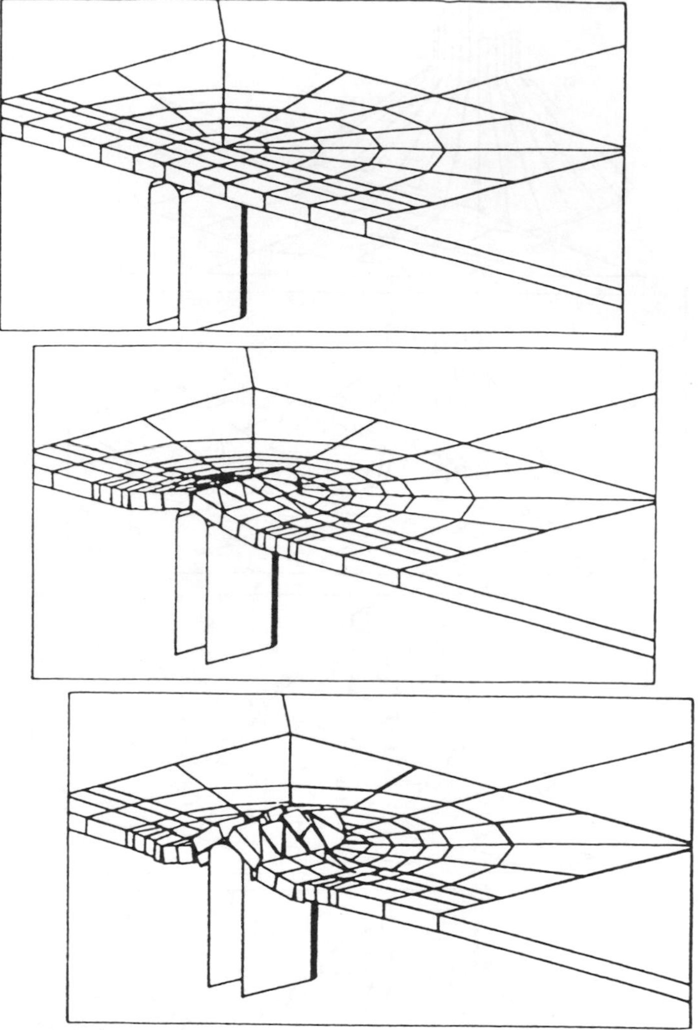

**Figure 12.19**
Simulation of underwater craft surfacing through ice

orientations. By exercising various plotting options, single blocks (elements) can be examined or cut away representations generated as illustrated in Figure 12.22. In this case regular basaltic columns are cut by subhorizontal joints. Once the Discrete Elements have been generated information on the mass, volume, moments of inertia, etc. of each block is automatically calculated.

The generation of irregular hexagonal columns can be accomplished using an algorithm based on the dual relationship between hexagonal and triangular tessellations. A tessellation of triangles for a random set of points is a well-known algorithm to those involved in contouring data. Once a set of triangles has been generated each side of every triangle is bisected (Figure 12.23).

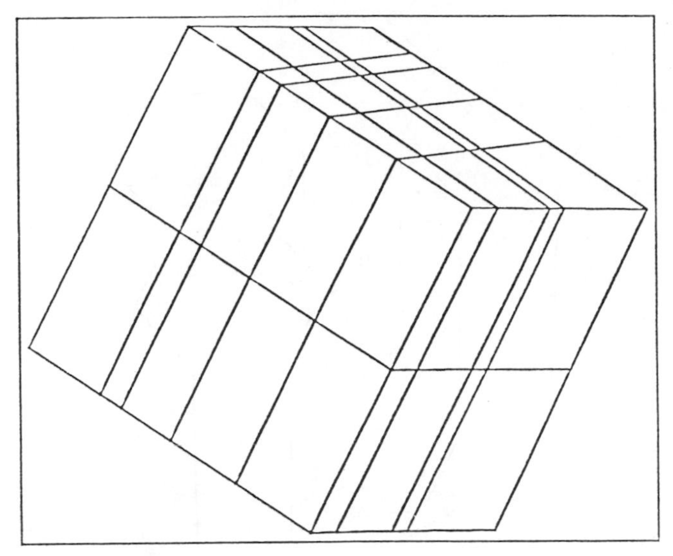

**Figure 12.20**
Joint pattern generated by RANJ03D with variation in size of blocks

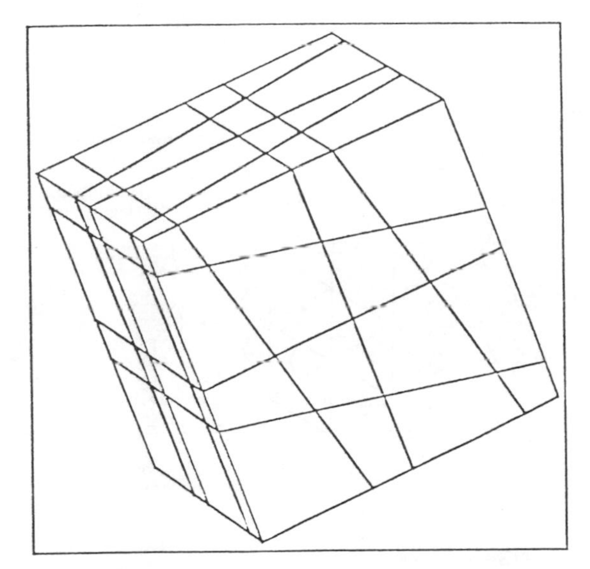

**Figure 12.21**
Joint pattern generated by RANJ03D with variation in orientation of joint planes

The point in each triangle where the bisectors meet is the centre of the circumscribing circle and is equidistant from each vertex. Hexagons are formed by the bisectors of the triangles. Figure 12.24 shows an irregular hexagonal mesh produced by such a procedure.

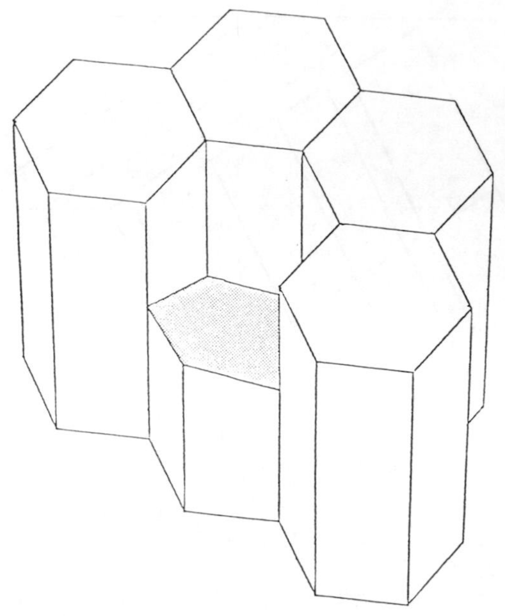

**Figure 12.22**
Cutaway representation of basaltic columns exposing a subhorizontal joint plant

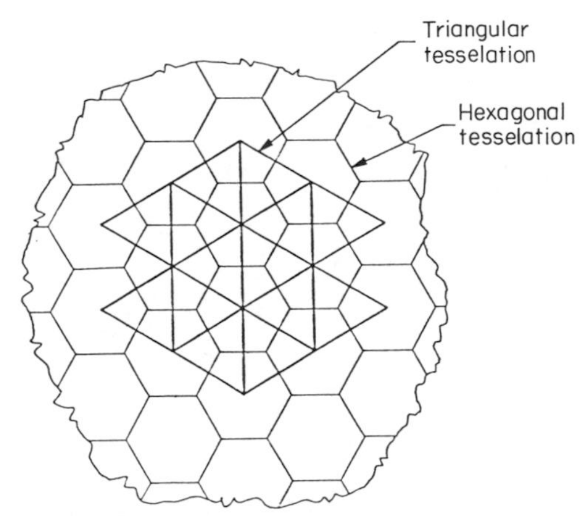


**Figure 12.23**
Illustration of dual nature of triangular and hexagonal tesselations of a plane

*Three-dimensional analysis of basalt response*

Three-dimensional Discrete Element analyses of jointed basalt have been conducted (Williams *et al.*, 1985c; Hocking *et al.*, 1987) as part of the Basalt Waste Isolation Project for Rockwell Hanford Operations. A number of

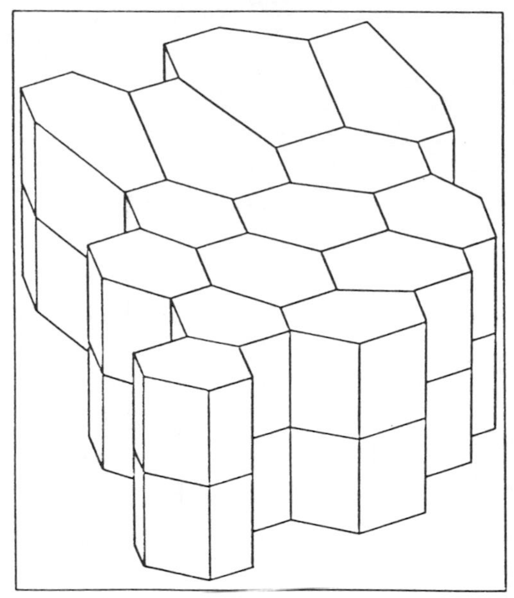

**Figure 12.24**
Example of generation of irregular hexagonal columns

large-scale field tests have been analysed including the stressing of a 2 m cube of basalt in a tunnel well (BWIP Staff, 1982). The test data was first analysed using a continuum Finite Element representation. This showed that the effect of the joints was significant and the response could not be represented accurately by a continuum technique. Therefore, a three-dimensional Discrete Element program ROCK3D was developed to analyse this specific test.

From the characterization studies of the Block Test volume and the results of the two-dimensional modelling the three-dimensional idealization illustrated in Figure 12.25(a) was adopted. Each column was idealized as having regular hexagonal cross-section with diameter 20 cm and average length between horizontal joints of 25 cm. Figure 12.25(b) shows part of the structure with elements removed to reveal the subhorizontal joints. Symmetry was used to reduce the number of elements so that a representative volume could be analyzed. The joint model utilized an elasto–plastic behaviour which was deduced from laboratory testing of single joints. A number of runs were executed in which the main loading direction was cycled between 2 and 12.5 MPa with a confining stress of 5 MPa being held constant in the other directions. A typical response is illustrated in Figure 12.26.

The agreement between modelled and measured response was relatively good, with response in the principal loading direction within approximately 20 percent of that measured. The three-dimensional coupling between

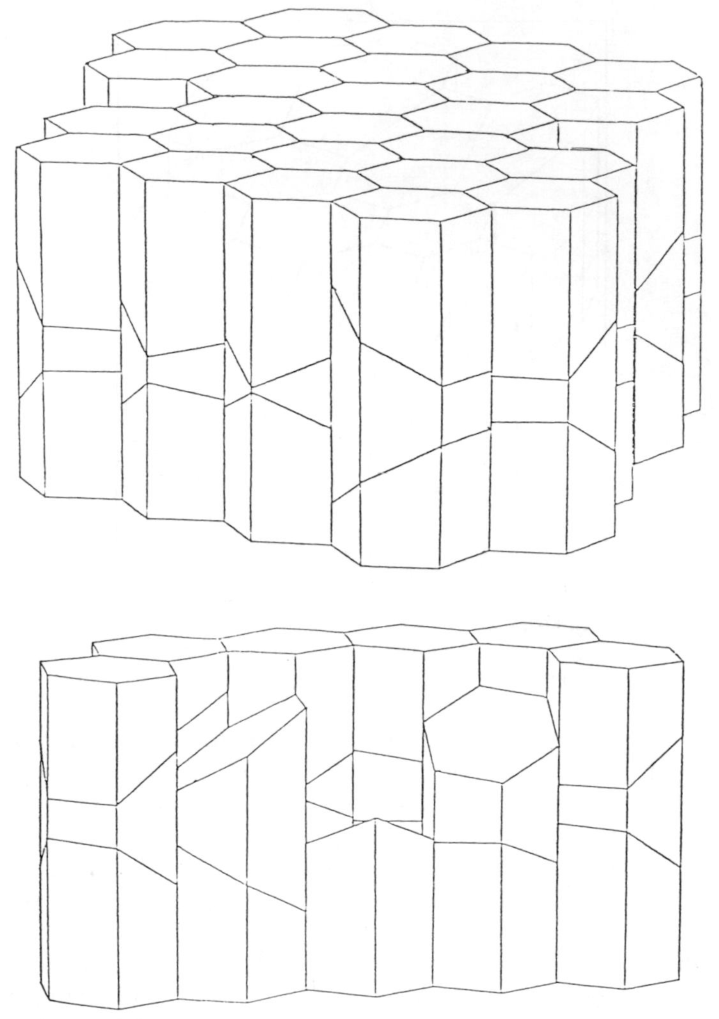

**Figure 12.25**
(a) Three-dimensional idealization for discrete element idealization of block test in basalt; (b) cutaway of block test idealization showing exposed subhorizontal joints

loading and out of loading directions was relatively good although some responses were too small to be quantitatively compared.

*Deformable Discrete Element models of rock joint behaviour*

The Discrete Element method is ideally suited to analysing rock joint behaviour since complex joint models are readily accommodated. As discussed in a previous chapter, the residual strength of a joint is often responsible for maintaining stability of the opening.

An example of a joint model exhibiting dilation, damage and residual

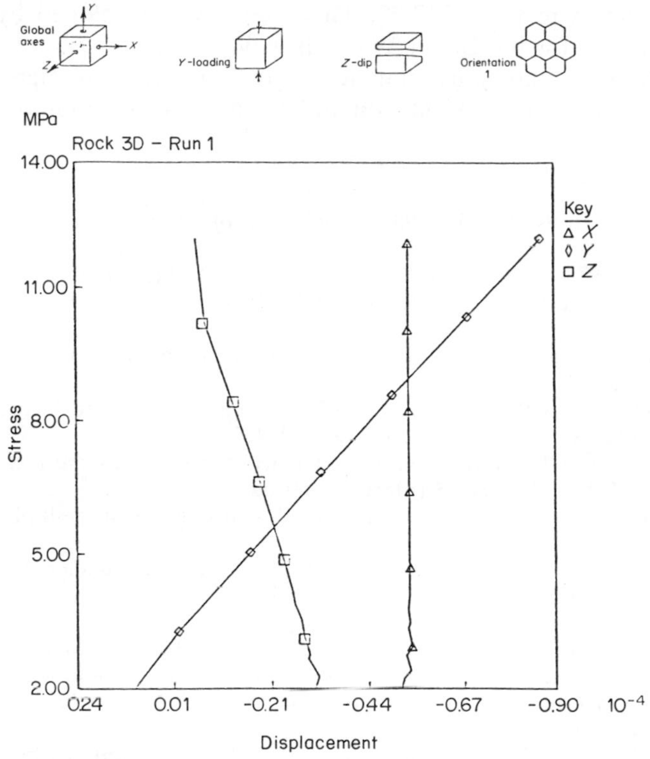


**Figure 12.26**
Results of vertical loading with continement in other directions with joint dip in 'Z' direction

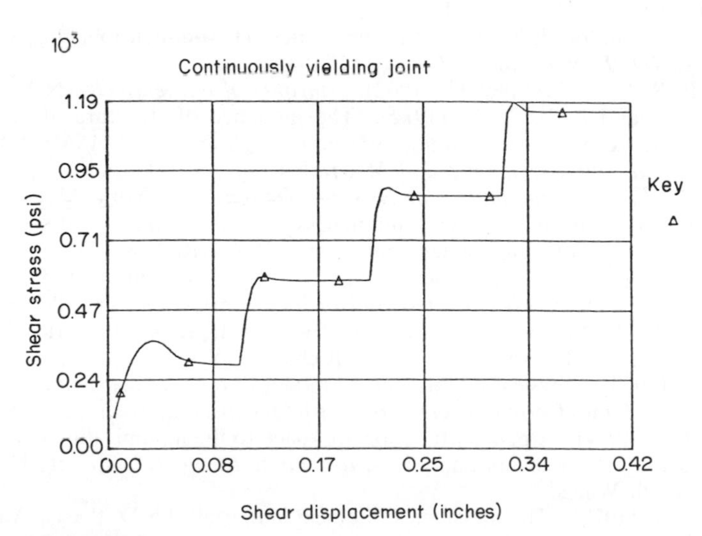


**Figure 12.27**
Example of the behaviour of the continuously yielding joint model

strength is shown in Figure 12.27. This model was developed by Cundall and Hart (1984), for the Basalt Waste Isolation project.

The model is a continuous, nonlinear, stress-displacement algorithm that exhibits continuous yielding, dilation and damage accumulation of the joint.

## REFERENCES


Ahmad, S., and Irons, B. M. (1980). *Technique of Finite Elements*, Chichester: Ellis–Horwood.

Aggson, J. R. (1978). 'Coal mine floor heave in the Beckley coalbed, an analysis'. Report of Investigations 8274, Denver Mining Research Center, Bureau of Mines, Denver, CO.

Arthur, J. R. F., and Menzies, B. K. (1972). 'Inherent anisotropy in a sand'. *Geotechnique*, **22**, 115–28.

Arthur, J. R. F., and Phillips, A. B. (1975). 'Homogeneous and layered sand in triaxial compression'. *Geotechnique*, **25**, 799–815.

Ashley, H. (1967). 'Observations on the dynamical behavior of large flexible bodies in orbit'. *A.I.A.A. Journal*, **5** (March), 460–69.

Babuska, I. (1973). 'The finite element method with Lagrangian multipliers'. *Num. Mat.* **20**.

Bandis, S. C., Lumsden, A. C., and Barton, N. R. (1983). 'Fundamentals of rock joint deformation'. *Int. J. Rock. Mech. and Min. Sci. & Geomech. Abstr.*, **20**(6), 249–68.

Barbosa, R., and Ghaboussi, J. (1989). 'Discrete Finite Element Method', 1st U.S. Conference on Discrete Element Methods, CSM, Golden, Colorado, Oct. 17–19.

Bathe, K., and Wilson, E. L. (1976). *Numerical Methods in Finite Element Analysis*. New Jersey: Prentice-Hall.

Belytschko, T., Yen, H. J., and Mullen, R. (1979). 'Mixed method for time integration'. *Computer Methods in Applied Mechanics and Engineering*, North-Holland Publishing Company.

Belytschko, T., Plesha, M., and Dowding, C. (1984). 'A computer method for stability analysis of caverns in jointed rock'. *Int'l Journal Numerical & Analytical Methods in Geomechanics*, **8**, 437–92.

Bicanic, N., and Hinton, E. (1979). 'Spurious modes in two-dimensional isoparametric elements'. *Int. J. Num. Meth. Engng*, **14**, 1545–57.

Bisplinghoff, R. L., and Ashley, H. (1962). *Principles of Aeroelasticity*. N.Y.: Dover.

Bodley, C. S., and Park, A. C. (1972). 'The influence of structural flexibility on the dynamic response of spinning spacecraft'. *Proc. AIAA/ASME/SAE 13th Structures, Structural Dynamics and Materials Conf.*, San Antonio, April.

Bowles, J. D. (1968). *Foundation Analysis and Design*. New York: McGraw-Hill.

Brezzi, F. (1974). 'On the existence, uniqueness and approximation of saddle-point problems arising from Lagrangian multipliers'. R.A.I.R.O. 8.

Burman, B. C. (1971). 'A numerical approach to the mechanics of discontinua'. Ph.D. Thesis, James Cook University of North Queensland, Townsville, Australia.

BWIP Staff (1982). 'Supporting document—Block test, Step 1 Report'. SD-BWI-TD-001, Rockwell Hanford Operations, Richland, WA.

Byrne, R. J. (1974). 'Physical and numerical model in rock and soil–slope stability'. Ph.D. thesis, James Cook University of North Queensland, Townsville, Australia.

Campbell, J. S. (1974). 'A penalty function approach to the minimization of quadratic functionals in finite element analysis'. *Proc. Finite Elements in Engng*, University of New South Wales.

Chappel, B. A. (1972). 'The mechanics of blocky material'. Ph.D. thesis, Australian National University, Canberra.

Chappel, B. A. (1974). 'Numerical and physical experiments with discontinua'. *Proc. 3rd Cong. Int. Soc. for Rock Mech.*, **2A**, 118, Denver, CO.
Chorlton, F. (1969). *Textbook of Dynamics*, London: Van Nostrand.
Clough, R. W., and Penzien, J. (1974). *Dynamics of Structures*. New York: McGraw-Hill.
Cormeau, I. (1975). 'Numerical stability in quasi-static elasto/viscoplasticity'. *Int. J. Num. Methods in Engineering*, **9**, 109–27.
Cundall, P. A. (1971). 'A computer model for simulating progressive, large-scale movements in block rock systems.' *Symp. Intl. Society of Rock Mechanics*, Nancy, France.
Cundall, P. A., and Hart (1984). 'Analysis of block test No. 1 inelastic rock mass behavior, Phase 2A Characterization of joint behavior'. Preliminary Report to Rockwell Hanford Operations, Richland, WA.
Cundall, P. A., and Strack, O. D. L. (1979a). 'A discrete numerical model for granular assemblies'. *Geotechnique*, **29**(1), 47–65.
Cundall, P. A., and Strack, O. D. L. (1979b). 'The development of constitutive Laws for soil using the distinct element method', *Numerical Methods in Geomechanics*, Aachen (ed. W. Wittke). 289–298.
Cundall, P. A., and Strack, O. D. L. (1979c). 'The distinct element method as a tool for research in granular media'. Report to NSF concerning grant ENG 76-20711, Part II, Dept. Civ. Min. Engng, University of Minnesota.
Cundall, P. A., and Strack, O. D. L. (1982). 'Modeling of microscopic mechanisms in granular material'. *Proc. US-Jap Sem. New Models Const. Rel. Mech. Gran. Mat.*, Itaca.
DeVeubeke, B. F. (1976). 'The dynamics of flexible bodies'. *Journal of Engineering Science*, **14**, 895–913.
Dowding, C. H., Belytschko, T. B., and Yen, H. S. (1983). 'Coupled finite element-rigid block method for transient analysis of rock caverns'. *Intl. Journal of Numerical and Analytical Methods in Geomechanics*, **1**(1).
Eringen, A. C. (1980). *Mechanics of Continua*, New York: R. E. Krieger Co.
Felippa, C. A. (1978). 'Iterative procedures for improving penalty function solutions of algebraic systems'. *Int. J. Numerical Methods Engineering*, **12**, 821–36.
Felippa, C. A. (1986). 'Penalty-function iterative procedures for mixed finite element formulations'. *Int. J. Numerical Methods Engng*, **22**, 267–79.
Finnie, I., and Heller, W. R. (1959). *Creep of Engineering Materials*, New York: McGraw-Hill Book Company, Inc.
Francavilla, A., and Zienkiewicz, O. C. (1975). 'A Note on Numerical Computation of Elastic Contact problems'. *Int. J. Num. Method. Engng*, **9**, 913–24.
Garofalo, F. (1965). *Fundamentals of Creep and Creep Rupture in Metals*, New York: Macmillan Company.
Goodman, R. E. (1970). 'The deformability of joints'. *ASTM Spec. Tech. Publ.* **477**, 174–96.
Goodman, R. E. (1976). *Methods of Geological Engineering in Discontinuous Rocks*, St. Paul, Minn.: West Publishing Co.
Goodman, R. E., and Bray, J. W. (1976). 'Toppling of rock slopes'. *Proc. ASCE Speciality Conf.*, 'Rock engineering for foundations and slopes', Boulder, Colorado.
Goodman, R. E., and Shi, Gen-hua (1984). *Block Theory and Its Application to Rock Engineering*, Englewood Cliffs, NJ: Prentice-Hall.
Hetenyi, M. (1946). *Beams on Elastic Foundations*, University of Michigan Press.
Hakuno, M., and Tarumi, Y. (1988). 'A granular assembly simulation for the seismic liquifaction of sand'. *Structural Engineering/Earthquake Engineering* **5**. No. 2, pp. 333s–342s (*Proc. of JSCE No. 398*).
Hakuno, M., Iwashita, K., and Uchida, Y. (1989). 'DEM simulation of cliff collapse

and debris flow'. 1st U.S. Conference of Discrete Element Methods, CSM, Golden, Colorado.

Hallquist, J. O., Goudrau, G. L., and Benson, D. J. (1985). 'Sliding interfaces with contact-impact in large-scale Lagrangian computations'. *Computer Methods in Applied Mechanics & Engineering*, **51**, 107–37.

Hamajima, R., and Kawai, T. (1981). 'On the discrete analysis of the jointed rock media'. *Journal of the Japan Society of Engineering Geology*, **22**(2), 217–24.

Hocking, G. (1977). 'Development and application of the boundary integral and rigid block methods for geotechnics', Ph.D. Thesis, Imperial College.

Hocking, G., Williams, J. R., and Mustoe, G. G. W. (1985). 'Dynamic global forces on an offshore structure from large ice floe impacts', ARCTIC '85 Conference, ASCE, San Francisco, March.

Hocking, G., Mustoe, G. G. W., and Williams, J. R. (1985). 'Influence of artificial island side-shapes and ice ride up and pile up', ARCTIC '85 Conference, ASCE, San Francisco, March.

Hocking, G., Mustoe, G. G. W., and Williams, J. R. (1987). 'Dynamic analysis for generalized three-dimensional contact and fracturing of multiple bodies', NUMETA '87, Swansea, UK, July 6–10, 1987. Balkema Publications.

Hughes, T. J. R., Taylor, R. L., and Kanoknukulchai, W. (1977). 'A finite element method for large displacement contact and impact problems', *Formulations and Computational Algorithms in Finite Element Analysis*' (eds) K. J. Bathe *et al.* M.I.T. Press.

Hughes, T. J. R., Taylor, R. L., and Kanoknukulchai, W. (1978). 'A finite element method for large displacement contact and impact problems', *Formulation in Finite Element Analysis* (eds) K. J. Bathe, J. T. Oden and W. Wunderlich, US–German Symposium on Finite Element Method.

Hughes, T. J. R., Taylor, R. L., Sackman, I. L., Curnier, A., and Kanoknukulchai, W. (1976). 'A finite element method for a class of contact–impact problems', *Comput. Meths. Appl. Mech. Engng*, **8**, 233–43.

Hughes, T. J. R., Taylor, R. L., Sackman, J. L., Curnier, A., and Kanoknukulchai, W. (1976). 'A finite element method for a class of contact–impact problems'. *Computer Methods in Applied Mechanics & Engineering*, **8**, 249–76.

Hult, J. A. H. (1966). *Creep in Engineering Structures*, Waltham, Massachusetts: Blaisdell Publishing Company.

Huston, R. L., and Passerello, C. E. (1979). 'On multi-rigid body system dynamics'. *Computers and Structures*, **10**, 439–46.

Huston, R. L., and Passerello, C. E. (1980). 'Multibody structural dynamics including translation between the bodies'. *Computers and Structures*, **12**, 713–20.

Jaeger, J. C. (1971). 'Friction of rocks and the stability of rock slopes'. Rankine Lecture, *Geotechnique*, **21**, 97–134.

Jaeger, C. (1972). *Rock Mechanics and Engineering*, London: Cambridge University Press.

Jaeger, J., and Cook, N. C. W. (1976). *Fundamentals of Rock Mechanics*, second ed., London: Chapman and Hall.

Johnson, W., and Mellor, P. B. (1973). *Engineering Plasticity*, London: Van Nostrand-Reinhold Company.

Juvinall, R. C. (1967). *Stress, Strain and Strength*, New York: McGraw-Hill Book Company. Inc.

Kawai, T. (1977). 'New discrete structural models and generalization of the method of limit analysis'. International Conference on Finite Elements in Nonlinear Solid and Structural Mechanics, Norway, **2**, G04.1–G04.20.

Kawai, T. (1977). 'New element models in discrete structural analysis'. *Japan Soc. Naval. Arch., Japan*, **141**, 174–80.

Kawai, T., Kawabata, K. Y., Kondou, I., and Kumagai, K. (1978). 'A new discrete model for analysis of solid mechanics problems'. *Proc. 1st Conf. Numerical Methods in Fracture Mech.*, Swansea, UK: pp. 26–7.

Kawai, T. (1979). 'Collapse load analysis of engineering structures by using new discrete element models'. IABSE Colloquium, Copenhagen.

Kraus, H. (1980). *Creep Analysis*, New York: John Wiley and Sons.

Krieg, R. D., and Krieg, D. B. (1977). 'Accuracies of numerical solution methods for the elastic-perfectly plastic mode', *Trans. ASME, Journal Pressure Vessel Technology*, **99**(4), 510–515.

Lee, I. K., and Herington, J. R. (1971). 'Stresses beneath granular embankments'. *Proc. 1st Australia–New Zealand Conf. Geomech.*, **1**, 291–97.

Likins, P. W. (1974). 'Analytical dynamics and nonrigid spacecraft simulation'. Jet Propulsion Lab. Tech. Report 32-1593, July 15.

Lorig, L. J., and Brady, B. H. G. (1982). 'A hybrid discrete element-boundary element method of stress analysis', *Proceedings, 23rd Symposium on Rock Mechanics*. Berkeley, CA.

Love, A. E. H. (1945). *A Treatise on the Mathematical Theory of Elasticity*. 4th Edition, New York: Dover.

Mahmood, A., and Mitchell, J. K. (1974). 'Fabric-property relationships in fine granular materials'. *Clays and Clay Minerals*, **22**, 397–408.

Maini, T., Cundall, P. A., Marti, J., Beresford, N. L., and Asgian, M. (1978). 'Computer modeling of jointed rock masses'. Tech. Report N-78-8, US Army Waterways Experiment Station, Vicksburg, MS, August.

Malkus, D. S., and Hughes, T. J. R. (1979). 'Mixed finite element methods—Reduced and selective integration techniques—a unification of concepts'. *Computer Meth. Appl. Mech. Engng*, **14**, 1785–804.

Manson, S. S. (1966). *Thermal Stress and Low Cycle Fatigue*, New York: McGraw-Hill Book Company, Inc.

McClintock, F. A., and Argon, A. S. (1966). *Mechanical Behavior of Materials*, Reading, Massachusetts: Addison Wesley Publishing Co., Inc.

McDonough, T. B. (1975). 'Formulation of the global equations of motion of a deformable body,' *A.I.A.A. Journal*, **14**(5), 656–60.

Meirovitch, L. (1972). 'Liapunov stability analysis of hybrid dynamical systems with multielastic domains'. *Intl. Journal Nonlinear Mechanics*, **7**, 425–43.

Milne, R. D. (1967). 'Some remarks on the dynamics of deformable bodies'. *A.I.A.A. Journal*, **6**(3), 556–8.

Mustoe, G. W. W., Williams, J. R., and Hocking, G. (1986). *The Discrete Element Method in Geotechnical Engineering*, Barking, Essex, UK: Elsevier Applied Science Publishers, (in press).

Mustoe, G. G. W., Williams, J. R., Hocking, G., and Worgan, K. J. (1987). 'Penetration and fracturing of brittle plates under dynamic impact'. NUMETA '87, Swansea, UK, July 6–10, 1978, Balkema Publications.

Mustoe, G. G. W. (1989). 'Special elements in discrete element analysis', 1st U.S. Conference on Discrete Element Methods, (Eds. Mustoe, Henriksen & Huttelmaier, CSM, Golden, Colorado.

Nakazawa, S., and T. Kawai (1978). 'A rigid element spring method with applications to non-linear problems'. *Proc. 1st Conf. Numerical Methods in Fracture Mechanics*, Swansea, UK: pp. 38–51.

Nour-Omid, B., and Wriggers, P. (1986). 'A two-level interation method for solution of contact problems'. *Comput. Meths. Appl. Mech. Engng*, **54**, 131–144, North-Holland.

Oda, M., Konishi, J., and Nemat-Nasser, S. 'Some experimentally based fundamental results on the mechanical behavior of granular materials'. *Geotechnique*, **30**(4), 479–95.

Oda, M. (1972). 'Initial fabrics and their relations to mechanical properties of the granular material'. *Soils and Foundations*, **12**(1), 17–37.

Oda, M. (1972). 'The mechanism of fabric changes during compressional deformation of sand'. *Soils and Foundations*, **12**(2), 1–18.

Oda, M. (1972). 'Deformation mechanism of sand in triaxial compression tests'. *Soils and Foundations*, **12**(4), 45–63.

Oden, J. T. (1982). 'Penalty-finite element methods for constrained problems in elasticity'. *Proc. Symposium on Finite Element Method*, New York: Science Press. Beijing, China. Gordon Breach, pp. 150–197.

Oden, J. T., and Campos, L. (1981). 'Some new results on finite element methods for contact problems with friction', In Hughes, Gartling and Spilker (eds), *New Concepts in Finite Element Analysis, ASME*, ADM—Vol. 44.

Oden, J. T., and Pires, E. B. (1983). 'Nonlocal and nonlinear friction laws and variational principles for contact problems in elasticity'. *J. Appl. Mech., ASME*, **50**, 67–76.

Odqvist, F. K. G. (1974). *Mathematical Theory of Creep and Creep Rupture*, Oxford: Clarendon Press.

Okamoto, N., and Nakazawa, M. (1979). 'Finite element incremental contact analysis with various frictional conditions'. *Int. J. Num. Meth. Engng*, **14**, 337–57.

Owen, D. R. J., and Hinton, E. (1980). *Finite Elements in Plasticity—Theory and Practice*, Swansea, UK: Pineridge Press.

Patton, F. D. (1966). 'Multiple Modes of Shear Failure in Rock'. *Proc. 1st Cong. ISRM* (Lisbon), **1**, 509–13.

Paul, B. (1961). 'Modification of the Coulomb–Mohr theory of fractive', *Journal of Applied Mechanics*, **28**, 259–68.

Penny, R. K., and Marriott, D. L. (1971). *Design for Creep*, London: McGraw-Hill Book Company, Ltd.

Pickering, D. J. (1970). 'Anisotropic elastic parameters for soil'. *Geotechnique*, **20**(3), 271–6.

Roark, R. J., and Young, W. C. (1975). *Formulas for Stress and Strain*. 5th ed. New York: McGraw-Hill.

Sachdeva, T. D., and Ramakrishnan, C. V. (1981). 'A finite element solution for the two-dimensional elastic contact problems with friction', *Int. J. Num. Meth. Engng*, **17**, 1257–71.

Sachdeva, T. D., Ramakrishnan, C. V., and Natarajan, R. (1981). 'A finite element method for the elastic contact problems'. *Trans. ASME, J. Engng Indus*, **103**, 456–61.

Schamaun, J. T. (1981). 'Lumped mass modeling of overburden motion during explosive blasting', Sandia National Laboratories Report, SAND 80-2413, February.

Schamaun, J. T. (1983). 'An engineering model for predicting rubble motion during blasting'. *Proceedings of 9th Conf. of Explosives and Blasting Techniques*, Soc. of Explosive Engineers, Dallas, TX, February.

Schreyer, H. L., Kulak, R. F., and Kramer, J. M. (1979). 'Accurate numerical solutions for elastic–plastic models'. *Trans. ASME, J. Pressure Vessel Technology*, **101**(3), 226–34.

Shi, Gen-hua, and Goodman, R. E. (1984). 'Discontinuous deformation analysis'. *Proceedings 25th U.S. Symposium on Rock Mechanics*, pp. 269–77.

Shi, Gen-hua (1988). 'Discontinuous Deformation Analysis—A new numerical model for statics and dynamics of block systems.' Ph.D. Dissertation, Dept. Civil Engineering, University of California, Berkeley.

Shi, Gen-hua (1989). 'Discontinuous Deformation Analysis—A new numerical model for the statics and dynamics of deformable block structures.' 1st U.S. Conference on Discrete Element Methods, CSM, Golden, Colorado.

Strack, O. D. L., and Cundall, P. A. (1978). 'The distinct element method as a tool for research in granular media'. Report to NSF concerning grant ENG 76-20711, Part I, Dept. Civ. Min. Engng. University of Minnesota.

Suh, N. P., and Turner, A. P. L. (1975). *Elements of the Mechanical Behavior of Solids*, New York: McGraw-Hill Book Company.

Taylor, L. M. (1983). 'BLOCKS, A block motion code for geomechanics studies'. Sandia National Laboratories Report, SAND 82-2373, March.

Taylor, L. M., and Preece, D. S. (1989). 'DMC—A rigid body motion code for determining the interaction of Multiple Spherical Particles'. Sandia National Laboratories, SAND-88-3482.

Thornton, C. (1989). 'Application of DEM to process engineering problems'. 1st US Conference on Discrete Element Methods, Golden, Colorado.

Timoshenko, S. P., and Woinowsky-Krieger, S. (1959). *Theory of Plates and Shells*, 2nd Edn, New York: McGraw-Hill.

Trollope, D. H., and Burman, B. C. (1980). 'Physical and numerical experiments with granular wedges'. *Geotechnique*, **30**(2), 137–57.

Trollope, D. H. (1957). 'The systematic arching theory applied to the stability analysis of embankments'. *Proc. 4th Int. Conf. S.M. and F.E.,* London, **2**, 383–88.

Tseng, J., and Olsen, M. D. (1981). 'The mixed finite element method applied to two-dimensional elastic contact problems'. *Int. J. Num. Meth. Engng*, **17**, 991–1014.

Walton, O. R. (1982). 'Explicit particle dynamics models for granular materials'. Numerical Methods in Geomechanics, Edmonton, Alberta. (A. A. Balkeona, Rotterdam).

Walton, O. R. (1983). 'Particle-dynamics calculations of shear flows'. In Mechanics of granular materials: new models and constitutive relations, Elsevier Science Publishers, Amsterdam, pp. 327–338.

Walton, O. R., Braun, R. L., Mallon, R. G., and Cervelli, D. M. (1988). 'Particle-dynamics calculations of gravity flows of inelastic, frictional spheres'. In Micromechanics of granular material, Elsevier Science Publishers, Amsterdam, pp. 153–161.

Watanabe, M., and Kawai, T. (1980). 'Simulation of the bending collapse of ice plates using a new discrete model'. *J. Soc. Naval Arch., Japan* (in Japanese), **147**, 306–15.

Webster, A. G. (1959). *The Dynamics of Particles and of Rigid, Elastic and Fluid Bodies*, 2nd edn, New York: Dover.

Wilkins, M. L. (1964). 'Calculation of elastic–plastic flow', *Methods in Computational Physics*, Vol. 3, New York: Academic Press.

Williams, J. R., Hocking, G., and Mustoe, G. G. W. (1985), 'The theoretical basis of the discrete element method', *NUMETA '85 Numerical Methods in Engineering, Theory and Application*, A. A. Balkema Publishers, Rotterdam, Conf. in Swansea, January 7–11.

Williams, J. R. (1985a). 'Task 1—Validation of UDEC-R thermal capabilities'. Report to Rockwell Handford Operations, Richland, WA.

Williams, J. R. (1985b). 'Task 4—Update of UDEC version 1.2 to UDEC-R version 1.0'. Report to Rockwell Hanford Operations, Richland, WA.

Williams, J. R., Hocking, G., and Hagan, R. (1985c). 'Task 5—UDEC-R baseline document (Volumes I and II)'. Report to Rockwell Hanford Operations, Richland, WA.

Williams, J. R., Mustoe, G. G. W., and Worgan, K. (1986). 'Force transfer and behavior of rubble piles'. Iowa: IAHR Ice Symposium, Iowa City.

Williams, J. R. (1987). 'A method for three-dimensional dynamic contact analysis of numerous deformable discrete bodies including automatic fracturing'. COMPLAS 87, Barcelona, Spain, April 6–10, U.K.: Pineridge Press.

Williams, J. R., and Mustoe, G. G. W. (1987). 'Modal methods for the analysis of

discrete systems'. *Computers and Geotechnics* (in press).
Williams, J. R., Mustoe, G. G. W., and Hocking, G. (1987). 'Dynamic soil analysis by inter-granular association', Report to US Air Force, Contract No. FO 8635–86–C–0314.
Williams, J. R., and Pentland, A. P., (1988). 'Superquadric object representation for dynamics of multibody structures'. Accepted for ASCE Structures. Congress '89.
Wriggens, P., Simo, J. C., and Taylor, R. L. (1985). 'Penalty and augmented Lagrangian formulations for contact problems'. *NUMETA '85 Numerical Methods in Engineering, Theory and Application*, A.A. Balkema Publishers, Rotterdam, Conf. in Swansea, January 7–11.
Yoshimura, N., and Kamesake, I. (1981). 'The estimation of crack pattern on ice by the new discrete model'. *Int. Symposium Ice, Quebec*, **11**, 663–72.
Zienkiewicz, O. C. (1977). *The Finite Element Method*, 3rd Edn, McGraw-Hill, UK.
Zienkiewicz, O. C., and Cormeau, I. C. (1974). 'Viscoplasticity–plasticity and creep in elastic solids–a unified numerical solution approach', *Int. J. Numerical Methods Engineering*, **8**, 821–45.
Zienkiewicz, O. C., Xi-Kui, Li., and Nakasawa, S. (1984). 'Iterative solution of mixed problems and the stress recovery procedures', Report C/R476/84, Inst. for Numerical Methods in Engineering, Univ. College of Swansea, Wales, U.K., June.

# Appendix I
## Invariants of Stress and Strain

Invariants of stress and strain tensor are scalar quantities which do not change with the transformation of coordinate axes. Stress tensor

$$[\boldsymbol{\sigma} = (\sigma_x, \sigma_y, \sigma_z, \tau_{xy}, \tau_{yz}, \tau_{zx})^T]$$

has three independent invariants as follows:
First invariant of stress

$$(I_1) = \sigma_1 + \sigma_2 + \sigma_3 \tag{AI.1}$$

where $\sigma_1$, $\sigma_2$, $\sigma_3$ are the principal stresses.
Second invariant of stress

$$(I_2) = \sigma_1\sigma_2 + \sigma_2\sigma_3 + \sigma_3\sigma_1 \tag{AI.2}$$

Third invariant of stress $= \sigma_1\sigma_2\sigma_3$ (AI.3)

It is sometimes convenient to divide the stress state into two parts, (a) the hydrostatic part, (b) the deviatoric part. The hydrostatic stress ($\sigma_m$) is defined as

$$\sigma_m = \frac{\sigma_x + \sigma_y + \sigma_z}{3} = \frac{\sigma_1 + \sigma_2 + \sigma_3}{3} \tag{AI.4}$$

The deviatoric stresses (**S**) are then given by

$$\begin{aligned} S_x &= \sigma_x - \sigma_m \\ S_y &= \sigma_y - \sigma_m \\ S_z &= \sigma_z - \sigma_m \end{aligned} \tag{AI.5}$$

$$S_{xy} = \tau_{xy}$$
$$S_{yz} = \tau_{yz}$$
$$S_{zx} = \tau_{zx}$$

The invariants of deviatoric stress tensor can also be defined.
First invariant of deviatoric stress

$$(J_1) = S_x + S_y + S_z = 0. \tag{AI.6}$$

Thus, the first invariant of deviatoric stress is identically equal to zero.
Second invariant of deviatoric stress

$$(J_2) = S_1S_2 + S_2S_3 + S_3S_1 \tag{AI.7}$$

where $S_1$, $S_2$, $S_3$ are principal deviatoric stresses and are equal to:

$$\begin{aligned} S_1 &= \sigma_1 - \sigma_m \\ S_2 &= \sigma_2 - \sigma_m \\ S_3 &= \sigma_3 - \sigma_m \end{aligned} \tag{AI.8}$$

Third invariant of deviatoric stress

$$(J_3) = S_1S_2S_3 = \tfrac{1}{3}[S_1^3 + S_2^3 + S_3^3) \tag{AI.9}$$

It can be shown that

$$J_2 = \frac{I_1^2}{3} - I_2 \tag{AI.10}$$

and

$$J_3 = I_3 - I_2\sigma_m + 2\sigma_m^3 \tag{AI.11}$$

The invariants of stress and deviatoric stress can also be written in terms of stress or deviatoric stress components. Thus,

$$I_1 = \sigma_x + \sigma_y + \sigma_z \tag{AI.12}$$

$$I_2 = \sigma_x\sigma_y + \sigma_y\sigma_z + \sigma_z\sigma_x - \tau_{xy}^2 - \tau_{yz}^2 - \tau_{zx}^2 \tag{AI.13}$$

$$I_3 = \sigma_x\sigma_y\sigma_z - \sigma_x \tau_{yz}^2 - \sigma_y \tau_{zx}^2 - \sigma_z\tau_{xy}^2 \tag{AI.14}$$

$$+2\tau_{xy}\tau_{yz}\tau_{zx}$$

$$J_1 = 0 \tag{AI.15}$$

$$J_2 = \tfrac{1}{6}[(\sigma_x - \sigma_y)^2 + (\sigma_y - \sigma_z)^2 + (\sigma_z - \sigma_x)^2] \qquad \text{(AI.16)}$$

$$+ \tau_{xy}^2 + \tau_{yz}^2 + \tau_{zx}^2$$

$$J_3 = S_x S_y S_z + 2\tau_{xy}\tau_{yz}\tau_{zx} - S_x\tau_{yz}^2 \qquad \text{(AI.17)}$$

$$-S_y\tau_{zx}^2 - S_z\tau_{xy}^2$$

Functions of stress or deviatoric stress are also invariants.

An invariant commonly used in geotechnical engineering is Lode's angle ($\theta_0$) and is defined as follows:

$$-\frac{\pi}{6} \leq \theta_0 = \tfrac{1}{3}\sin^{-1}\left[-\frac{3\sqrt{3}}{2}\frac{J_3}{(J_2)^{3/2}}\right] \leq \frac{\pi}{6} \qquad \text{(AI.18)}$$

The angle $\theta_0$ varies between $-30°$ to $+30°$ and is also invariant. The choice of this invariant is not without reasons. In the principal stress space, $\sigma_1 + \sigma_2 + \sigma_3 =$ constant represents a plane, known as the $\pi$ plane.
The trace of the yield function on the $\pi$ plane, for an isotropic material must have a six-fold symmetry. Thus, a 60° sector is sufficient to describe the yield conditions of an isotropic material. Many configurations for testing of materials have a fixed Lode's angle as given below:

| Test | Lode's angle |
|---|---|
| Uniaxial compression | 30° |
| Triaxial compression | 30° |
| Triaxial extension | −30° |
| Uniaxial extension | −30° |
| Simple shear | 0° |

It should be noted that the principal stresses also are invariants of the stress tensor. They are related to other invariants by

$$\begin{Bmatrix} \sigma_1 \\ \sigma_2 \\ \sigma_3 \end{Bmatrix} = \sigma_m \begin{Bmatrix} 1 \\ 1 \\ 1 \end{Bmatrix} + 2\sqrt{\frac{J_2}{3}} \begin{Bmatrix} \sin\left(\theta_0 + \frac{2\pi}{3}\right) \\ \sin\theta_0 \\ \sin\left(\theta_0 + \frac{4\pi}{3}\right) \end{Bmatrix} \qquad \text{(AI.19)}$$

Invariants of strain [$\boldsymbol{\epsilon} = \{\epsilon_x, \epsilon_y, \epsilon_z, \gamma_{xy}, \gamma_{yz}, \gamma_{zx}\}^T$] are:

First invariant

$$(E_1) = \epsilon_x + \epsilon_y + \epsilon_z = \epsilon_v \text{ (volumetric strain)}$$

Second invariant

$$(E_2) = \epsilon_x\epsilon_y + \epsilon_y\epsilon_z + \epsilon_z\epsilon_x - \tfrac{1}{4}(\gamma_{xy}^2 + \gamma_{yz}^2 + \gamma_{zx}^2)$$

Third invariant

$$(E_3) = \epsilon_x\epsilon_y\epsilon_z - \tfrac{1}{4}(\epsilon_x\gamma_{yz}^2 + \epsilon_y\gamma_{zx}^2 + \epsilon_z\gamma_{xy}^2) + \tfrac{1}{4}\cdot\gamma_{xy}\gamma_{yz}\gamma_{zx}$$

It is possible to define invariants of deviatoric strains as well but they are not of much practical value.

# Appendix II
## Transformation of Stresses, Strains and Compliances

### (a) Stresses

In rock mechanics one frequently encounters the problem of transformation of stresses, strains and compliances, etc. Let $x$, $y$, $z$ represent a set of Cartesian coordinate axes. Stresses and strains in this system are denoted by $\boldsymbol{\sigma}$ and $\boldsymbol{\epsilon}$, respectively. Let $x'$, $y'$, $z'$ represent another Cartesian coordinate system, in which the corresponding stresses and strains are represented by $\boldsymbol{\sigma}'$ and $\boldsymbol{\epsilon}'$. The coordinates of a point in the $x'$, $y'$, $z'$ system are related to the coordinates of the same point in $x$, $y$, $z$ system by

$$\begin{Bmatrix} x' \\ y' \\ z' \end{Bmatrix} = \begin{bmatrix} l_1 & m_1 & n_1 \\ l_2 & m_2 & n_2 \\ l_3 & m_3 & n_3 \end{bmatrix} \begin{Bmatrix} x \\ y \\ z \end{Bmatrix} \tag{AII.1}$$

where $l_1$, $m_1$, $n_1$ are direction cosines of $x'$ axis, $l_2$, $m_2$, $n_2$ are direction cosines of $y'$ axis and $l_3$, $m_3$, $n_3$ are direction cosines of $z'$ axis.

The stresses $\boldsymbol{\sigma}'$ in the $x'$, $y'$, $z'$ system are given by

$$\boldsymbol{\sigma}' = T_\sigma \boldsymbol{\sigma}$$

where $T_\sigma$ is a $6 \times 6$ transformation matrix given by

$$T_\sigma = \begin{bmatrix} l_1^2 & m_1^2 & n_1^2 & 2l_1m_1 & 2m_1n_1 & 2l_1n_1 \\ l_2^2 & m_2^2 & n_2^2 & 2l_2m_2 & 2m_2n_2 & 2l_2n_2 \\ l_3^2 & m_3^2 & n_3^2 & 2l_3m_3 & 2m_3n_3 & 2l_3n_3 \\ l_1l_2 & m_1m_2 & n_1n_2 & l_1m_2 + l_2m_1 & m_1n_2 + m_2n_1 & l_1n_2 + l_2n_1 \\ l_2l_3 & m_2m_3 & n_2n_3 & l_2m_3 + l_3m_2 & m_2n_3 + m_3n_2 & l_2n_3 + l_3n_2 \\ l_1l_3 & m_1m_3 & n_1n_3 & l_1m_3 + l_3m_1 & m_1n_3 + m_3n_1 & l_1n_3 + l_3n_1 \end{bmatrix} \quad \text{(AII.2)}$$

Stresses, in the above equation, are arranged in the order $\sigma_x$, $\sigma_y$, $\sigma_z$, $\tau_{xy}$, $\tau_{yz}$, $\tau_{zx}$. The transformation matrix $T_\sigma$ above can be condensed if only certain components of stress are required. For example, to obtain normal and shear stresses on a joint plane, one can write:

$$\begin{Bmatrix} \sigma'_z \\ \tau'_{yz} \\ \tau'_{zx} \end{Bmatrix} = \begin{bmatrix} l_3^2 & m_3^2 & n_3^2 & 2l_3m_3 & 2m_3n_3 & 2l_3n_3 \\ l_2l_3 & m_2m_3 & n_2n_3 & l_2m_3 + l_3m_2 & m_2n_3 + m_3n_2 & l_2n_3 + l_3n_2 \\ l_1l_3 & m_1m_3 & n_1n_3 & l_1m_3 + l_3m_1 & m_1n_3 + m_3n_1 & l_1n_3 + l_3n_1 \end{bmatrix} \boldsymbol{\sigma} \quad \text{(AII.3)}$$

where $z'$ axis is normal to the joint, $x'$, $y'$ axes lie on the joint plane but can be chosen arbitrarily. The resultant shear stress ($\tau$) on the joint plane is given by

$$\tau = \sqrt{(\tau'_{yz})^2 + (\tau'_{zx})^2} \quad \text{(AII.4)}$$

## (b) Strains

Strains in $x'$, $y'$, $z'$ system ($\boldsymbol{\epsilon}'$) are related to strains in $x$, $y$, $z$ system ($\boldsymbol{\epsilon}$) by

$$\boldsymbol{\epsilon}' = T_\epsilon \boldsymbol{\epsilon} \quad \text{(AII.5)}$$

where $T_\epsilon$ is given by

$$T_\epsilon = \begin{bmatrix} l_1^2 & m_1^2 & n_1^2 & l_1m_1 & m_1n_1 & l_1n_1 \\ l_2^2 & m_2^2 & n_2^2 & l_2m_2 & m_2n_2 & l_2n_2 \\ l_3^2 & m_3^2 & n_3^2 & l_3m_3 & m_3n_3 & l_3n_3 \\ 2l_1l_2 & 2m_1m_2 & 2n_1n_2 & l_1m_2 + l_2m_1 & m_1n_2 + m_2n_1 & l_1n_2 + l_2n_1 \\ 2l_2l_3 & 2m_2m_3 & 2n_2n_3 & l_2m_3 + l_3m_2 & m_2n_3 + m_3n_2 & l_2n_3 + l_3n_2 \\ 2l_1l_3 & 2m_1m_3 & 2n_1n_3 & l_1m_3 + l_3m_1 & m_1n_3 + m_3n_1 & l_1n_3 + l_3n_1 \end{bmatrix} \boldsymbol{\epsilon} \quad \text{(AII.6)}$$

Strains in the above equation are arranged in the order $\epsilon_{x'}$, $\epsilon_{y'}$, $\epsilon_{z'}$, $\gamma_{x'y'}$, $\gamma_{y'z'}$, $\gamma_{z'x'}$.

### (c) Compliances

The work done in the two coordinate systems must be equal. This implies

$$\boldsymbol{\sigma}'^T\boldsymbol{\epsilon}' = \boldsymbol{\sigma}^T\boldsymbol{\epsilon} \tag{AII.7}$$

also

$$\boldsymbol{\epsilon}'^T\boldsymbol{\sigma}' = \boldsymbol{\epsilon}^T\boldsymbol{\sigma}$$

The compliance matrix ($c'$) defined by

$$\boldsymbol{\epsilon}' = c'\boldsymbol{\sigma}' \tag{AII.8}$$

in $x'$, $y'$, $z'$ system can be related to compliance (c) in $x$, $y$, $z$ system by making use of equation (AII.7). It can be shown that

$$c' = T_\epsilon c T_\epsilon^{-1}$$

where $T_\epsilon$ has been defined in equation (AII.6).

# Appendix III Fundamental Solutions

TWO-DIMENSIONAL KELVIN SOLUTION
(Banerjee and Butterfield, 1981)

$$\mathbf{U}(Q,P) = C_1 \begin{bmatrix} r_x^2 - C_2 ln\, r\,, & r_x r_y \\ r_x r_y & r_y^2 - C_2 ln\, r \end{bmatrix}$$

where

$$C_1 = \frac{1}{8\pi G(1-\nu)} \qquad \text{for plane strain}$$

$$C_1 = \frac{1+\nu}{8\pi G} \qquad \text{for plane stress}$$

and

$$C_2 = 3 - 4\nu \qquad \text{for plane strain}$$

$$C_2 = 3 - 4\frac{\nu}{1+\nu} \qquad \text{for plane stress}$$

$$r = \sqrt{(x_Q - x_P)^2 + (y_Q - y_P)^2}$$

$$r_x = (x_Q - x_P)/r$$

$$r_y = (y_Q - y_P)/r$$

$$\mathbf{S}(Q,P) = \frac{C_3}{r}\begin{bmatrix} r_x(C_4 + 2r_x^2) & r_y(2r_x^2 - C_4) \\ r_x(2r_y^2 - C_4) & r_y(C_4 + 2r_y^2) \\ r_y(C_4 + 2r_x^2) & r_x(C_4 + 2r_y^2) \end{bmatrix}$$

where

$$C_3 = \frac{-1}{4\pi(1-\nu)} \qquad \text{for plane strain}$$

$$C_3 = \frac{-(1+\nu)}{4\pi} \qquad \text{for plane stress}$$

and

$$C_4 = 1 - 2\nu \qquad \text{for plane strain}$$

$$C_4 = \frac{1-\nu}{1+\nu} \qquad \text{for planc stress}$$

## TWO-DIMENSIONAL INTEGRATED KELVIN SOLUTIONS (Crouch and Starfield, 1983) (refer to Figure 9.8)

$$\Delta\bar{\mathbf{U}} = -C_1\begin{bmatrix} C_2 r_a + 4C_5\,\bar{y}(\theta_1 - \theta_2) & \bar{y}\ln\frac{r_1}{r_2} \\ \bar{y}\,\ln\frac{r_1}{r_2} & C_2 r_a + 2C_4\bar{y}(\theta_I - \theta_2) \end{bmatrix}$$

where $C_1$, $C_4$ and $C_2$ are as previously defined for the Kelvin solution and

$$C_5 = (1-\nu) \qquad \text{for plane strain}$$

$$C_5 = \frac{1}{(1+\nu)} \qquad \text{for plane stress}$$

$$r_a = (x+a)\ln r_2 - (x-a)\ln r_1$$

$$\theta_1 = \arctan\frac{\bar{y}}{\bar{x}-a}$$

$$\theta_2 = \arctan\frac{\bar{y}}{\bar{x}+a}$$

$$r_1 = \sqrt{(\bar{x}-a)^2 + \bar{y}^2}$$

$$r_2 = \sqrt{(\bar{x}+a)^2 + \bar{y}^2}$$

$$\Delta\bar{\mathbf{S}} = -C_3 \begin{bmatrix} C_6 \ln \dfrac{r_1}{r_2} + \bar{y}^2 r_b & -2\nu(\theta_1 - \theta_2) - \bar{y} r_c \\ -C_4 \ln \dfrac{r_1}{r_2} - \bar{y}^2 r_b & -2C_5(\theta_1 - \theta_2) + \bar{y} r_c \\ -2C_3(\theta_1 - \theta_2) - \bar{y} r_c & -C_4 \ln \dfrac{r_1}{r_2} - \bar{y}^2 r_b \end{bmatrix}$$

where $C_4$ and $C_5$ are as discussed previously and

$$C_6 = 3 - 2\nu \qquad \text{for plane strain}$$

$$C_6 = 3 - 2 \frac{\nu}{1+\nu} \qquad \text{for plane stress}$$

$$r_b = \frac{1}{r_1^2} - \frac{1}{r_2^2}$$

$$r_c = \frac{\bar{x} - a}{r_1^2} - \frac{\bar{x} + a}{r_2^2}$$

## TWO-DIMENSIONAL DISPLACEMENT DISCONTINUITY SOLUTION (Crouch and Starfield, 1983)

$$\mathbf{U}_\Delta = C_3 \begin{bmatrix} 2C_5 f_y - y f_{xx} & -C_4 f_x - y f_{xy} \\ C_4 f_x - y f_{xy} & 2C_5 f_y - y f_{yy} \end{bmatrix}$$

$$f_x = -\ln k \frac{r_2}{r_1}$$

$$f_y = \theta_1 - \theta_2$$

$$f_{xy} = -y\left(\frac{1}{r_2^2} - \frac{1}{r_1^2}\right)$$

$$f_{xx} = -f_{yy} = \frac{x+a}{r_1^2} - \frac{x-a}{r_2^2}$$

# Appendix IV
## Choosing Eigenmodes of a Rectangular Element

The stiffness matrix $[K]$ for a rectangular element of length $b$ and height $c$ is given by Cook [14] as follows:

$$[K] = \frac{Qt}{12(1-m^2)} \begin{bmatrix} A_1 & C_1 & A_2 & -C_2 & A_4 & -C_1 & A_3 & C_2 \\ C_1 & B_1 & C_2 & B_3 & -C_1 & B_4 & -C_2 & B_2 \\ A_2 & C_2 & A_1 & -C_1 & A_3 & -C_2 & A_4 & C_1 \\ -C_2 & B_3 & -C_1 & B_1 & C_2 & B_2 & C_1 & B_4 \\ A_4 & -C_1 & A_3 & C_2 & A_1 & C_1 & A_2 & -C_2 \\ -C_1 & B_4 & C_2 & B_2 & C_1 & B_1 & C_2 & B_3 \\ A_3 & -C_2 & A_4 & C_1 & A_2 & C_2 & A_1 & -C_1 \\ C_2 & B_2 & C_1 & B_4 & -C_2 & B_3 & -C_1 & B_1 \end{bmatrix}$$

In plane stress:

$$Q = E, m = \nu$$

In plane strain:

$$Q = \frac{E}{1-\nu^2}, \qquad m = \frac{\nu}{1-\nu}$$

$$A_1 = (4 - m^2)c/b + 1.5(1 - m)b/c$$

$$A_2 = -(4 - m^2)c/b + 1.5(1 - m)b/c$$

$$A_3 = (2 + m^2)c/b - 1.5(1 - m)b/c$$

$$A_4 = -(2 + m^2)c/b - 1.5(1 - m)b/c$$

$$C_1 = 1.5(1 + m)$$

$$C_2 = 1.5(1 - 3m)$$

$B_1 - B_4$ are obtained from $A_1 - A_4$ by interchanging $b$ and $c$.

If $(\phi]$ is the matrix of eigenvectors then $[\phi]^T[K][\phi]$ should be diagonal. Taking $[\phi]$ as below

$$[\phi]^T = \begin{bmatrix} 1 & 0 & 1 & 0 & 1 & 0 & 1 & 0 \\ 0 & 1 & 0 & 1 & 0 & 1 & 0 & 1 \\ c & -b & c & b & -c & b & -c & b \\ b & 0 & -b & 0 & -b & 0 & b & 0 \\ 0 & c & 0 & c & 0 & -c & 0 & -c \\ -b & -c & -b & c & b & c & b & -c \\ 1 & 0 & -1 & 0 & 1 & 0 & -1 & 0 \\ 0 & -1 & 0 & 1 & 0 & -1 & 0 & 1 \end{bmatrix}$$

and calculating $[\phi]^T[K][\phi]$ the following is derived

$$[K] = 16 \begin{bmatrix} 0 & 0 & 0 & 0 & 0 & 0 & 0 & 0 \\ & 0 & 0 & 0 & 0 & 0 & 0 & 0 \\ & & 0 & 0 & 0 & 0 & 0 & 0 \\ & & & 3bc & 3mbc & 0 & 0 & 0 \\ & & & 3mbc & 3bc & 0 & 0 & 0 \\ & & & & & Q & 0 & 0 \\ \text{Symmetric} & & & & & & (1 - m^2)c/b & 0 \\ & & & & & & & (1 - m^2)b/c \end{bmatrix}$$

where

$$Q = b^2(A_1 + A_2 - A_3 - A_4)/4 + c^2(B_1 + B_2 - B_3 - B_4)/4 + bc(C_1 + C_2)$$

It is easily shown that the above equation can be totally decoupled by choosing the following combined strain modes

$$\epsilon_{xx} + \epsilon_{yy} \Rightarrow (\phi_4)^T = \{b, c, -b, c, -b, -c, b, -c\}$$

$$\epsilon_{xx} - \epsilon_{yy} \Rightarrow (\phi_5)^T = \{b, -c, -b, -c, -b, c, b, c\}$$

It can be checked that these new modes do indeed diagonalize the full stiffness matrix $[K]$ to give

$$[K] = 16 \begin{bmatrix} 0 & & & & & & & \\ & 0 & & & & & & \\ & & 0 & & & 0 & & \\ & & & 6(1 + m)bc & & & & \\ & & & & 6(1 - m)bc & & & \\ & 0 & & & & Q & & \\ & & & & & & (1 - m^2)c/b & \\ & & & & & & & (1 - m^2)b/c \end{bmatrix}$$

Figure IV.1 shows the eigenmodes and eigenvalues for a square element and Figure IV.2 for a rectangular element both with nonzero Poisson's ratio.

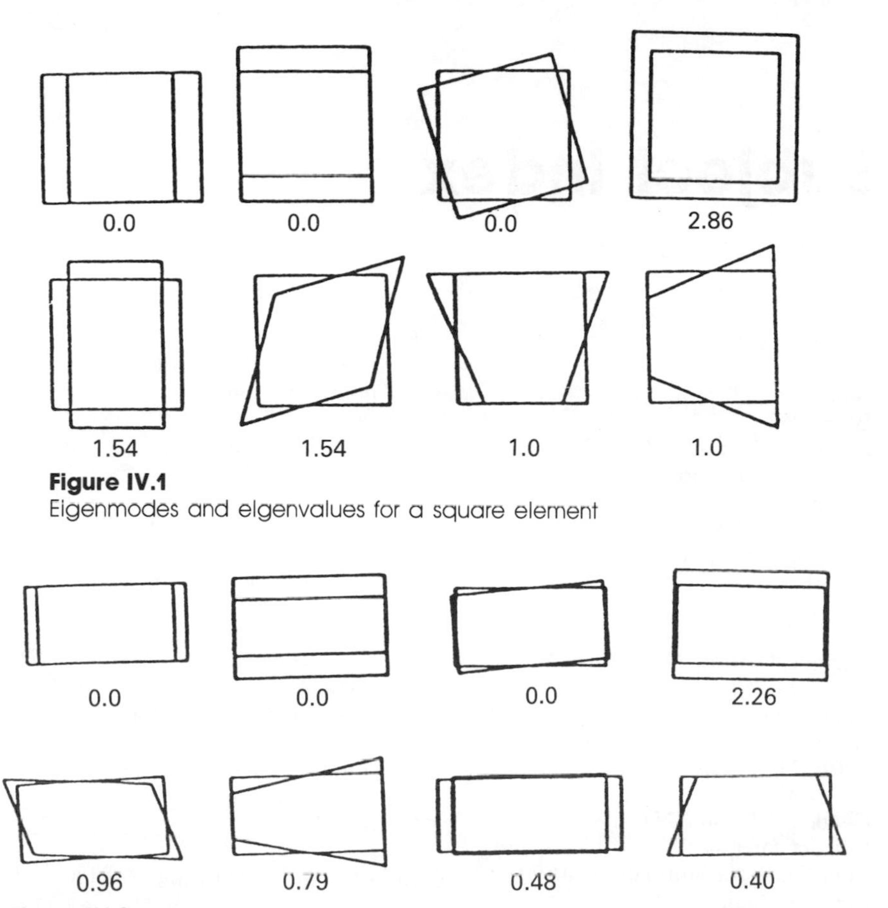


**Figure IV.1**
Eigenmodes and eigenvalues for a square element

**Figure IV.2**
Eigenmodes and eigenvalues for a rectangular element

# Subject Index